# Multivariate Kernel Smoothing and Its Applications

# MONOGRAPHS ON STATISTICS AND APPLIED PROBABILITY



**Semialgebraic Statistics and Latent Tree Models**
*Piotr Zwiernik 146*

**Inferential Models**
Reasoning with Uncertainty
*Ryan Martin and Chuanhai Liu 147*

**Perfect Simulation**
*Mark L. Huber 148*

**State-Space Methods for Time Series Analysis**
Theory, Applications and Software
*Jose Casals, Alfredo Garcia-Hiernaux, Miguel Jerez, Sonia Sotoca, and A. Alexandre Trindade 149*

**Hidden Markov Models for Time Series**
An Introduction Using R, Second Edition
*Walter Zucchini, Iain L. MacDonald, and Roland Langrock 150*

**Joint Modeling of Longitudinal and Time-to-Event Data**
*Robert M. Elashoff, Gang Li, and Ning Li 151*

**Multi-State Survival Models for Interval-Censored Data**
*Ardo van den Hout 152*

**Generalized Linear Models with Random Effects**
Unified Analysis via H-likelihood, Second Edition
*Youngjo Lee, John A. Nelder, and Yudi Pawitan 153*

**Absolute Risk**
Methods and Applications in Clinical Management and Public Health
*Ruth M. Pfeiffer and Mitchell H. Gail 154*

**Asymptotic Analysis of Mixed Effects Models**
Theory, Applications, and Open Problems
*Jiming Jiang 155*

**Missing and Modified Data in Nonparametric Estimation With R Examples**
*Sam Efromovich 156*

**Probabilistic Foundations of Statistical Network Analysis**
*Harry Crane 157*

**Multistate Models for the Analysis of Life History Data**
*Richard J. Cook and Jerald F. Lawless 158*

**Nonparametric Models for Longitudinal Data**
with Implementation in R
*Colin O. Wu and Xin Tian 159*

**Multivariate Kernel Smoothing and Its Applications**
*José E. Chacón and Tarn Duong 160*

*For more information about this series please visit:*
*https://www.crcpress.com/Chapman--HallCRC-Monographs-on-Statistics--Applied-Probability/book-series/CHMONSTAAPP*

# Multivariate Kernel Smoothing and Its Applications

José E. Chacón
Tarn Duong

CRC Press
Taylor & Francis Group
Boca Raton London New York

CRC Press is an imprint of the
Taylor & Francis Group, an **informa** business
A CHAPMAN & HALL BOOK

The cover figure is the heat map corresponding to the kernel density estimator of the distribution of lighting locations in the Paris city council.

CRC Press
Taylor & Francis Group
6000 Broken Sound Parkway NW, Suite 300
Boca Raton, FL 33487-2742

First issued in paperback 2020



ISBN-13: 978-0-367-57173-3 (pbk)
ISBN-13: 978-1-4987-6301-1 (hbk)

Version Date: 20180311



**Visit the Taylor & Francis Web site at**
**http://www.taylorandfrancis.com**

**and the CRC Press Web site at**
**http://www.crcpress.com**

*Para Anabel, Mario y Silvia, en los que empleé toda la suerte que me tocaba en esta vida*

*À mon bien-aimé éternel, Christophe*

# Contents

# Preface

Kernel smoothers, for their ability to render masses of data into useful and interpretable summaries, form an integral part of the modern computational data analysis toolkit. Their success is due in a large part to their intuitive appeal which makes them accessible to non-specialists in mathematics.

Our goal in writing this monograph is to provide an overview of kernel smoothers for multivariate data analysis. In keeping with their intuitive appeal, for each different data analytic situation, we present how a kernel smoother provides a solution to it, illustrating it with statistical graphics created from the experimental data, and supporting it with minimal technical mathematics. These technical details are then gradually filled in to guide the reader towards a more comprehensive understanding of the underlying statistical, mathematical and computational concepts of kernel smoothing. It is our hope that the book can be read at different levels: for data scientists who wish to apply kernel smoothing techniques to their data analysis problems, for undergraduate students who aim to understand the basic statistical properties of these methods, and for post-graduate students/specialist researchers who require details of their technical mathematical properties. An expanded explanation of how to read this monograph is included in Section 1.5.

The most well-known kernel smoothers are kernel density estimators, which convert multivariate point clouds into a smooth graphical visualisation, and so can be considered as an important improvement over data histograms. Interest in them is not solely restricted to their visual appeal, as their mathematical form leads to the quantification of key statistical characteristics which greatly assist in drawing meaningful conclusions from the observed data. From this base case of density estimation, kernel smoothers can be extended to a wide range of more complex statistical data analytic situations, including the identification of data-rich regions, clustering (unsupervised learning), classification (supervised learning) and hypothesis testing.

Many of these complex data analysis problems are closely related to derivatives of the density function, especially the first (gradient) and second (Hessian) derivatives. An important contribution of this monograph is to provide a systematic treatment of multivariate density derivative estimation. Rather than being an arcane theoretical problem with limited practical applications, as it has been usually historically viewed, density derivative estimation is now recognised as a key component in the toolkit for exploratory data analysis and statistical inference.

We can cover only a small selection of the possible data analysis situations in this monograph, and we focus on those with a sufficient level of maturity to be able to offer meaningful analyses of experimental data which are informed by solid mathematical justifications. This has lead to the exclusion of important and interesting problems such as density estimation in non-Euclidean spaces (e.g., simplices, spheres), mode estimation, receiver operating characteristic (ROC) curve analysis, conditional estimation, censored data analysis, or regression. Rather than being a pessimistic observation, this augurs well for the continuing potential of kernel smoothing methods to contribute to the solution of a wide range of complex multivariate data analysis problems in the future.

In our voyage through the field of kernel smoothing techniques, we have been fortunate to benefit from the experience of all of our research collaborators, who are too numerous to mention individually here. Though it would be remiss of us to omit that we have been greatly influenced by our mentors: Antonio Cuevas, Agustín García Nogales, Martin Hazelton, Berwin Turlach and Matt Wand. A special thank you also goes to Larry Wasserman for his encouragement at the initial stages of this project, during the stay of J.E.C. at the Carnegie Mellon University in 2015: probably this book would not have been written without such initial push. Moreover, regarding the writing of this monograph itself, we are grateful to all those who have reviewed the manuscript drafts and whose feedback have much contributed to improving them. We thank all the host and funding organisations as well, which are also too numerous to list individually, for supporting our research work over the past two decades.

And last, but of course not least, J.E.C. would like to thank his family for their loving and support, and T.D. would like to thank his family and most of all, Christophe.

José E. Chacón
*Badajoz, Spain*

Tarn Duong
*Paris, France*

# List of Figures

# List of Tables

# List of Algorithms

Chapter 1

# Introduction

The late twentieth century witnessed the arrival of an omnipresence of data, ushering in the so-called Information Age (Castells, 2010). Statistics plays a key role in this age, as it can be broadly characterised as a meta-science which draws meaningful conclusions from data. This role has become increasingly critical with the advent of Open Data, which professes the unrestricted access to data as a source material, as well as similar freedoms to produce and to disseminate any suitable analyses from them (OKI, 2016). To tackle the analysis of the widely different forms of data, there are correspondingly a wide range of statistical techniques.

## 1.1 Exploratory data analysis with density estimation

Exploratory data analysis (EDA), introduced by Tukey (1977), is a fundamental class of statistical methods for understanding the structure of data sets. As its name suggests, its relationship with more formal inferential methods is sometimes cast with the former solely serving as a preliminary role to the ultimate goal of inference. This belies its importance in its own right, especially given that increasingly accessible computing power has rendered it increasingly informative.

Graphical techniques occupy a central place in EDA, as appropriate visual summaries have the ability to convey concisely the deep structure of the data which would otherwise be difficult to discern from purely numerical summaries. The usual starting point is that we suppose that the observed data set consists of the realisations of a continuous multi-dimensional random variable $\boldsymbol{X}$, whose underlying unknown probabilistic mechanism is fully determined by its probability density function $f$. The density function provides a wealth of information about the data and its distribution, so considerable efforts, historically and currently, have been expended to develop its statistical estimation. If each data point is represented by a multi-dimensional vector, then the most basic form of a density estimator is the scatter plot which visualises the data as points in this multi-dimensional space.

A motivating data set is the `tempb` data. In Figure 1.1 is the scatter plot of the $n = 21908$ pairs of the daily minimum and maximum temperatures (°C) in the GHCN-D v2 time series (Menne et al., 2012) from 1 Jan 1955 to 31 Dec 2015 recorded at the weather station in Badajoz, Spain. It is located at (38.88N, 6.83W) and has GHCN-D station code SP000004452. As this is a dense data set in the central regions, its standard scatter plot results in solid coloured regions which obscure the underlying data structure. To enhance the scatter plot we use alpha blending, where the data points are represented by translucent dots, and where overlapping shapes reinforce rather than obscure each other as the overlapping regions lead to an increase in the displayed opacity (Porter & Duffle, 1984). For alpha blended scatter plots, darker regions thus indicate regions of higher data density.

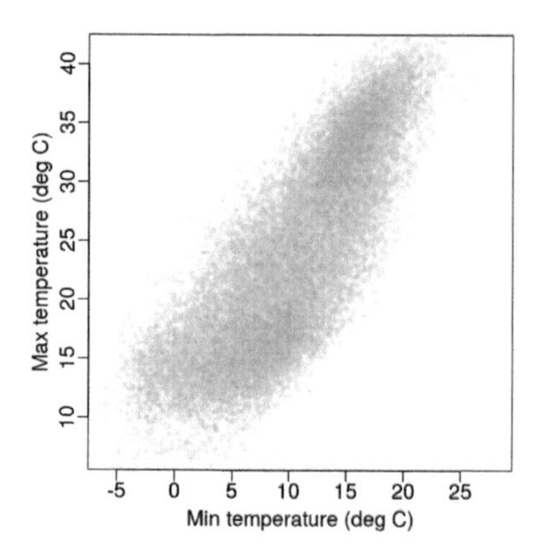


Figure 1.1 *Scatter plot of minimum and maximum daily temperatures (°C) of the* $n = 21908$ *days between 1 Jan 1955 to 31 Dec 2015, recorded at the weather station in Badajoz, Spain.*

Alpha blended scatter plots can be considered as a low level form of data smoothing of the density function, though there exist more sophisticated data smoothing techniques. Data smoothers in general replace each infinitesimal data point with a smooth function over its local neighbourhood in the data space. With this smooth representation of the data sample, we are able to proceed to more suitable visualisations. We will highlight kernel smoothers in this monograph as they form a class of methods which are highly applicable for the practical analysis of experimental data, as well as residing within a solid theoretical framework which assures the statistical rigour of their construction and software implementation. Since the introduction of kernel estimators of density functions, or more concisely kernel density estimators, in the pioneering period of the 1950s–60s (Fix & Hodges, 1951; Rosenblatt, 1956; Parzen, 1962), they have become indispensable within the EDA toolkit.

The daily temperature data are suited to the classical kernel density estimation case, as there are few extreme outlying values or thick tails or a data support with a rigid boundary. Figure 1.2 illustrates the `worldbank` data, in which such unfavourable features are present, and so require the modification of the classical estimator.

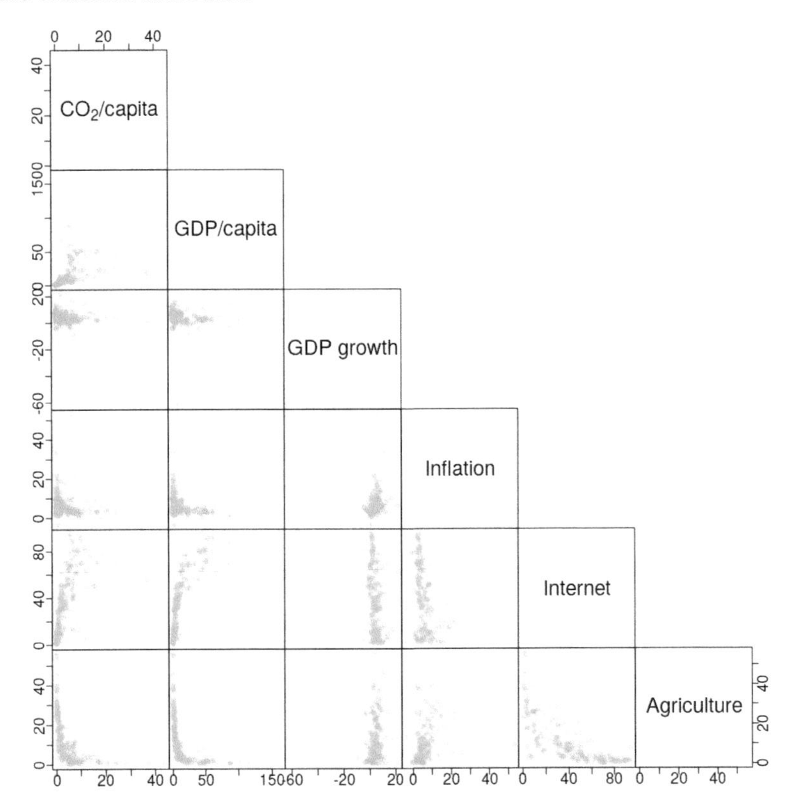


Figure 1.2 *Scatter plot matrix of the World Bank development indicators for the $n = 218$ national entities. There are six variables: $CO_2$ emissions per capita (thousands Kg), GDP per capita (thousands of current USD), annual GDP growth rate (%), annual inflation rate (%), number of internet users in the population (%) and the added value of agricultural production as a ratio of the GDP (%).*

This is a subset of six development indicators (World Bank Group, 2016) for $n = 218$ national entities for the year 2011, which is the latest year for which they are consistently available. These indicators are the carbon dioxide ($CO_2$) emissions per capita (thousands Kg), the gross domestic product

(GDP) per capita (thousands of current US$), the annual GDP growth rate (%), the annual inflation rate (%), the number of internet users in the population (%) and the added value of agricultural production as a ratio of the total GDP (%). The first pair of variables ($CO_2$ emissions, GDP per capita) exhibit semi-infinite values on $[0, \infty)$ as well as thick tails. The second pair (GDP growth, inflation) can be both positive and negative, and have some highly extreme values. The third pair of the percentages of internet users and the added value of agricultural production are strictly bounded in [0%, 100%]. We will demonstrate the modifications of standard kernel estimators in response to these more challenging density estimation cases.

## 1.2 Exploratory data analysis with density derivatives estimation

The data density gradient (first derivative) and the density curvature (second derivative) provide additional information about the structure of the data set which are not evident from investigating only the density function itself. This supplementary information from the derivatives provides a suitable analysis of data which follow filamentary structures.

The `quake` data are the locations of the $n = 2646$ severe earthquakes in the years 100 to 2016 inclusive within the Circum-Pacific belt seismic zone (more commonly known as the Pacific Ring of Fire) as approximately delimited by the United States Geological Survey (USGS, 2017).

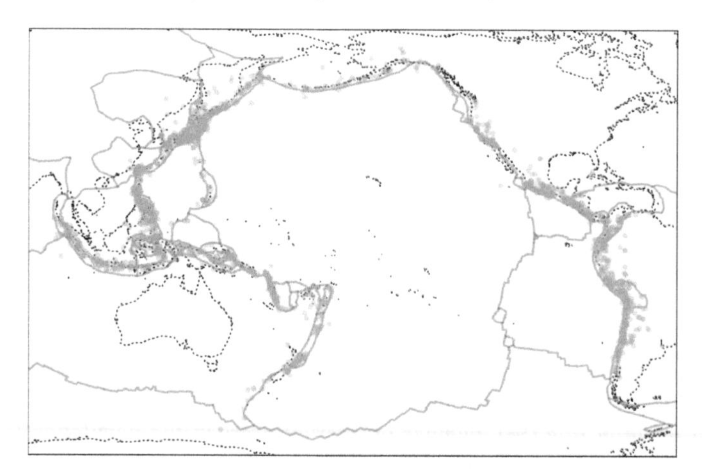

Figure 1.3 *Scatter plot of the geographical locations (longitude, latitude) of the* $n =$ 2646 *major earthquakes in the Circum-Pacific belt seismic zone. The locations are the green points. The boundaries of the land masses are the dotted grey curves, and the boundaries of the tectonic plates are the solid blue curves.*

This is the most important seismic zone in the world, in terms of the number of severe earthquakes, as well as the size of the geographical area. The locations are catalogued in the Global Significant Earthquake Database (NGDC/WDS, 2017), and are plotted as the green points in Figure 1.3, with the longitude ranging from 0 to 360°. The boundaries of the land masses are the dotted grey curves, and the boundaries of the tectonic plates are the solid blue curves (Bird, 2003). We will demonstrate the utility of kernel estimators of the density gradient and curvature to summarise this filamentary data set of earthquake locations.

## 1.3 Clustering/unsupervised learning

Density estimation can be considered as the base case for all other data smoothing problems for EDA. So the success in practical and theoretical terms of kernel methods here has seen kernel estimators being applied to more complex data analysis situations, e.g., clustering (also known as unsupervised learning) where observations within a cluster are similar and observations from different clusters are dissimilar. Clustering benefits greatly when extra information beyond the data density values is supplied.

The development of biotechnologies is one of the major generators of experimental data. The availability of these data, in terms of both quality and quantity, is the driver of the corresponding development of quantitative data analysis in the biological sciences. In cellular biology, the properties of cells are studied via staining with fluorochrome markers. These markers attach to a particular type of cell, and fluoresce at a known range of wavelengths when exposed to a laser light. This fluorescent signal therefore allows the identification of the cell type amongst a mixed cell population. One such biotechnology is a flow cytometer. In this machine, the stained cell sample is directed in a thin tube where the cells pass singly before a laser light, and their fluorescent signal is captured by a series of sensors. A flow cytometer is capable of automatically gathering information from much larger numbers of cells than a manual experiment/analysis (see Givan, 2001 for an overview).

In the hematopoietic stem cell transplant (HSCT) experiments (Aghaeepour et al., 2013), cell samples are collected from 30 mice which have received a graft transplant. In the `hsct` data set, there are 278005 observations with the fluorescent levels (normalised to 0 to 1023) of the fluorochromes: FITC-CD45.1, APC.CD45.2 and PE-Ly65/Mac1. A successful graft can be established by the presence of monocytes/granulocytes amongst the recipient cell population. In Figure 1.4 is the scatter plot of $n = 6236$ measurements for subject mouse #12. When coupled with biological knowledge on characterising (a) recipient versus donor cells and (b) monocytes/granulocytes from other cell types in terms of the fluorescence measurements, we will highlight the role of kernel estimators of density derivatives in their role in facilitating more accurate clustering results for complex data.

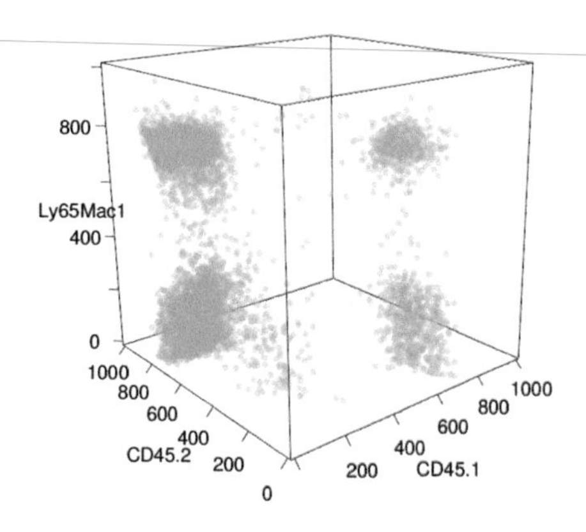


Figure 1.4 *Scatter plot of the fluorescent signals of the $n = 6236$ cells of subject 12 from a hematopoietic stem cell transplant operation. The fluorochrome on the x-axis is FITC-CD45.1, y-axis is APC-CD45.2, z-axis is PE-Ly65/Mac1.*

## 1.4 Classification/supervised learning

Classification/supervised learning is a similar data analysis problem to clustering/unsupervised learning in that we search for groups with similar members. The main difference is that for the former, the training data are augmented by an auxiliary variable for the group label. This auxiliary variable label gives us the number and location of the groups (which is one of the main goals of cluster analysis), so we are not required to carry out this task. On the other hand, we have the test data which are drawn from the same population as the training data for which no auxiliary group label is observed. The main goal of classification is to determine if these groups in the training data are able to yield useful estimated group labels for the test data.

A cardiotocogram is a biotechnology which measures the instantaneous foetal heart rate and uterine contraction signals during pregnancy, and is an important tool for evaluating the health of a foetus. The `cardio` data consist of measurements taken from 2126 foetal cardiotocograms, collected by Ayres-de Campos et al. (2000) and are available from the UCI Machine Learning Repository (Lichman, 2013). This complete data set is divided into a random 25% subset ($n = 532$) to be the training data sample and the remaining 75% ($m = 1594$) to be the test data sample. For the training sample, the cardiotocograms are classified into three groups according to the foetal state: normal ($n_1 = 412$, green), suspect ($n_2 = 83$, orange), and pathological ($n_3 = 37$, purple) in the scatter plot in Figure 1.5(a). These group labels were determined by three expert obstetricians and a consensus group label finally assigned to each cardiotocogram. From the wide range of variables derived from the cardiotocogram, we analyse the abnormal short term variability (percentage) and the mean level of the cardiotocographic signal histogram (0–255). The aim is to assign a group label for the test data in Figure 1.5(b) based on the groups in the training data. We will highlight the role of kernel estimators for classification for these training/test data sets.

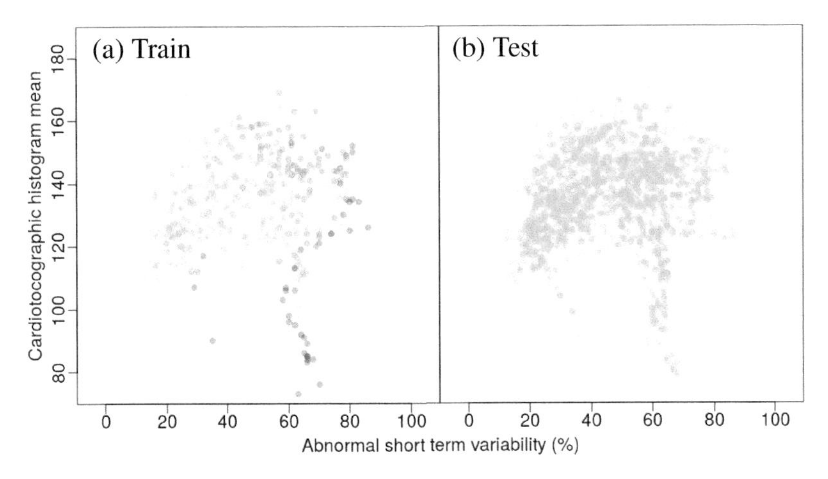


Figure 1.5 *Scatter plots of the foetal cardiotocographic data. The x-axis is the abnormal short term variability (percentage), and the y-axis is the mean of the cardiotocographic histogram (0–255). (a) Training data ($n = 532$) with three groups: normal ($n_1 = 412$, green), suspect ($n_2 = 83$, orange), pathological ($n_3 = 37$, purple). (b) Test data ($m = 1594$) with no group labels.*

## 1.5 Suggestions on how to read this monograph

For the data analyst who wishes to gain a broad understanding of the state of the art of practical multivariate data analysis with kernel smoothers using a minimal mathematical theoretical framework, the following chapters illustrate the different kernel smoothers with experimental data sets by highlighting their intuitive, visual approach to data analysis.

Sections 2.1–2.4 serve as an introduction into kernel smoothers by presenting kernel density estimators as an improvement on histograms. These sections deal with this basic case of kernel smoothing, and the subsequent sections explore their extensions to more complex data analytic cases.

Sections 4.1–4.3 introduce some modified density estimators, e.g., for data with rigid boundaries, where these standard density estimators are not sufficient.

Sections 5.1–5.3 focus on the estimation of the first derivative (gradient) and the second derivative (curvature) of the density function. Whilst it has been recognised for some time now that these derivatives provide crucial information that is not found in the density itself, practical algorithms for estimating them are less well-developed. These sections set up a usable framework that includes the previously treated density estimation as a special case, and provide the foundations for topics, e.g. machine learning, treated in the subsequent chapters.

Section 6.1 tackles level sets of the density function to identify data-rich regions. Section 6.2 (excluding 6.2.3) deals with clustering/unsupervised learning, and in particular the mean shift clustering which implements the density gradient estimators to improve clustering accuracy. Section 6.3 explores density ridges, which implement density curvature estimators, as a high-content extension of principal components. Section 6.4 is the first foray into statistical hypothesis testing via significant modal regions (based on the density curvature) as regions of interest.

Chapter 7 extends from these single data sample analysis situations to 2- and $k$-sample comparison problems. Section 7.1 highlights the case for the difference between two density functions. Section 7.2 tackles classification/supervised learning for comparing $k$ density estimators. Section 7.3 (excluding 7.3.1) introduces density estimation for data measured with error. Section 7.4 provides a nearest neighbour mean shift clustering (suitable for high-dimensional data), by exploiting the connections between kernel and nearest neighbour estimators.

Section 8.1 outlines the `R` commands to produce the displayed figures from the accompanying `ks` package and associated scripts.

For a university undergraduate level, the remaining additional sections elaborate the mathematical framework of kernel smoothers. They provide the underlying mathematical reasoning for the widespread applicability of kernel smoothers for data analysis.

Sections 2.5–2.8 concern the squared error analysis for density estimators. Sections 3.1–3.9 formalise the crucial problem of bandwidth (or smoothing parameter) selection for density estimators. Section 4.4 examines the secondary question of kernel function choice. Sections 5.5–5.6 formalise the bandwidth selection problem for density derivative estimators, with a unified framework that includes as a special case the bandwidths for density estimation. Section 6.2.3 completes the description of the mean shift clustering algorithm. Section 7.3.1 describes classical deconvolution density estimation.

For a university postgraduate level, the additional sections, i.e., Sections 2.9, 3.10, 4.5–4.6, 5.9, 7.5.1–7.5.2, 8.2–8.5, contain the detailed mathematical derivations and other specialised topics which require a certain level of mathematical sophistication to comprehend. When read in conjunction with the other chapters, they provide an overview of many sub-fields in kernel smoothing, as well as in-depth knowledge of the approaches to the bandwidth selection problems.

These suggestions are summarised in Table 1.1 below. They should not be be considered as strict recommendations, but rather as loose guidance from the authors, especially for novices to multivariate kernel smoothing.

| **Topic** | **Data analyst** | **Under-graduate** | **Post-graduate** |
|---|---|---|---|
| Density estimation | 2.1–2.4 | 2.1–2.8 | 2.1–2.9 |
| Bandwidth selection | – | 3.1–3.9 | 3.1–3.10 |
| Modified density estimation | 4.1–4.3 | 4.1–4.4 | 4.1–4.6 |
| Density derivative estimation | 5.1–5.3 (ex. 5.1.3) | 5.1–5.6 (ex. 5.1.3) | 5.1–5.9 |
| Level set estimation | 6.1 | 6.1 | 6.1 |
| Density-based clustering | 6.2 (ex. 6.2.3) | 6.2 | 6.2 |
| Density ridge estimation | 6.3 | 6.3 | 6.3 |
| Feature significance | 6.4 | 6.4 | 6.4 |
| Density difference estimation | 7.1 | 7.1 | 7.1 |
| Classification | 7.2 | 7.2 | 7.2 |
| Data measured with error | 7.3 (ex. 7.3.1) | 7.3 | 7.3, 7.5.1 |
| Nearest neighbour estimation | 7.4 | 7.4 | 7.4, 7.5.2 |
| Computation | 8.1 | 8.1 | 8.1–8.5 |

Table 1.1 *Suggestions on how to read this monograph.*

This monograph can also be considered as complementary to those which exist already on kernel and other non-parametric smoothing methods, such as Devroye & Györfi (1985), Silverman (1986), Scott (1992, 2015), Wand & Jones (1995), Simonoff (1996), Bowman & Azzalini (1997), Devroye & Lugosi (2001*a*), Wasserman (2006), Klemelä (2009), and Horová et al. (2012).

Chapter 2

# Density estimation

In the introduction, we saw the scatter plot visualisation of the density of the daily temperature data, which can be interpreted as a low level smoothing of the data density. In this chapter, we pursue more advanced statistical estimators of the density function. Section 2.1 begins with histograms, as they were the first non-parametric density estimators. Section 2.2 motivates kernel density estimators as improvements over histograms. Sections 2.3–2.4 begin to tackle the choice of the bandwidth matrix, by offering practical insights into the choice amongst the most common data-based algorithms. Sections 2.5–2.8 set up a rigorous mathematical framework for optimal bandwidth selection. Section 2.9 fills in the mathematical details that were previously omitted in order to facilitate freely flowing text.

Unless stated otherwise, the data $\boldsymbol{X}_1,\ldots,\boldsymbol{X}_n$ are assumed to be a $d$-dimensional random sample drawn from a common density $f$. This is shorthand for a collection of independent and identically distributed random variables, where each of the $\boldsymbol{X}_i$ is a $d$-dimensional random vector drawn from the target density $f$. If required, its coordinates are written as $\boldsymbol{X}_i = (X_{i1},\ldots,X_{id})$, and a vector $\boldsymbol{x} \in \mathbb{R}^d$ is similarly denoted as $\boldsymbol{x} = (x_1,\ldots,x_d)$. An $m \times n$ matrix

$$\mathbf{A} = \begin{bmatrix} a_{11} & \cdots & a_{1n} \\ \vdots & \ddots & \vdots \\ a_{m1} & \cdots & a_{mn} \end{bmatrix} \in \mathcal{M}_{m\times n}$$

can be written in its vectorised form as $\mathbf{A} = [a_{11},\ldots,a_{m1};\cdots;a_{1n},\ldots,a_{mn}]$ to maintain a compact horizontal notation.

## 2.1 Histogram density estimation

The histogram was the first non-parametric density estimator, though its exact date of invention is not known for certain, but likely to have been in the 17th century, according to the summary of its historical genesis (Scott, 2015, pp.16–17). Its chronological primacy is due to its simple computation.

To construct a histogram, the preliminary step is to discretise the sample space in subregions known as bins. This discretisation usually consists of an equidistant rectangular grid, starting at the anchor point (the marginal minima) and containing $M$ bins, $B_1, \ldots, B_M$, each with width $b_i$ along the $i$-th coordinate. A histogram density estimator is a step function with a constant value within each of the bins, where the constant is given by the proportion of data points $\boldsymbol{X}_i$ which fall in the bin divided by the bin volume. Writing $\boldsymbol{b} = (b_1, \ldots, b_d)$ for the vector of bin widths, this is expressed as

$$\hat{f}_{\text{hist}}(\boldsymbol{x}; \boldsymbol{b}) = \frac{N_j}{nb_1 \cdots b_d} \quad \text{for all } \boldsymbol{x} \in B_j$$

where $N_j = \sum_{i=1}^n \mathbf{1}\{\boldsymbol{X}_i \in B_j\}$ is the random variable that counts the number of data points that fall in the $j$-th bin $B_j$. Constructing a histogram involves the choice of two tuning parameters: the anchor point and the binwidth vector $\boldsymbol{b}$.

**Example 2.1** The influence of the anchor point with the daily temperature data introduced in Section 1.1 is illustrated in Figure 2.1. A histogram with anchor point $(-5.54, 6.52)$ is shown in Figure 2.1(a) and with an anchor point $(-4.71, 7.71)$ in Figure 2.1(b), which is a translation by one-half bin of the left histogram.

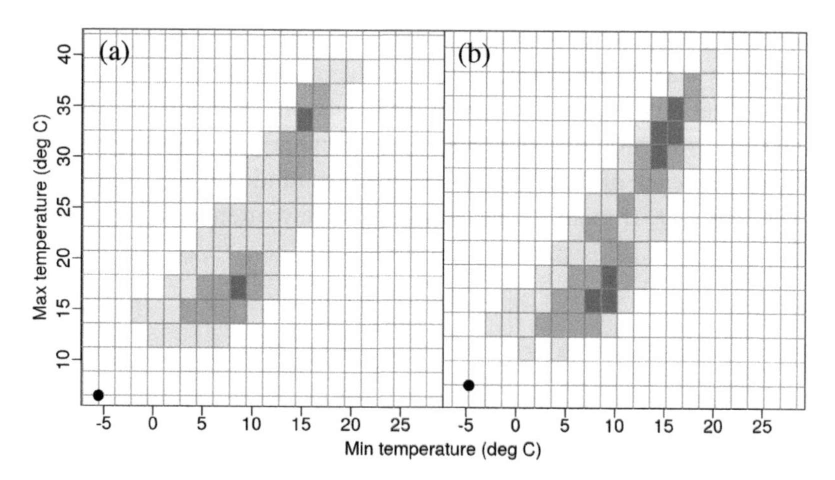


Figure 2.1 *Anchor point placement for the histograms of the daily temperature data. (a) Anchor point* $(-5.54, 6.52)$. *(b) Anchor point* $(-4.71, 7.71)$ *(black circles). The height of the histogram is represented by a colour gradient: white/grey for low density through to purple for high density. The common binwidth is* $\boldsymbol{b} = (1.66, 2.37)$.

The colour scale indicates density: starting with white for low density, increasing through grey and ending with purple for high density. The left histogram is clearly bimodal, whereas the right histogram is less clearly so. This change in the visual appearance is induced solely by a translation in anchor point and the resulting bin interval boundaries.

The binwidth utilised in Figure 2.1 is $\hat{\boldsymbol{b}}_{\text{NS}}$, given by the normal scale binwidth for the $i$-th dimension by, as noted in Simonoff (1996, p. 98),

$$\hat{b}_{\text{NS},i} = 2 \cdot 3^{1/(d+2)} \pi^{d/(2d+4)} s_i n^{-1/(d+2)} \tag{2.1}$$

where $s_i$ is the $i$-th marginal standard deviation. Equation (2.1) gives $\boldsymbol{b} = \hat{\boldsymbol{b}}_{\text{NS}} = (1.66, 2.37)$ for the daily temperature data. This binwidth is optimal when the target density is a normal density. As the data are manifestly non-normal, we explore histograms with different binwidths.

The influence of the binwidth is illustrated in Figure 2.2. The histogram with binwidth equal to $\boldsymbol{b}/3$ is Figure 2.2(a), and can be considered to be undersmoothed, as it appears to reduce insufficiently the data as presented in the scatter plot. The bin counts tend to be low and similar to each other as the bins are small, and so the light grey colours dominate. On the other hand, the right histogram in Figure 2.2(b) with a binwidth equal to $3\boldsymbol{b}$ is oversmoothed since much of the structure of the data are obscured by bins which are too large and contain counts which are too aggregated. So whilst the normal scale binwidth is not always optimal, it nonetheless provides a reasonable starting point to avoid extreme under- and oversmoothing. □

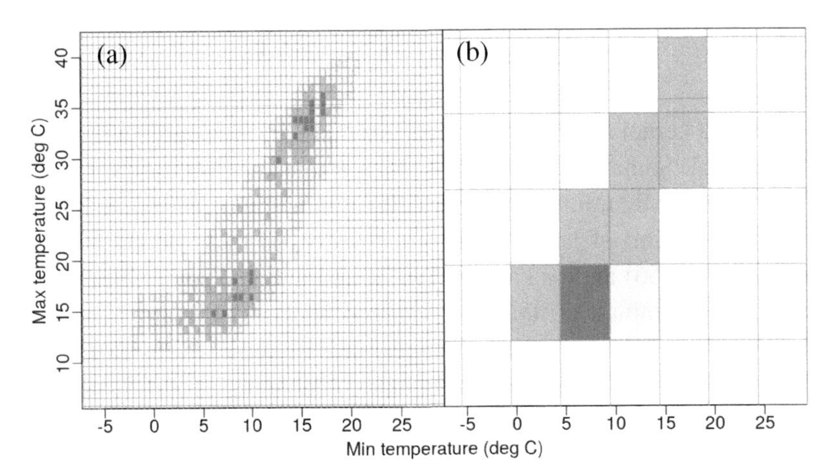


Figure 2.2 *Binwidths for the histograms of the daily temperature data. (a) Undersmoothed histogram, with binwidth* $\boldsymbol{b}/3$. *(b) Oversmoothed histogram, with binwidth* $3\boldsymbol{b}$. *The base binwidth is* $\boldsymbol{b} = (1.66, 2.37)$.

In addition to these tuning parameters, histograms display a discontinuous appearance as they have a constant value within each rectangular bin.

This can be attenuated, whilst not being entirely eliminated, by various approaches: (i) replacing rectangular with hexagonal bins which have more obtuse corners (Carr et al., 1987), or (ii) replacing the constant count by a linear interpolation (frequency polygons), or (iii) averaging several histograms with translated anchor points (average shifted histograms), which also attenuate the anchor point placement problem. Options (ii) and (iii) are explored in detail in Simonoff (1996, Ch. 4) and Scott (2015, Ch. 4–5).

In the latter reference, it is also shown that the limiting case of averaging an infinite number of shifted histograms coincides with the kernel density estimator. For references about histograms, the interested reader is referred to the papers (and the references therein) of Lugosi & Nobel (1996) for their consistency properties, Beirlant et al. (1994) for the asymptotic distribution of its error, or Scott (1979) and Wand (1997) for data-based rules for automatic binwidth selection.

Further details about histograms will be covered only sporadically as we turn our focus onto kernel smoothers.

## 2.2 Kernel density estimation

Apart from the histogram, the most popular density estimator is most likely to be the kernel density estimator, which is defined as

$$\hat{f}(\boldsymbol{x};\mathbf{H}) = n^{-1}\sum_{i=1}^{n} K_{\mathbf{H}}(\boldsymbol{x}-\boldsymbol{X}_i). \tag{2.2}$$

Here, $K$ is a kernel function, which means that it is an integrable function with unit integral. Sometimes additional requirements are imposed on $K$, such as that it is a smooth, unimodal, spherically symmetric density function.

The invention of kernel density estimators is traditionally attributed to Rosenblatt (1956) and/or Parzen (1962), and they are sometimes referred to as Parzen-Rosenblatt estimators or Parzen window estimators (but see also Akaike, 1954). However, the first appearance of kernel estimators is likely to be Fix & Hodges (1951): as the original technical report is difficult to find, it has been re-published as Fix & Hodges (1989).

The form of $\hat{f}$ in Equation (2.2) can be interpreted from a couple of different points of view: (i) from an estimation point $\boldsymbol{x}$, it is a local weighted averaging estimator where the weight of $\boldsymbol{X}_i$ decreases as its distance to $\boldsymbol{x}$ increases, or (ii) from a data point $\boldsymbol{X}_i$, its probability mass is smoothed in the local neighbourhood according to the scaled kernel to represent the unobserved data points.

The crucial tuning parameter is the bandwidth $\mathbf{H}$, also called the window width matrix. It is a symmetric, positive definite, $d \times d$ matrix of smoothing parameters. The bandwidth controls the orientation and the extent of the smoothing applied via the scaled kernel $K_{\mathbf{H}}(\boldsymbol{x}) = |\mathbf{H}|^{-1/2}K(\mathbf{H}^{-1/2}\boldsymbol{x})$, where $|\mathbf{H}|$ is the determinant of $\mathbf{H}$ and $\mathbf{H}^{-1/2}$ is the inverse of its matrix square root. A scaled kernel is positioned so that its mode coincides with each data point $\boldsymbol{X}_i$, which is expressed mathematically as $K_{\mathbf{H}}(\boldsymbol{x} - \boldsymbol{X}_i)$. The scaled kernels are summed and the division by $n$ ensures that the overall probability mass of $\hat{f}$ remains one. As the kernels are placed on each data point, the anchor point placement problem that the histogram suffers from is thus eliminated. The kernel density estimator also inherits the smoothness of the individual kernels. The increased smoothness of kernel estimators in comparison to histograms is not solely an aesthetic improvement, as it also leads to improved statistical properties, which is elaborated in the subsequent sections.

The most widely used multivariate kernel function is the normal kernel

$$K(\boldsymbol{x}) = (2\pi)^{-d/2} \exp(-\tfrac{1}{2}\boldsymbol{x}^\top \boldsymbol{x})$$

which is the standard $d$-variate normal density function. The scaled, translated normal kernel is

$$K_{\mathbf{H}}(\boldsymbol{x} - \boldsymbol{X}_i) = (2\pi)^{-d/2}|\mathbf{H}|^{-1/2} \exp\left\{ -\tfrac{1}{2}(\boldsymbol{x} - \boldsymbol{X}_i)^\top \mathbf{H}^{-1}(\boldsymbol{x} - \boldsymbol{X}_i)\right\}$$

which is a normal density centred at $\boldsymbol{X}_i$ and with variance matrix $\mathbf{H}$. This is one of the main reasons that we parametrise $\mathbf{H}$ as a variance matrix, i.e., on the squared data scale, rather than the square root of a variance matrix to be comparable on the data scale. For this, and other reasons, the normal kernel is almost universally preferred for multivariate data, in contrast to the univariate case where other kernels can be preferred. Consequently we use the normal kernel in numerical calculations, unless otherwise indicated, throughout this monograph.

**Example 2.2** A graphical illustration of the construction of Equation (2.2) on a subset of the daily temperature data is given in Figure 2.3. The data points are the solid green circles. On the left in Figure 2.3(a) are the 10 individual scaled normal kernels, each with variance $\mathbf{H} = [6.71, 6.04; 6.04, 10.42]$, and centred on each of the data points. On the right in Figure 2.3(b) is the contour plot of the kernel density estimate. The contours are the solid black curves, which are level sets of the density estimate, and the regions between two consecutive contours are filled with a colour gradient as used previously.

Following the same procedure, Figure 2.4 shows the kernel density estimate of the complete data set of the $n = 21908$ observations with $\mathbf{H} =$

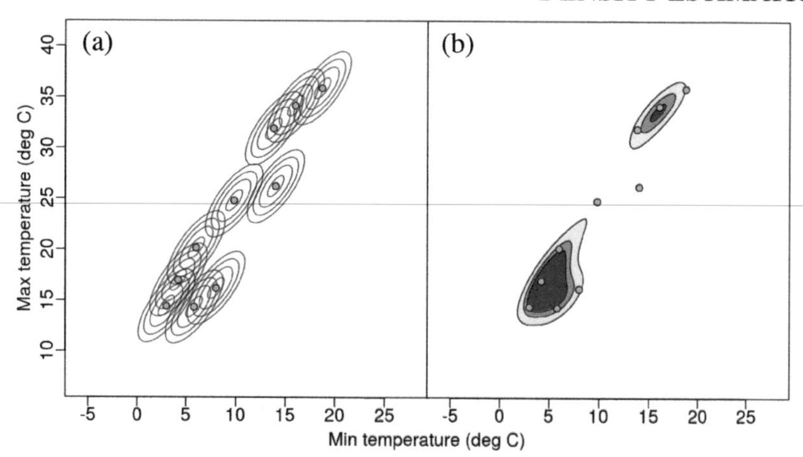


Figure 2.3 *Construction of a density estimate on a subset of the daily temperature data. Data points are the solid green circles. (a) Individual scaled kernels centred on the data points. (b) Kernel density estimate.*

$[0.67, 0.60; 0.60, 1.04]$. In Figure 2.4(a), there are 3 probability contours corresponding to the quartiles 25%, 50%, 75% of the probability mass (a more precise definition is given below). The wire-frame or perspective plot in Figure 2.4(b) is a more direct 3-dimensional visualisation. Whilst it avoids a potential visual bias, as it does not require the choice of a finite number of contours since it is able to display a continuum of density heights, the trade-off is that a 3-dimensional display on a 2-dimensional page is not always easy to interpret, e.g., it is not straightforward to establish the relative heights of the two modes due to the perspective distortion. □

### 2.2.1 *Probability contours as multivariate quantiles*

The levels of the contours play an important role in conveying visual information in contour plots of kernel density estimates.

**Example 2.3** In Figure 2.5(a), there are 9 contours (0.0010, 0.0017, 0.0023, 0.0028, 0.0032, 0.0034, 0.0038, 0.0044, 0.0049) which are linear in a probability space, in the sense that each contour encloses 10% of the sample space, e.g. $\mathbb{P}(\boldsymbol{X} \in \{\boldsymbol{x} : 0.0023 < f(\boldsymbol{x}) \leq 0.0028\}) = 0.1$ for $\boldsymbol{X} \sim f$. A linear scale of the density heights, e.g., in geographical or atmospheric pressure maps, is also widely used, as illustrated in Figure 2.5(b). There are 9 contours spaced linearly from 0.0009 to 0.0059, which cover the density range. Figures 2.4(a) and 2.5(a) show that the number of contours subtly influences the

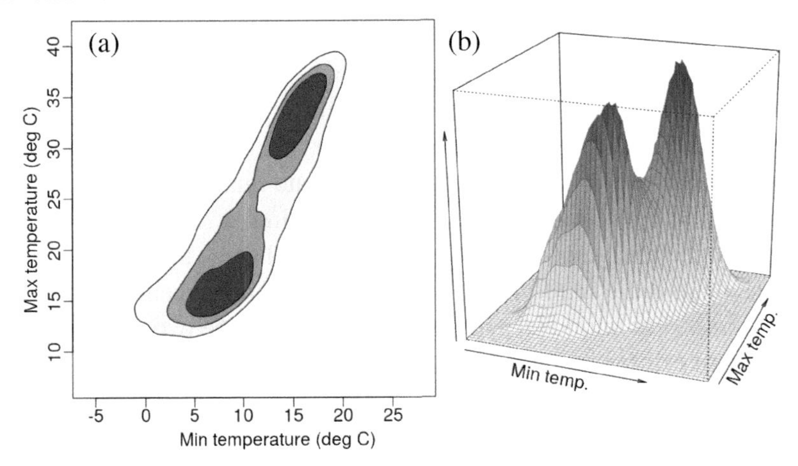


Figure 2.4 *Plots of the density estimate of the daily temperature data, with sequential colour scale. (a) Filled contour plot with quartile (25%, 50%, 75%) contours. (b) Wire frame/perspective plot.*

interpretation of the density estimator $\hat{f}(\boldsymbol{x};\mathbf{H})$. With the 3 contours located at the quartiles, Figure 2.4(a) emphasises the larger scale features. Whereas Figure 2.5(a), the inclusion of smaller, larger and more intermediate deciles, present a higher resolution map. An appropriate number of contours depends largely on the goal of the visual analysis; some practical recommendations about the contour levels can be found in Delicado & Vieu (2015). □

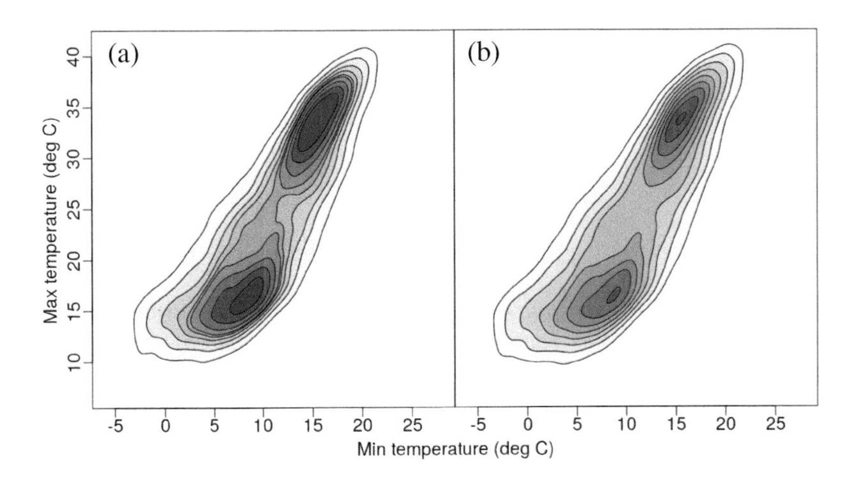


Figure 2.5 *Choice of contour levels for the density estimate of the daily temperature data. (a) Probability (decile) contour levels. (b) Linear contour levels. Both plots have 9 contour levels each.*

The probability contours in Figures 2.4(a) and 2.5(a) are defined as follows: for any $\tau \in (0,1)$, the $100\tau\%$ contour is the region with the smallest area which encloses $100\tau\%$ of the probability mass of the density function (Bowman & Foster, 1993*b*; Hyndman, 1996), and hence can be interpreted as a multivariate equivalent to the usual univariate quantiles. There is a difference, even in the univariate case: quantiles delimit a *central* region containing some percentage of the probability mass, whereas these highest density regions delimit the *smallest* region containing some percentage of the probability mass. For unimodal symmetric distributions the two entities coincide, but they give different answers for multimodal distributions.

**Example 2.4** The multivariate median corresponding to the 50% contour region of the temperature data in Figure 2.6(a) does not occupy a uniform half of the data space but is concentrated in the central regions, indicating the inhomogeneity of this distribution of the temperature measurements. Figure 2.6(b) is the 95% contour region, which can be interpreted as a conservative estimate of the effective support of the density function (we revisit the topic of density support estimation in Section 6.1). □

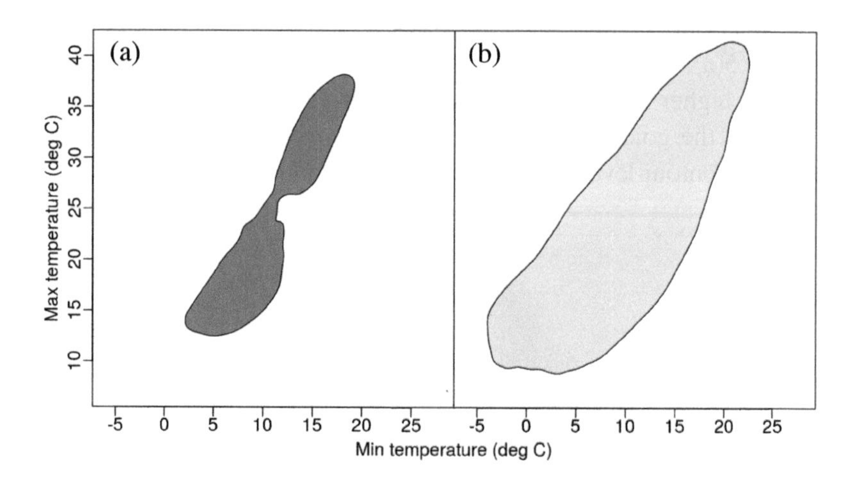


Figure 2.6 *Probability contours contour levels for the density estimate of the daily temperature data. (a) 50% contour as a multivariate median. (b) 95% contour as an effective data support.*

There are several advantages of these probability contours: (i) the minimum, the maximum, and the step size of the contours can be decided before any calculation, and (ii) they allow a probabilistic interpretation of defined contours. For example, the linear contour levels for Figure 2.5(b) may have to be adjusted if new data are added or the data scale is changed. On the other hand, the probability contours are scale invariant and are well-defined for any changes to the data set.

Formally, a $100\tau\%$ region of a density $f$ is defined as the level set $\mathcal{L}(f_\tau) = \{\boldsymbol{x} : f(\boldsymbol{x}) \geq f_\tau\}$ with its corresponding contour level $f_\tau$ such that $\mathbb{P}(\boldsymbol{X} \in \mathcal{L}(f_\tau)) = 1 - \tau$ and that $\mathcal{L}(f_\tau)$ has a minimal hypervolume. This implicit definition somewhat belies their importance, so Hyndman (1996) elucidates them with an alternative interpretation. Let $Y = f(\boldsymbol{X})$ be the random variable $\boldsymbol{X}$ evaluated at its own density function, and $f_\tau$ be the $\tau$-quantile of $Y$, i.e., $F_Y(f_\tau) = \tau$ where $F_Y$ is the cumulative distribution function of $Y$. Then $\mathbb{P}(\boldsymbol{X} \in \mathcal{L}(f_\tau)) = \int_{\mathbb{R}^d} f(\boldsymbol{x}) \mathbf{1}\{\mathcal{L}(f_\tau)\}\, d\boldsymbol{x} = \mathbb{E}\{\mathbf{1}\{f(\boldsymbol{X}) \geq f_\tau\}\} = \mathbb{P}(f(\boldsymbol{X}) \geq f_\tau) = \mathbb{P}(Y \geq f_\tau) = 1 - F_Y(f_\tau) = 1 - \tau$. The minimal hypervolume property is more complicated to establish concisely, and for brevity we leave the reader to peruse it in Hyndman (1996). It follows that an easy-to-compute estimate of the probability contour level $f_\tau$ is given by the $\tau$-th quantile $\hat{f}_\tau$ of $\hat{f}(\boldsymbol{X}_1; \mathbf{H}), \ldots, \hat{f}(\boldsymbol{X}_n; \mathbf{H})$. This estimator is valid for any consistent estimator of $f$, and not only for a kernel estimator.

### 2.2.2 *Contour colour scales*

Along with the contour levels, the colour scheme of these contour regions also plays an important role in graphical visualisation. We have used a sequential colour scheme in Figures 2.1–2.6 which feature a white background (low density), progressing to grey (medium) to purple (high). All the colour scales used in this monograph take their cue from Zeileis et al. (2009). The visualisation of kernel density estimators require some care as it depends on the goal of the data analysis. It should not detract though that kernel density estimators strike a happy medium between displaying the intricate data structures, whilst still reducing the visual complexity of a scatter plot. A deeper study of the visualisation of density estimators can be found in Klemelä (2009).

**Example 2.5** In Figure 2.7 are two alternative sequential colour schemes. Figure 2.7(a) is a heat colour scale with pink (low), violet (medium), blue (high). Figure 2.7(b) is the terrain colour scale, commonly used in geographical land maps where the green represents sea level (low) to yellow/beige (medium) and through to white for the snow-capped mountains (high). □

## 2.3 Gains from unconstrained bandwidth matrices

The bandwidth matrix in the definition of the kernel density estimator $\hat{f}(\boldsymbol{x}; \mathbf{H}) = n^{-1} \sum_{i=1}^{n} K_{\mathbf{H}}(\boldsymbol{x} - \boldsymbol{X}_i)$ was taken to be a symmetric, positive definite matrix. In the univariate case the bandwidth matrix $\mathbf{H}$ reduces to a pos-

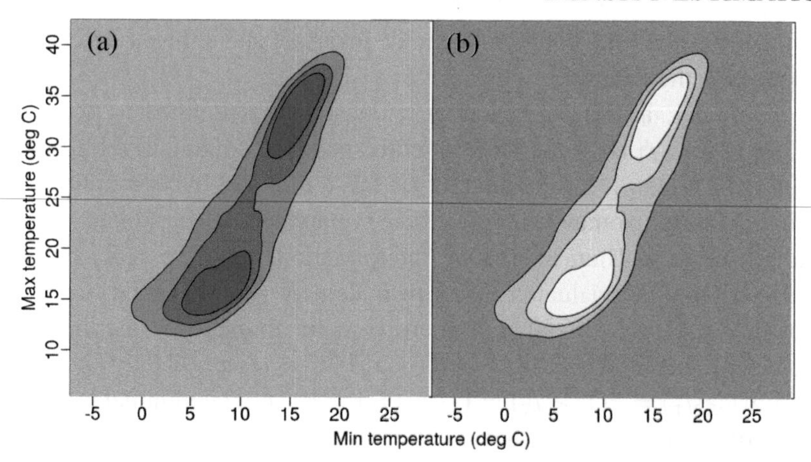


Figure 2.7 *Filled contour plots of the density estimate of the daily temperature data, with alternative colour scales. (a) Heat colours: pink (low), violet (medium), blue (high). (b) Terrain colours: green (low), yellow/beige (medium), white (high).*

itive scalar, so writing $\mathbf{H} = h^2$ for some $h > 0$ the kernel estimator in Equation (2.2) can be written more simply, since the kernel scaling simplifies to $K_h(x) = K(x/h)/h$, as

$$\hat{f}(x;h) = n^{-1} \sum_{i=1}^{n} K_h(x - X_i).$$

For multivariate data, the bandwidth matrix offers a richer class of the possibilities for smoothing. Historically, the bandwidth matrix was initially restricted to the class $\mathcal{A} = \{h^2\mathbf{I}_d : h > 0\}$ of a positive scalar multiplying the identity matrix $\mathbf{I}_d$ (Cacoullos, 1966), resulting in a simple multivariate kernel density estimator

$$\hat{f}(\boldsymbol{x};h) = (nh^d)^{-1} \sum_{i=1}^{n} K\big((\boldsymbol{x} - \boldsymbol{X}_i)/h\big),$$

or to the class $\mathcal{D} = \{\text{diag}(h_1^2, \ldots, h_d^2) : h_1, \ldots, h_d > 0\}$ of positive definite diagonal matrices (Epanechnikov, 1969), leading to

$$\hat{f}(\boldsymbol{x};h_1,\ldots,h_d) = (nh_1 \cdots h_d)^{-1} \sum_{i=1}^{n} K\big((x_1 - X_{i1})/h_1, \ldots, (x_d - X_{id})/h_d\big).$$

The most general class of unconstrained matrices (i.e., symmetric positive definite matrices) $\mathcal{F} = \{\mathbf{H} \in \mathcal{M}_{d\times d} : \mathbf{H} > 0, \mathbf{H} = \mathbf{H}^\top\}$ was introduced in Deheuvels (1977), although it remained mostly unconsidered until the 1990s, partly due to computational limitations but also because the kernel estimator based on simpler parametrisations were substantially easier to analyse given the then-available mathematical tools.

A comprehensive treatment of the different classes of bandwidth matrix parametrisations was carried out by Wand & Jones (1993), who considered more classes than these three, e.g., $\mathcal{V} = \{h^2\mathbf{S} : h > 0\}$ where $\mathbf{S}$ is the sample variance matrix. These authors strongly advocate against the use of the scalar-based classes like $\mathcal{A}$ or $\mathcal{V}$, because $\mathcal{A}$ applies the same amount of smoothing for all coordinate directions and the sample variance matrix in $\mathcal{V}$ does not usually capture the density curvature and its orientation: a practical illustration of this issue is given in Chacón & Duong (2010, Section 3.1.3). Wand & Jones (1993) recommend the unconstrained class $\mathcal{F}$ for most data analysis cases, followed by the diagonal class $\mathcal{D}$. Since the publication of this influential study, the question of the bandwidth matrix parametrisation is usually cast as a choice between the diagonal and unconstrained classes. Furthermore, a data-based procedure to make this choice is outlined in Chacón (2009).

**Example 2.6** The `grevillea` data is an experimental data example that validates the utilisation of unconstrained matrices. It contains the geographical locations (in decimal degrees) of the specimens of *Grevillea uncinulata*, more commonly known as the Hook leaf *Grevillea*, which is an endemic floral species to south Western Australia. This region of south Western Australia is one of the 25 'biodiversity hotspots' which are 'areas featuring exceptional concentrations of endemic species and experiencing exceptional loss of habitat' identified in Myers et al. (2000) to assist in formulating priorities in biodiversity conservation policies. These data are available from the open data platform of the Australian Living Atlas (CSIRO, 2016).

Figure 2.8(a) is the scatter plot of the $n = 222$ observations of the (longitude, latitude) measurements in decimal degrees. The data points are plotted in green, the ocean is blue, and the land mass is white. The kernel density estimate with an unconstrained bandwidth $[0.058, -0.045; -0.045, 0.079]$ is shown in Figure 2.8(b), and with a diagonal bandwidth $\text{diag}(0.035, 0.040)$ in Figure 2.8(c). Both density estimates yield multi-modal densities. For the diagonal bandwidth matrix, the modes are all clearly separated. For the unconstrained matrix, the two largest modes are less clearly delimited from each other, and the oblique contours suggest a directionality in the geographical distribution. In the upper right corners of Figure 2.8(b)–(c) are the contours of the kernel with the given bandwidth matrices: the unconstrained matrix gives obliquely oriented kernels whereas the diagonal matrix gives kernels which are oriented parallel to the coordinate axes. □

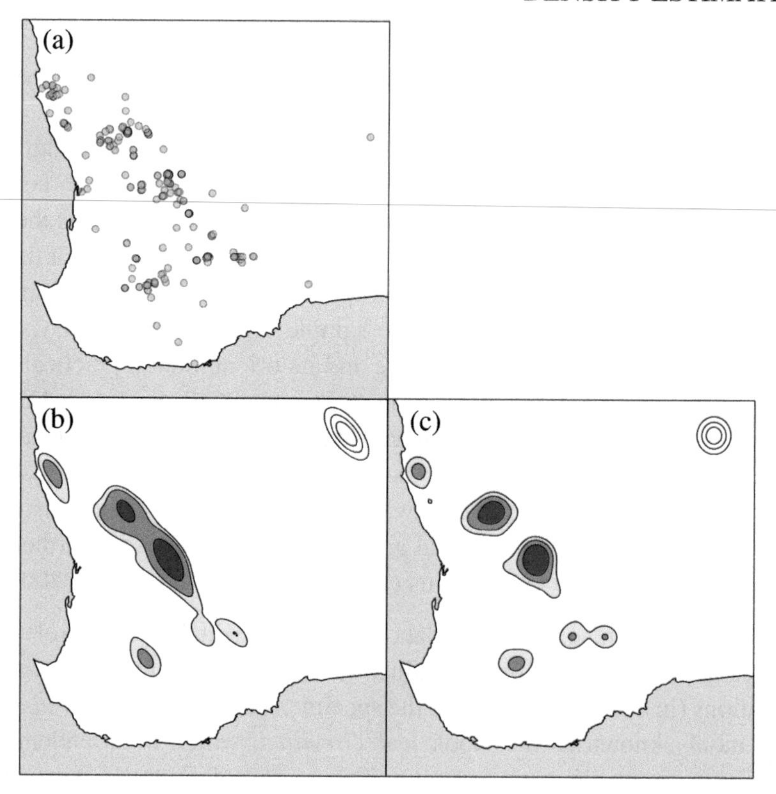


Figure 2.8 *(a) Scatter plot of the geographical locations (longitude, latitude) of the* $n = 222$ Grevillea uncinulata *specimens in south Western Australia. Observed locations are the green points, the ocean is blue, and the land mass is white. (b) Density estimate with an unconstrained bandwidth* $[0.058, -0.045; -0.045, 0.079]$. *(c) Density estimate with diagonal bandwidth* $\text{diag}(0.035, 0.040)$. *In the upper right corner in (b)–(c) is the kernel function with the given bandwidth.*

**Example 2.7** As the target density remains unknown for the experimental data example above, we analyse a synthetic example drawn from the dumbbell normal mixture $\frac{4}{11}N((-2,2),\mathbf{I}_2) + \frac{3}{11}N((0,0),[0.8,-0.72;-0.72,0.8]) + \frac{4}{11}N((2,-2),\mathbf{I}_2)$ introduced by Duong & Hazelton (2005*b*). Its contour plot is given in Figure 2.9(a). The overall mixture is unimodal with steep, oblique contours in the centre, with two shoulder regions which lead into more circular contours with a shallower gradient in the tails. For an $n = 1000$ random sample, density estimators were computed using an unconstrained bandwidth $[0.23, -0.19; -0.19, 0.23]$ and a diagonal bandwidth $\text{diag}(0.093, 0.089)$, shown in Figure 2.9(b)–(c). The unconstrained matrix gives a unimodal density estimate as it orients the kernels appropriately, and which also gives elliptical tail contours. The diagonal matrix gives more circular tail contours, but it is unable to apply appropriate smoothing in the oblique direction. This causes the upper shoulder region to be separated from the central mode, and thence the appearance of a multimodal estimate. □

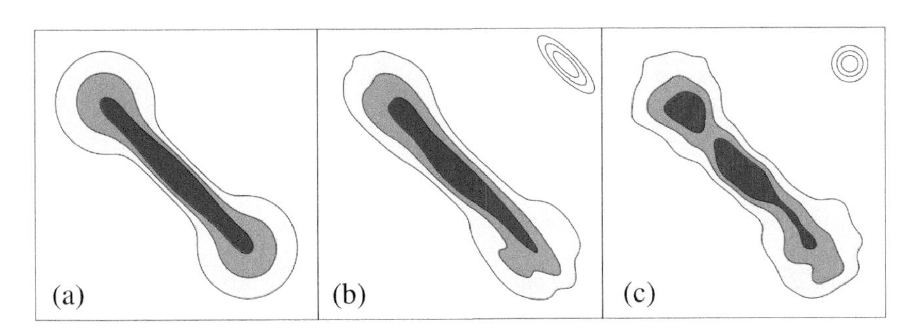


Figure 2.9 *Potential gains of an unconstrained over a diagonal bandwidth matrix for the dumbbell density. (a) Unimodal target dumbbell density. (b) Unimodal density estimate with unconstrained bandwidth* $[0.23, -0.19; -0.19, 0.23]$. *(c) Multimodal density estimate with diagonal bandwidth* $\text{diag}(0.093, 0.089)$. *For (b)–(c), in the upper right corner in each panel is the kernel function with the given bandwidth.*

There is much to gain from using unconstrained bandwidth matrices if there is an important probability mass oriented away from the coordinate axes. A number of studies (e.g., Wand & Jones, 1993, 1994; Duong & Hazelton, 2003; Chacón & Duong, 2010; Chacón et al., 2011; Chacón & Duong, 2011) confirm this good theoretical performance for a wide range of target density shapes and for a wide range of experimental data, including multivariate but non-spatial data. For this reason, we recommend the general use of unconstrained bandwidth matrices for density estimation.

## 2.4 Advice for practical bandwidth selection

For the *Grevillea* data, the key question is how the bandwidth [0.058, –0.045; –0.045, 0.079] is obtained. It is a trade-off between under- and oversmoothing in kernel estimators, analogous to that for histograms in Figure 2.2.

**Example 2.8** In Figure 2.10(a), a kernel density estimate with a bandwidth $\mathbf{H}^2$ is undersmoothed, whereas in Figure 2.10(b), a bandwidth of $\mathbf{H}^{1/2}$ leads to oversmoothing. Since the determinant of $\mathbf{H}$ is less than 1, then powers greater than 1 lead to less smoothing, and powers less than 1 lead to more smoothing. The bandwidth is the equivalent to the binwidth for histograms, so whilst kernel estimators resolve many of problems of associated with histograms, the key remaining one is the dependence on the smoothing parameters. □

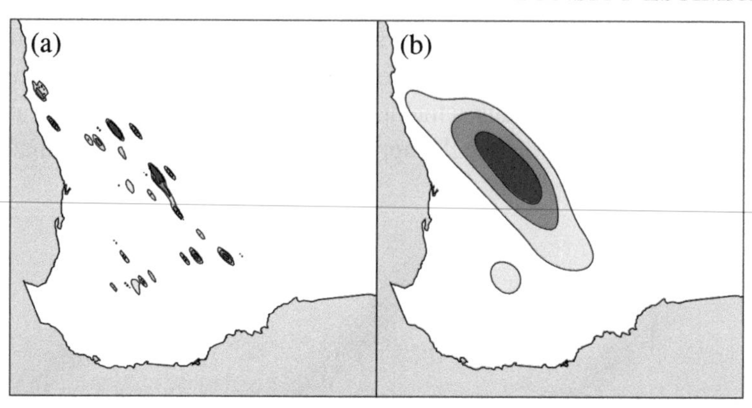


Figure 2.10 *Under- and oversmoothing of the density estimate of the* Grevillea *data. (a) Undersmoothed density estimate with bandwidth* $\mathbf{H}^2$. *(b) Oversmoothed density estimate with bandwidth* $\mathbf{H}^{1/2}$. *The base bandwidth is* $\mathbf{H} = [0.058, -0.045; -0.045, 0.079]$.

Selecting an optimal amount of smoothing can be cast into a more formal mathematical framework as an example of a bias-variance trade-off. A small bandwidth/undersmoothing gives low bias as the corresponding density estimate closely follows the data set at hand, but on a different data set would give a substantially different density estimate, thus presenting a high variability. A large bandwidth/oversmoothing has low variance as it tends to give the same features from different data sets, but where each density estimate would have high bias as it does not sufficiently take into account the data set at hand.

The question remains, which is/are the most suitable bandwidth(s) for the given data set? These will be exposited in detail in Chapter 3 after we have presented a sufficient mathematical framework, so we only briefly outline them here. The normal scale $\hat{\mathbf{H}}_{\text{NS}}$, maximal smoothing $\hat{\mathbf{H}}_{\text{MS}}$ and normal mixture $\hat{\mathbf{H}}_{\text{NM}}$ selectors refer to those bandwidths that minimise the squared estimation error of $\hat{f}$ for the cases in which the unknown density $f$ is replaced, respectively, by a single normal, a beta, and a normal mixture density with their parameters suitably estimated from the data. The main cross validation selectors are unbiased $\hat{\mathbf{H}}_{\text{UCV}}$, biased $\hat{\mathbf{H}}_{\text{BCV}}$ and smoothed $\hat{\mathbf{H}}_{\text{SCV}}$, which are variations of the usual leave-one-out cross validation approaches to estimate the squared estimation error of $\hat{f}$. The main competitor to cross validation is the plug-in selector $\hat{\mathbf{H}}_{\text{PI}}$, which is based on an asymptotic approximation of the squared estimation error.

**Example 2.9** In Figure 2.11 are the density estimates of the *Grevillea* data for these different bandwidth selectors (except for the maximal smoothing as it is usually similar to the normal scale selector).

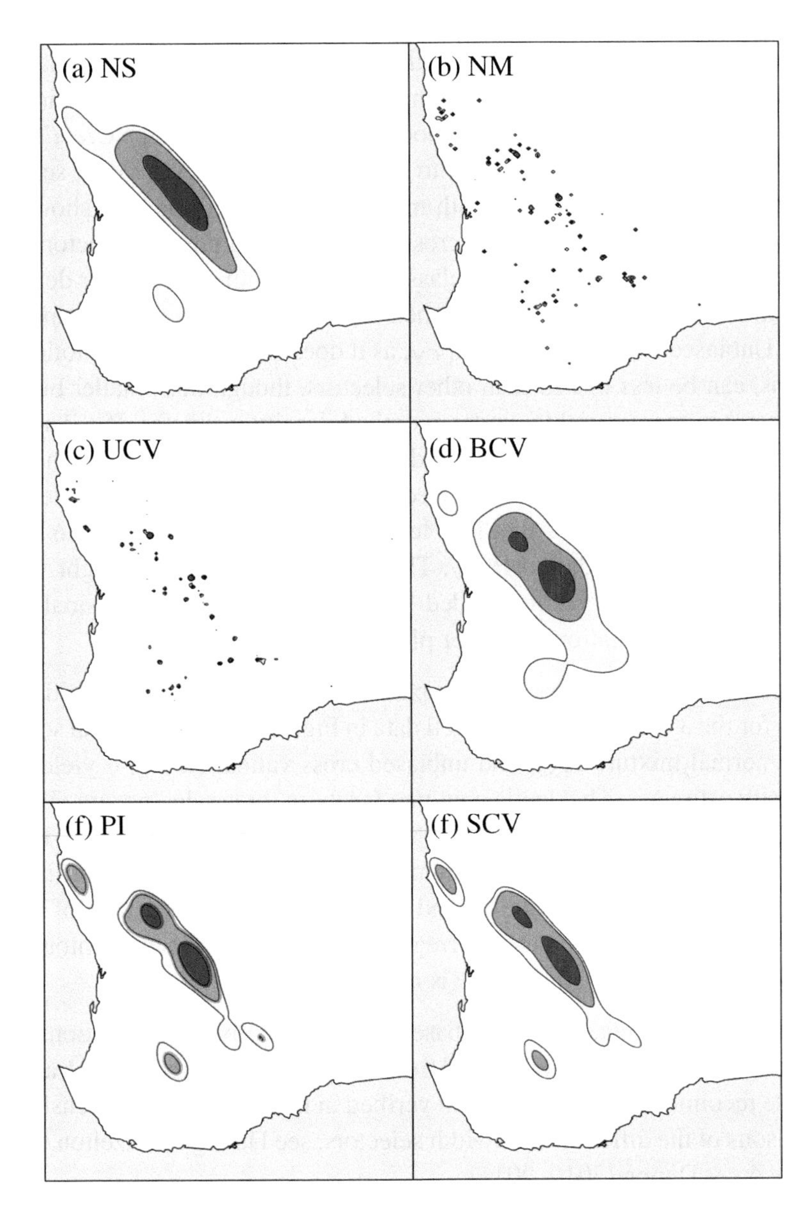


Figure 2.11 *Different bandwidth selectors for the density estimates of the* Grevillea *data. (a) Normal scale* $\hat{\mathbf{H}}_{\text{NS}}$. *(b) Normal mixture* $\hat{\mathbf{H}}_{\text{NM}}$. *(c) Unbiased cross validation* $\hat{\mathbf{H}}_{\text{UCV}}$. *(d) Biased cross validation* $\hat{\mathbf{H}}_{\text{BCV}}$. *(e) Plug-in* $\hat{\mathbf{H}}_{\text{PI}}$. *(f) Smoothed cross validation* $\hat{\mathbf{H}}_{\text{SCV}}$.

The normal scale $\hat{\mathbf{H}}_{\text{NS}}$ and maximal smoothing $\hat{\mathbf{H}}_{\text{MS}}$ selectors are efficient to compute but tend to produce oversmoothed density estimates, so they are useful for a quick visualisation of the overall trends in the data. The normal mixture selector $\hat{\mathbf{H}}_{\text{NM}}$ offers a richer class of parametric distributions than the normal scale and maximal smoothing selectors, but for the *Grevillea* data, the normal mixture fit is unable to appropriately model the data set and so gives an inappropriate bandwidth matrix. In general, it has not shown sufficient promise compared to the cross validation and plug-in selectors. These latter are the most widely used classes of bandwidth selectors for density estimation, due to their well-established theoretical and empirical properties.

Unbiased cross validation $\hat{\mathbf{H}}_{\text{UCV}}$, as it does not rely on asymptotic expansions, can be less biased than other selectors, though this smaller bias tends to result in more variable, undersmoothed density estimates. If a density has many isolated modes then this undersmoothing may be useful. The biased cross validation $\hat{\mathbf{H}}_{\text{BCV}}$ is not advised in general as its lack of an independent pilot bandwidth implies that it is less competitive than the plug-in $\hat{\mathbf{H}}_{\text{PI}}$ and smoothed cross validation $\hat{\mathbf{H}}_{\text{SCV}}$. These latter two (and their slight variants) are the most widely recommended bandwidth selectors, with a small advantage to the computationally faster plug-in methods. □

**Example 2.10** We repeat this comparison of these different bandwidth selectors for the 3-dimensional stem cell data in Figure 2.12. The normal scale $\hat{\mathbf{H}}_{\text{NS}}$ and normal mixture $\hat{\mathbf{H}}_{\text{NM}}$ and unbiased cross validation $\hat{\mathbf{H}}_{\text{UCV}}$ yield similar density estimates. The decile contours for these three selectors are ellipsoidal and evenly spaced, and indicating an oversmoothed estimate. The BCV selector has been omitted, since a computer implementation is not available for 3-dimensional data. For the PI and SCV selectors, more details of the data structure are visible due to the irregularly shaped and spaced contour shells, and because the upper left mode is more peaked. □

Whilst it is inadvisable to base these general recommendations on the sole analyses of the *Grevillea* and the stem cell data in Figures 2.11 and 2.12, these recommendations have been verified in numerous, comprehensive comparisons of the different bandwidth selectors, see Duong & Hazelton (2005*b*); Chacón & Duong (2010, 2011).

## 2.5 Squared error analysis

So far we have kept the mathematics as minimal as possible in our exposition, to make it an amenable introduction for a wider audience. In this section we start to address the most crucial factor for kernel estimation, i.e., the op-

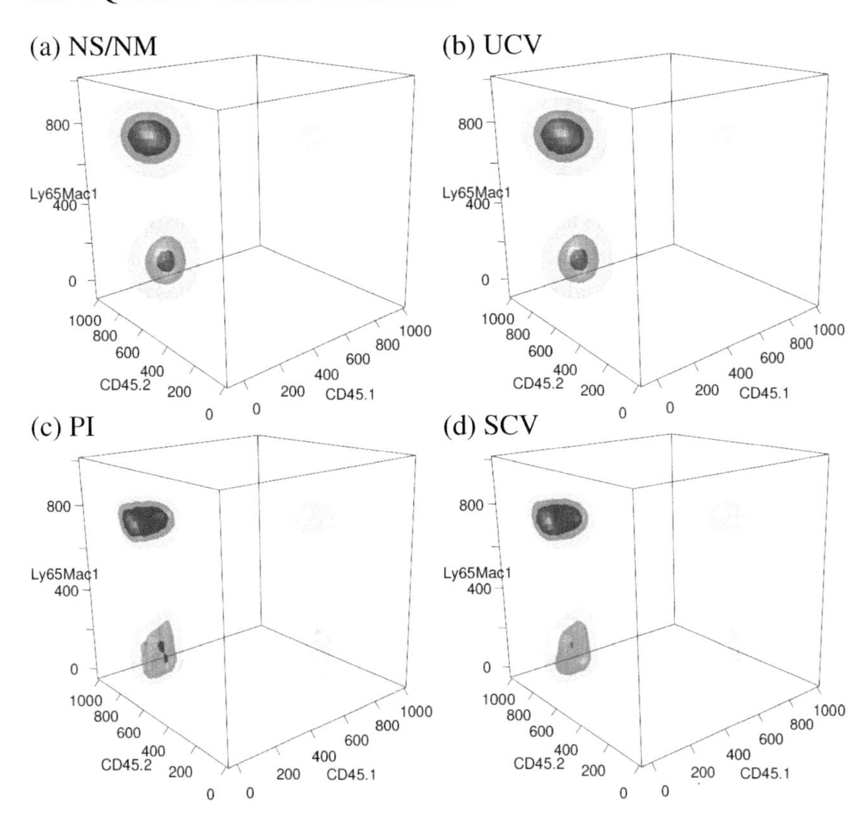


Figure 2.12 *Different bandwidth selectors for the density estimates of the stem cell data. (a) Normal scale* $\hat{\mathbf{H}}_{\rm NS}$*/normal mixture* $\hat{\mathbf{H}}_{\rm NM}$*. (b) Unbiased cross validation* $\hat{\mathbf{H}}_{\rm UCV}$*. (c) Plug-in* $\hat{\mathbf{H}}_{\rm PI}$*. (d) Smoothed cross validation* $\hat{\mathbf{H}}_{\rm SCV}$*.*

timal choice of smoothing for the class of unconstrained matrices. Despite the apparent complexity of this class, once suitable mathematical tools are employed its analysis is not more complicated than for the class of diagonal bandwidth matrices.

To establish a notion of an optimal bandwidth, we require a discrepancy measure. For density estimation at a fixed point $\boldsymbol{x}$, since the target density $f(\boldsymbol{x})$ is a real number, the most common way to measure the performance of the kernel density estimator $\hat{f}(\boldsymbol{x};\mathbf{H})$ is through the mean squared error (MSE), which allows for the variance plus squared bias decomposition

$$\text{MSE}\{\hat{f}(\boldsymbol{x};\mathbf{H})\} = \mathbb{E}\{[\hat{f}(\boldsymbol{x};\mathbf{H}) - f(\boldsymbol{x})]^2\} = \text{Var}\{\hat{f}(\boldsymbol{x};\mathbf{H})\} + \text{Bias}^2\{\hat{f}(\boldsymbol{x};\mathbf{H})\}$$

where

$$\text{Var}\{\hat{f}(\boldsymbol{x};\mathbf{H})\} = \mathbb{E}\left\{[\hat{f}(\boldsymbol{x};\mathbf{H}) - \mathbb{E}\{\hat{f}(\boldsymbol{x};\mathbf{H})\}]^2\right\}$$
$$\text{Bias}\{\hat{f}(\boldsymbol{x};\mathbf{H})\} = \mathbb{E}\{\hat{f}(\boldsymbol{x};\mathbf{H})\} - f(\boldsymbol{x}).$$

The MSE$\{\hat{f}(\boldsymbol{x};\mathbf{H})\}$ is a local, point-wise, discrepancy measure. In density estimation, the interest usually lies in the global behaviour of $\hat{f}$ as an estimator of $f$. To measure the global performance, a possibility is to integrate the MSE with respect to $\boldsymbol{x}$, to obtain (after an application of Fubini's theorem) the mean integrated squared error (MISE),

$$\text{MISE}\{\hat{f}(\cdot;\mathbf{H})\} = \mathbb{E}\int_{\mathbb{R}^d}\{\hat{f}(\boldsymbol{x};\mathbf{H}) - f(\boldsymbol{x})\}^2 d\boldsymbol{x}.$$

The MISE is the expected (squared) $L_2$ distance between $\hat{f}$ and $f$. The $L_2$ error is the most studied approach because it affords a simple mathematical treatment. Other $L_p$ distances are possible here, with some appealing reasons for using the $L_1$ distance (Devroye & Györfi, 1985).

The variance-squared bias decomposition of the MSE now leads to a decomposition of the MISE into integrated variance (IV) and integrated squared bias (ISB), namely

$$\text{MISE}\{\hat{f}(\cdot;\mathbf{H})\} = \text{IV}\{\hat{f}(\cdot;\mathbf{H})\} + \text{ISB}\{\hat{f}(\cdot;\mathbf{H})\}$$

where

$$\text{IV}\{\hat{f}(\cdot;\mathbf{H})\} = \int_{\mathbb{R}^d}\text{Var}\{\hat{f}(\boldsymbol{x};\mathbf{H})\}\,d\boldsymbol{x}$$
$$\text{ISB}\{\hat{f}(\cdot;\mathbf{H})\} = \int_{\mathbb{R}^d}\text{Bias}^2\{\hat{f}(\boldsymbol{x};\mathbf{H})\}\,d\boldsymbol{x}.$$

The MISE is a non-stochastic quantity that describes the performance of the kernel density estimator with respect to a typical sample from the true density. This is due to the expected value included in its definition. However, in some situations it is of interest to measure how the kernel estimator performs, not for an average sample, but for the data that we have at hand. This is the case, for instance, when the goal is to compare the kernel estimator based on different data-based methods to select the bandwidth matrix. To that end, a stochastic discrepancy measure depending on the data at hand is the integrated squared error (ISE), defined as

$$\text{ISE}\{\hat{f}(\cdot;\mathbf{H})\} = \int_{\mathbb{R}^d}\{\hat{f}(\boldsymbol{x};\mathbf{H}) - f(\boldsymbol{x})\}^2 d\boldsymbol{x}.$$

We emphasise that, whereas the MISE is a real number, the ISE is a random variable. Fundamental central limit theorems for the ISE were obtained by Hall (1984) while Jones (1991) explored the respective advantages and disadvantages of the MISE and ISE as the target discrepancy measure.

Convolutions play a key role in the analysis of kernel estimators. The convolution of two integrable functions $f$ and $g$ is a new function $f * g$, defined as $(f * g)(\boldsymbol{x}) = \int_{\mathbb{R}^d} f(\boldsymbol{x}-\boldsymbol{y})g(\boldsymbol{y})\,d\boldsymbol{y}$. If $\boldsymbol{X}$ and $\boldsymbol{Y}$ are independent random variables with densities $f$ and $g$, respectively, then $\boldsymbol{X}+\boldsymbol{Y}$ has density $f * g$. The notion of convolution of functions can be extended to the convolution of two probability distributions. A kernel density estimator can be considered to be the convolution of the probability measure induced by the scaled kernel $K_{\mathbf{H}}$ with the empirical distribution of the data. This is an alternative mathematical statement of the smoothing action of the kernel on the data, for it reveals the kernel density estimator as the density of the distribution resulting from perturbing the sample data points with independent random variables distributed according to $K_{\mathbf{H}}$.

Then, the expected value and the variance of the kernel density estimator can be written as

$$\begin{aligned} \mathbb{E}\{\hat{f}(\boldsymbol{x};\mathbf{H})\} &= (K_{\mathbf{H}} * f)(\boldsymbol{x}) \\ \operatorname{Var}\{\hat{f}(\boldsymbol{x};\mathbf{H})\} &= n^{-1}\{(K_{\mathbf{H}}^2 * f)(\boldsymbol{x}) - (K_{\mathbf{H}} * f)(\boldsymbol{x})^2\} \end{aligned} \tag{2.3}$$

as shown in the seminal papers of Rosenblatt (1956) and Parzen (1962). Here, $K_{\mathbf{H}}^2$ is short for $(K_{\mathbf{H}})^2$, that is, squaring takes place after scaling. Combining the two equalities in Equation (2.3) we obtain a more explicit formula for the MSE, namely

$$\operatorname{MSE}\{\hat{f}(\boldsymbol{x};\mathbf{H})\} = n^{-1}\{(K_{\mathbf{H}}^2 * f)(\boldsymbol{x}) - (K_{\mathbf{H}} * f)(\boldsymbol{x})^2\} + \{(K_{\mathbf{H}} * f)(\boldsymbol{x}) - f(\boldsymbol{x})\}^2.$$

Integrating over $\boldsymbol{x}$, it follows that

$$\begin{aligned} \operatorname{MISE}\{\hat{f}(\cdot;\mathbf{H})\} = &\{n^{-1}|\mathbf{H}|^{-1/2}R(K) - n^{-1}R(K_{\mathbf{H}} * f)\} \\ &+ \left\{R(K_{\mathbf{H}} * f) - 2\int_{\mathbb{R}^d}(K_{\mathbf{H}} * f)(\boldsymbol{x})f(\boldsymbol{x})\,d\boldsymbol{x} + R(f)\right\} \end{aligned} \tag{2.4}$$

where $R(a) = \int_{\mathbb{R}^d} a(\boldsymbol{x})^2\,d\boldsymbol{x}$ for a square integrable function $a\colon \mathbb{R}^d \to \mathbb{R}$, see Wand & Jones (1993). The first set of braces in Equation (2.4) contains the expression for the IV, the second the ISB.

Whilst Equation (2.4) is an exact closed form expression, its dependence on the bandwidth is not fully elucidated as the latter enters implicitly via the integrals involving the scaled kernel. In order to render the effect of the bandwidth more apparent, it is useful to derive an asymptotic approximation to the MISE, which is called the asymptotic mean integrated squared error (AMISE), and satisfies $\text{MISE}\{\hat{f}(\cdot;\mathbf{H})\} \sim \text{AMISE}\{\hat{f}(\cdot;\mathbf{H})\}$ as $n \to \infty$. Here, the tilde is used as a binary operator $a \sim b$ to indicate that $a = a_n$ and $b = b_n$ are asymptotically equivalent sequences, i.e., that $a_n/b_n \to 1$ as $n \to \infty$.

## 2.6 Asymptotic squared error formulas

We present the arguments leading to the AMISE formula. In order to make the exposition easier to follow for non-experts, and to hint at the tools involved, some details are treated here in a slightly heuristic way. For the interested reader, a rigorous mathematical proof is given in Section 2.9.

The main mathematical tool used to obtain the desired asymptotic approximation is Taylor's theorem. This is a well-known result, but we require a particular formulation that fits our needs as the infinitesimal entity involved in our multivariate expansions is not a real number or a vector, as is usual, but a positive definite matrix $\mathbf{H}$, either on its own or in the matrix product $\mathbf{H}^{1/2}\boldsymbol{z}$ for some fixed $\boldsymbol{z} \in \mathbb{R}^d$.

First we require a systematic way of organising the set of all the partial derivatives of a given order. Starting with the differential operator with respect to $\boldsymbol{x}$, which we denote as $\mathsf{D} = (\partial/\partial x_1, \ldots, \partial/\partial x_d)$, the $r$-th order differential operator $\mathsf{D}^{\otimes r}$ is defined as the formal $r$-fold Kronecker product of $\mathsf{D}$ with itself, using the convention that a multiplication of two first order differentials is understood as being a second order differential: $(\partial/\partial x_i)(\partial/\partial x_j) = \partial^2/(\partial x_i \partial x_j)$. For example, the second order differential operator, containing all the mixed partial derivatives of order two, can be expressed in a vector as $\mathsf{D} \otimes \mathsf{D}$, and this is related to the usual $d \times d$ Hessian matrix $\mathsf{H}$ via the identity $\mathsf{D} \otimes \mathsf{D} = \text{vec}\,\mathsf{H}$, where vec is the vectorization operator which acts on a matrix by stacking its columns on top of one another. So the two forms of derivatives contain exactly the same partial derivatives albeit arranged in a different way.

Moreover, this notation is consistent with the aforementioned multiplication convention, since the Hessian matrix can be expressed as $\mathsf{H} = \mathsf{D}\mathsf{D}^\top$ and one of the properties of the Kronecker product is that $\text{vec}(\boldsymbol{a}\boldsymbol{b}^\top) = \boldsymbol{b} \otimes \boldsymbol{a}$ for any $\boldsymbol{a}, \boldsymbol{b} \in \mathbb{R}^d$. See Appendix B for a brief summary of the properties of the vec operator, the Kronecker product and other special matrices, which we will reuse often in the sequel.

There does not appear (yet) to be much gain in rewriting the Hessian matrix in this way. In order to establish the theoretical properties of kernel estimators with unconstrained bandwidth matrices, the usual techniques (e.g., Wand, 1992; Wand & Jones, 1993, 1994) involved element-wise partial derivative analyses, which then required complex indexing operations to map the individual elements to its matrix form. These authors thus concentrated on diagonal bandwidth matrices in order to reduce the complexity of the calculations. This algebraic complexity explains, in large part, the lack of results for kernel estimators with unconstrained bandwidths. With vectorised derivatives, we are able to analyse the unconstrained case without requiring more complexity than the constrained cases.

Using these vectorised derivatives it is possible to write the Taylor expansion of an $r$-times continuously differentiable function $f$ at a point $\boldsymbol{x}+\boldsymbol{a}$ for a small perturbation $\boldsymbol{a}$ as

$$f(\boldsymbol{x}+\boldsymbol{a}) = \sum_{j=0}^{r} \frac{1}{j!} \mathsf{D}^{\otimes j} f(\boldsymbol{x})^\top \boldsymbol{a}^{\otimes j} + Re(\boldsymbol{a}), \tag{2.5}$$

where $\boldsymbol{x}, \boldsymbol{a} \in \mathbb{R}^d$ and the remainder $Re(\boldsymbol{a})$ is of order smaller than $\|\boldsymbol{a}\|^r$ as $\boldsymbol{a} \to 0$, which we express as $o(\|\boldsymbol{a}\|^r)$, meaning that $Re(\boldsymbol{a})/\|\boldsymbol{a}\|^r \to 0$ as $\boldsymbol{a} \to 0$. Here $\|\boldsymbol{a}\| = (\boldsymbol{a}^\top \boldsymbol{a})^{1/2}$ denotes the Euclidean norm of $\boldsymbol{a}$.

Finally, in order to use the Taylor expansions to find an asymptotic expression for the MISE we require the following assumptions:

**Conditions A**

**(A1)** The density function $f$ is square integrable and twice differentiable, with all of its second order partial derivatives bounded, continuous and square integrable.

**(A2)** The kernel $K$ is square integrable, spherically symmetric and with a finite second order moment; this means that $\int_{\mathbb{R}^d} \boldsymbol{z} K(\boldsymbol{z}) d\boldsymbol{z} = \boldsymbol{0}$ and that $\int_{\mathbb{R}^d} \boldsymbol{z}^{\otimes 2} K(\boldsymbol{z}) d\boldsymbol{z} = m_2(K) \operatorname{vec} \mathbf{I}_d$ with $m_2(K) = \int_{\mathbb{R}^d} z_i^2 K(\boldsymbol{z}) d\boldsymbol{z}$ for all $i = 1, \ldots, d$.

**(A3)** The bandwidth matrices $\mathbf{H} = \mathbf{H}_n$ form a sequence of positive definite, symmetric matrices such that $\operatorname{vec} \mathbf{H} \to 0$ and $n^{-1} |\mathbf{H}|^{-1/2} \to 0$ as $n \to \infty$.

Once these tools have been established, we start by developing an asymptotic form for the bias. After a change of variables, the expected value of $\hat{f}(\boldsymbol{x}; \mathbf{H})$ can be written as

$$\mathbb{E}\{\hat{f}(\boldsymbol{x}; \mathbf{H})\} = \int_{\mathbb{R}^d} K_{\mathbf{H}}(\boldsymbol{x}-\boldsymbol{y}) f(\boldsymbol{y}) d\boldsymbol{y} = \int_{\mathbb{R}^d} K(\boldsymbol{z}) f(\boldsymbol{x} - \mathbf{H}^{1/2}\boldsymbol{z}) d\boldsymbol{z}. \tag{2.6}$$

What is required to proceed is a Taylor expansion of $f(\boldsymbol{x} - \mathbf{H}^{1/2}\boldsymbol{z})$ around $f(\boldsymbol{x})$. By applying Equation (2.5) we obtain

$$f(\boldsymbol{x} - \mathbf{H}^{1/2}\boldsymbol{z}) = f(\boldsymbol{x}) - \mathsf{D} f(\boldsymbol{x})^\top \mathbf{H}^{1/2} \boldsymbol{z} + \tfrac{1}{2} \mathsf{D}^{\otimes 2} f(\boldsymbol{x})^\top (\mathbf{H}^{1/2}\boldsymbol{z})^{\otimes 2} + o(\|\operatorname{vec} \mathbf{H}\|).$$

Noting that $(\mathbf{H}^{1/2}\boldsymbol{z})^{\otimes 2} = (\mathbf{H}^{1/2})^{\otimes 2}\boldsymbol{z}^{\otimes 2}$ and taking into account the condition (A2) on $K$, substituting this expansion of $f(\boldsymbol{x}-\mathbf{H}^{1/2}\boldsymbol{z})$ into Equation (2.6) yields

$$\begin{aligned}\mathbb{E}\{\hat{f}(\boldsymbol{x};\mathbf{H})\} &= f(\boldsymbol{x}) + \tfrac{1}{2}\mathsf{D}^{\otimes 2}f(\boldsymbol{x})^\top(\mathbf{H}^{1/2})^{\otimes 2}m_2(K)\operatorname{vec}\mathbf{I}_d + o(\|\operatorname{vec}\mathbf{H}\|)\\ &= f(\boldsymbol{x}) + \tfrac{1}{2}m_2(K)\mathsf{D}^{\otimes 2}f(\boldsymbol{x})^\top\operatorname{vec}\mathbf{H} + o(\|\operatorname{vec}\mathbf{H}\|),\end{aligned}$$

where the formula $\operatorname{vec}(\mathbf{ABC}) = (\mathbf{C}^\top \otimes \mathbf{A})\operatorname{vec}\mathbf{B}$, for conformable matrices $\mathbf{A},\mathbf{B},\mathbf{C}$, was used for the second equality (see Appendix B). From this, squaring and integrating with respect to $\boldsymbol{x}$ we obtain

$$\operatorname{ISB}\{\hat{f}(\cdot;\mathbf{H})\} = \tfrac{1}{4}m_2(K)^2\{\operatorname{vec}^\top\mathbf{R}(\mathsf{D}^{\otimes 2}f)\}(\operatorname{vec}\mathbf{H})^{\otimes 2} + o(\|\operatorname{vec}\mathbf{H}\|^2) \quad (2.7)$$

where, for an arbitrary vector-valued function $\boldsymbol{a}\colon \mathbb{R}^d \to \mathbb{R}^p$, we denote $\mathbf{R}(\boldsymbol{a}) = \int_{\mathbb{R}^d}\boldsymbol{a}(\boldsymbol{x})\boldsymbol{a}(\boldsymbol{x})^\top d\boldsymbol{x} \in \mathcal{M}_{p\times p}$, that is,

$$\mathbf{R}(\mathsf{D}^{\otimes 2}f) = \int_{\mathbb{R}^d}\mathsf{D}^{\otimes 2}f(\boldsymbol{x})\mathsf{D}^{\otimes 2}f(\boldsymbol{x})^\top d\boldsymbol{x} \in \mathcal{M}_{d^2\times d^2}.$$

Equation (2.7) presents the dominant term of the ISB as a (Kronecker) square in $\operatorname{vec}\mathbf{H}$. Using again the properties of the vec operator and the Kronecker product it is possible to write this term in alternative forms, like

$$\begin{aligned}\tfrac{1}{4}m_2(K)^2\{\operatorname{vec}^\top\mathbf{R}(\mathsf{D}^{\otimes 2}f)\}(\operatorname{vec}\mathbf{H})^{\otimes 2} &= \tfrac{1}{4}m_2(K)^2(\operatorname{vec}^\top\mathbf{H})\mathbf{R}(\mathsf{D}^{\otimes 2}f)\operatorname{vec}\mathbf{H}\\ &= \tfrac{1}{4}m_2(K)^2\int_{\mathbb{R}^d}\operatorname{tr}^2\{\mathbf{H}\mathsf{H}f(\boldsymbol{x})\}\,d\boldsymbol{x}\end{aligned}$$

(see Chacón & Duong, 2010). The first equality shows the nature of the ISB as a quadratic form in $\operatorname{vec}\mathbf{H}$, and the second equality gives the expression provided for the first time in Wand (1992).

Regarding the IV, starting from Equation (2.3) and integrating with respect to $\boldsymbol{x}$ we have, for the first term,

$$\begin{aligned}n^{-1}\int_{\mathbb{R}^d}(K_{\mathbf{H}}^2 * f)(\boldsymbol{x})\,d\boldsymbol{x} &= n^{-1}\int_{\mathbb{R}^d}\int_{\mathbb{R}^d}K_{\mathbf{H}}(\boldsymbol{x}-\boldsymbol{y})^2 f(\boldsymbol{y})\,d\boldsymbol{y}d\boldsymbol{x}\\ &= n^{-1}|\mathbf{H}|^{-1/2}\int_{\mathbb{R}^d}\int_{\mathbb{R}^d}K(\boldsymbol{z})^2 f(\boldsymbol{x}-\mathbf{H}^{1/2}\boldsymbol{z})\,d\boldsymbol{z}d\boldsymbol{x}\\ &= n^{-1}|\mathbf{H}|^{-1/2}R(K) \qquad (2.8)\end{aligned}$$

where the second equality follows from the change of variables $\boldsymbol{z} = \mathbf{H}^{-1/2}(\boldsymbol{x}-\boldsymbol{y})$ and the third one from Fubini's theorem. For the second term in Equation (2.3) we can take advantage of the previous calculations for $\mathbb{E}\{\hat{f}(\boldsymbol{x};\mathbf{H})\}$

to immediately obtain $n^{-1}\int_{\mathbb{R}^d}(K_{\mathbf{H}}*f)(\boldsymbol{x})^2 d\boldsymbol{x} = n^{-1}R(f) + o(n^{-1})$. Since our assumption $\operatorname{vec}\mathbf{H} \to 0$ implies $|\mathbf{H}| \to 0$ by Hadamard's inequality (see Magnus & Neudecker, 1999, Chapter 11), in view of Equation (2.8) it follows that this second term in the IV is of a smaller order than the first one. Therefore,

$$\mathrm{IV}\{\hat{f}(\cdot;\mathbf{H})\} = n^{-1}|\mathbf{H}|^{-1/2}R(K) + o(n^{-1}|\mathbf{H}|^{-1/2}). \tag{2.9}$$

Combining Equations (2.7) and (2.9), it follows that an asymptotic approximation to the MISE can be written as

$$\begin{aligned}\mathrm{AMISE}\{\hat{f}(\cdot;\mathbf{H})\} &= n^{-1}|\mathbf{H}|^{-1/2}R(K) + \tfrac{1}{4}m_2(K)^2\{\operatorname{vec}^\top \mathbf{R}(\mathsf{D}^{\otimes 2}f)\}(\operatorname{vec}\mathbf{H})^{\otimes 2}\\ &= n^{-1}|\mathbf{H}|^{-1/2}R(K) + \tfrac{1}{4}m_2(K)^2(\operatorname{vec}^\top \mathbf{H})\mathbf{R}(\mathsf{D}^{\otimes 2}f)\operatorname{vec}\mathbf{H}\\ &= n^{-1}|\mathbf{H}|^{-1/2}R(K) + \tfrac{1}{4}m_2(K)^2\int_{\mathbb{R}^d}\operatorname{tr}^2\{\mathbf{H}\mathsf{H}f(\boldsymbol{x})\}\,d\boldsymbol{x}.\end{aligned} \tag{2.10}$$

Comparing the AMISE to the MISE in Equation (2.4), we note that the leading term $n^{-1}|\mathbf{H}|^{-1/2}R(K)$ of the IV is virtually indistinguishable asymptotically from its exact form $n^{-1}|\mathbf{H}|^{-1/2}R(K) - n^{-1}R(K_{\mathbf{H}}*f)$ for any density $f$. This excellent approximation is difficult to improve on, and there is no effect on bandwidth selection, at least in an asymptotic sense, in changing the exact variance for its leading term (see Chacón & Duong, 2011, Theorem 1). In contrast, the asymptotic ISB is less uniformly accurate and its accuracy depends on the structure of $f$. This implies that the methods which we subsequently explore rely on refining this ISB approximation.

In deriving Equation (2.10), only the condition $\operatorname{vec}\mathbf{H} \to 0$ on the sequence of bandwidth matrices was needed. If we additionally impose $n^{-1}|\mathbf{H}|^{-1/2} \to 0$ we obtain that $\mathrm{AMISE}\{\hat{f}(\cdot;\mathbf{H})\} \to 0$. These two conditions on the bandwidth are usually assumed to guarantee consistency of the kernel density estimator. In fact, reasoning as in Theorem II.2 in Chapter 4 of Bosq & Lecoutre (1987), it can be shown that condition (A3) on the bandwidth is necessary and sufficient so that the MISE converges to zero for any density, with the sole assumption that $f$ is square integrable. And even this assumption on $f$ is not necessary to ensure consistency in the $L_1$ context (Devroye, 1983).

In comparison to the exact MISE in Equation (2.4), in the AMISE, the variance-bias trade-off is more apparent. When the bandwidth has small entries, the bias tends to be small as well, whilst the variance is inflated. When the bandwidth is large, in the sense of having a large determinant, the bias tends also to be large (again due to Hadamard's inequality) with a correspondingly diminishing variance. This mathematically expresses the effect of the bandwidth on the kernel density estimator in the empirically observed under- and oversmoothing in Figure 2.11.

Equation (2.10) represents a general formulation for the asymptotic approximation of the MISE, since it is valid with unconstrained bandwidths and for an arbitrary dimension. By reducing it to specific cases it is possible to recover some of the more usual AMISE formulas. In the univariate case with $\mathbf{H} = h^2$, Equation (2.10) gives the more familiar expression

$$\mathrm{AMISE}\{\hat{f}(\cdot;h)\} = n^{-1}h^{-1}R(K) + \tfrac{1}{4}m_2(K)^2R(f'')h^4$$

(Rosenblatt, 1956, 1971). For comparison, the role that $R(f'') = \int_{\mathbb{R}} f''(x)^2dx$ traditionally plays as a measure of the roughness/curvature of the density is now played in the multivariate case by the (symmetric) curvature matrix $\mathbf{R}(\mathsf{D}^{\otimes 2}f) \in \mathcal{M}_{d^2\times d^2}$, which includes as its entries all the terms of the form

$$\psi_{ij,k\ell} = \int_{\mathbb{R}^d} \frac{\partial^2 f(\boldsymbol{x})}{\partial x_i \partial x_j} \frac{\partial^2 f(\boldsymbol{x})}{\partial x_k \partial x_\ell} d\boldsymbol{x}$$

for all possible choices of $i, j, k, \ell \in \{1, 2, \ldots, d\}$.

In the bivariate case, an explicit formula for $\mathbf{R}(\mathsf{D}^{\otimes 2}f)$ is

$$\mathbf{R}(\mathsf{D}^{\otimes 2}f) = \left[\begin{array}{c|c} \mathbf{R}_{11} & \mathbf{R}_{12} \\ \hline \mathbf{R}_{21} & \mathbf{R}_{22} \end{array}\right] = \left[\begin{array}{cc|cc} \psi_{11,11} & \psi_{11,12} & \psi_{11,21} & \psi_{11,22} \\ \psi_{12,11} & \psi_{12,12} & \psi_{12,21} & \psi_{12,22} \\ \hline \psi_{21,11} & \psi_{21,12} & \psi_{21,21} & \psi_{21,22} \\ \psi_{22,11} & \psi_{22,12} & \psi_{22,21} & \psi_{22,22} \end{array}\right] \in \mathcal{M}_{2^2\times 2^2}$$

where each block $\mathbf{R}_{ij}$ is a $2\times 2$ matrix, given by

$$\mathbf{R}_{ij} = \begin{bmatrix} \psi_{i1,j1} & \psi_{i1,j2} \\ \psi_{i2,j1} & \psi_{i2,j2} \end{bmatrix} \in \mathcal{M}_{2\times 2}.$$

More generally, it can be shown that $\mathbf{R}(\mathsf{D}^{\otimes 2}f)$ can be written using $d\times d$ blocks, of dimension $d \times d$ each, where the $(i,j)$th block is the matrix $\mathbf{R}_{ij} = [\psi_{ik,j\ell}]_{k,\ell=1}^d \in \mathcal{M}_{d\times d}$.

Still in the single-parameter case where $\mathbf{H} = h^2\mathbf{I}_d \in \mathcal{A}$, but in the multivariate context, observe that $(\mathsf{D}\otimes\mathsf{D})^\top \operatorname{vec}\mathbf{I}_d = \mathsf{D}^\top\mathsf{D} = \operatorname{tr}\mathsf{H} \equiv \triangle$, with $\triangle = \sum_{i=1}^d \partial^2/(\partial x_i^2)$ denoting the Laplacian operator, i.e., the trace of the Hessian operator. Then, the AMISE formula in Equation (2.10) simplifies to

$$\mathrm{AMISE}\{\hat{f}(\cdot;h)\} = n^{-1}h^{-d}R(K) + \tfrac{1}{4}m_2(K)^2R(\triangle f)h^4. \tag{2.11}$$

So the effect of using a constrained $\mathbf{H} \in \mathcal{A}$, as opposed to an unconstrained one, is mainly reflected in the ISB. Since $\operatorname{tr}\{(\mathsf{D}\otimes\mathsf{D})(\mathsf{D}\otimes\mathsf{D})^\top\} = (\mathsf{D}^\top\mathsf{D})^2 = \triangle^2$, it follows that $R(\triangle f) = \operatorname{tr}\mathbf{R}(\mathsf{D}^{\otimes 2}f)$. Instead of taking into account the full curvature matrix $\mathbf{R}(\mathsf{D}^{\otimes 2}f)$, only its trace contributes to the ISB.

For a diagonal bandwidth $\mathbf{H} = \operatorname{diag}(h_1^2,\dots,h_d^2) \in \mathcal{D}$, writing $\boldsymbol{h} = (h_1^2,\dots,h_d^2) \in \mathbb{R}^d$ and reasoning as above, it is straightforward to verify that

$$\operatorname{AMISE}\{\hat{f}(\cdot;\boldsymbol{h})\} = (nh_1\cdots h_d)^{-1}R(K) + \tfrac{1}{4}m_2(K)^2\{\operatorname{diag}\mathbf{R}(\mathsf{D}^{\otimes 2}f)\}^\top\boldsymbol{h}^{\otimes 2} \tag{2.12}$$

where $\operatorname{diag}\mathbf{R}(\mathsf{D}^{\otimes 2}f) \in \mathbb{R}^{d^2}$ is the vector comprising the elements from the main diagonal of the curvature matrix $\mathbf{R}(\mathsf{D}^{\otimes 2}f)$. Using a diagonal bandwidth matrix implies that, asymptotically, only the diagonal of the curvature matrix (as opposed to the full curvature matrix) influences the ISB of the kernel density estimator.

## 2.7 Optimal bandwidths

The optimal bandwidth is often defined as the minimiser of the MISE, i.e., $\mathbf{H}_{\text{MISE}} = \operatorname{argmin}_{\mathbf{H}\in\mathcal{F}} \operatorname{MISE}\{\hat{f}(\cdot;\mathbf{H})\}$, where the minimisation is taken over the class $\mathcal{F}$. As the MISE is complicated to compute due to the presence of the integrals and convolutions, a simpler proxy would be $\mathbf{H}_{\text{AMISE}} = \operatorname{argmin}_{\mathbf{H}\in\mathcal{F}} \operatorname{AMISE}\{\hat{f}(\cdot;\mathbf{H})\}$ as an asymptotically optimal target bandwidth, since it can be shown that $\mathbf{H}_{\text{MISE}}$ and $\mathbf{H}_{\text{AMISE}}$ are asymptotically equivalent as $n \to \infty$. In any case, it is worth pointing out that these optimal bandwidths depend on the unknown density $f$, as they are defined in terms of the true estimation error, and therefore it is not possible to compute them from the data. This is why they are also known as oracle bandwidths.

If the optimal bandwidth is sought among the class $\mathcal{A}$, then a further advantage of the asymptotic surrogate is that it has an explicit form, since the AMISE approximation in Equation (2.11) is minimised at

$$h_{\text{AMISE}} = \left[dR(K)/\{m_2(K)^2R(\triangle f)\}\right]^{1/(d+4)} n^{-1/(d+4)}.$$

If a diagonal bandwidth $\mathbf{H} = \operatorname{diag}(\boldsymbol{h})$ is employed, where $\boldsymbol{h} = (h_1^2,\dots,h_d^2) \in \mathbb{R}^d$, then Wand & Jones (1994) showed that in the multivariate case an explicit form for the optimal $\boldsymbol{h}_{\text{AMISE}}$ exists only for $d = 2$, where

$$\begin{aligned}\boldsymbol{h}_{\text{AMISE}} = {} & \left(\{\psi_{22,22}/\psi_{11,11}\}^{1/8}, \{\psi_{11,11}/\psi_{22,22}\}^{1/8}\right) \\ & \times \left[R(K)/\{m_2(K)^2(\psi_{11,11}^{1/2}\psi_{22,22}^{1/2} + \psi_{11,22})\}\right]^{1/6} n^{-1/6}.\end{aligned}$$

Unfortunately, there exists no explicit formula for the unconstrained asymptotic oracle bandwidth $\mathbf{H}_{\text{AMISE}}$ either. However, it can be shown (see Section 2.9.2 below) that $\mathbf{H}_{\text{AMISE}}$ is also of order $n^{-2/(d+4)}$, as its constrained counterparts.

## 2.8 Convergence of density estimators

From a data analysis point of view, the closeness of the density estimate $\hat{f}$ to the target density $f$ is an important feature to take into account. It is natural then to consider the rate at which $\text{MISE}\{\hat{f}(\cdot;\mathbf{H})\}$ tends to zero. This MISE convergence implies that the density estimator has various desirable properties, among which the most important are that $\hat{f}$ is asymptotically unbiased and is consistent. This assures that, at least in the limiting case, kernel density estimators perform well.

Under (A1)–(A3) in Conditions A, it has been asserted above that the optimal bandwidth is of order $n^{-2/(d+4)}$. This implies that the minimal MISE rate is $\inf_{\mathbf{H}\in\mathcal{F}} \text{MISE}\{\hat{f}(\cdot;\mathbf{H})\} = O(n^{-4/(d+4)})$. In comparison, the minimal MISE rate of convergence for a histogram density estimator $\hat{f}_{\text{hist}}$ is $O(n^{-2/(d+2)})$, which is uniformly slower than the $n^{-4/(d+4)}$ rate for a kernel density estimator for all $d$, see Simonoff (1996, p. 97, Equation (4.2)). This is a formal statement of the asymptotic improvement of the multivariate kernel estimators over histograms. The discreteness of the latter, apart from providing a less aesthetically pleasing visualisation, is also the main source of the slowness in their convergence to the target density in comparison to the former.

This $O(n^{-4/(d+4)})$ rate for kernel density estimators is uniformly slower than the parametric rate $O(n^{-1})$ for all $d$, implying that non-parametric kernel estimation is a more difficult problem than its parametric counterpart and hence (i) requires larger sample sizes to achieve the same accuracy or (ii) for the same sample size, achieves a lower level of accuracy. The fast parametric rate of convergence is valid only if a correctly specified parametric form is used. In the case of a misspecified parametric form, the rate is not necessarily faster than for non-parametric estimators, and the former can even be shown to be inconsistent under certain misspecifications. One of the major advantages of the non-parametric approach is its consistency under minimal assumptions.

The $O(n^{-4/(d+4)})$ rate implies that as the dimension $d$ increases, kernel estimation becomes increasingly difficult. This is usually referred to as the *curse of dimensionality*, a term originally coined by Bellman (1961). The problem is not that the rate becomes slower for higher dimensions, because this can be avoided by using superkernels or infinite-order kernels (see Politis & Romano, 1999, Remark 5). The intrinsic difficulty in high dimensions relies on the fact that kernel estimators are based on local neighbourhoods, where the extent of such neighbourhoods is determined by the fixed bandwidth matrix. The sparsity of data in higher dimensions implies that increasingly larger neighbourhoods are required to include a substantial enough effective sample size to produce a reliable non-parametric density estimate.

Nevertheless, kernel smoothing plays an important role for low-to-moderate dimensional data. Examples include spatial analysis for microscopy images (2D/3D spatial + 1 grey level) and the analysis of geographical trends/clusters/filaments, where the data is inherently low dimensional, and where increasing the number of dimensions does not necessarily lead to better statistical analysis since it implies that visualisation is then no longer easy to interpret. Indeed, the visual component of kernel smoothing for exploratory data analysis is crucial for many non-mathematician users.

Moreover, in some current advanced applications of kernel smoothing (which are described in Chapters 6–7), the main end goal is not to obtain a precise estimation of the entire density function, but to use it as an intermediate tool to discover the regions of interest, e.g. modal regions, density ridges and clusters. In these cases, kernel smoothers have been demonstrated to be useful for high-dimensional data, and even for functional/infinite-dimensional data, e.g., Ciollaro et al. (2016).

## 2.9 Further mathematical analysis of density estimators

### 2.9.1 *Asymptotic expansion of the mean integrated squared error*

We provide a formal technical proof for the asymptotic expansion of the MISE. The ‘heuristic’ component in the presentation provided in Section 2.6 concerns mainly the issue of why the smaller order terms in the pointwise expansions can be discarded after several integration steps, and this is treated in a more careful and rigorous way here. All the asymptotic arguments in this monograph could be formalised in a similar way, but we only give the details for the base case of density estimation.

**Theorem 1** *Suppose that (A1)–(A3) in Conditions A hold.*
*(i) The integrated squared bias of the kernel density estimator can be expanded as*

$$\mathrm{ISB}\{\hat{f}(\cdot;\mathbf{H})\} = \tfrac{1}{4}m_2(K)^2\{\mathrm{vec}^\top \mathbf{R}(\mathsf{D}^{\otimes 2}f)\}(\mathrm{vec}\,\mathbf{H})^{\otimes 2} + o(\|\mathrm{vec}\,\mathbf{H}\|^2).$$

*(ii) The integrated variance of the kernel density estimator can be expanded as*

$$\mathrm{IV}\{\hat{f}(\cdot;\mathbf{H})\} = n^{-1}|\mathbf{H}|^{-1/2}R(K) + o(n^{-1}|\mathbf{H}|^{-1/2}).$$

**Proof** Since $K$ is symmetric, a Taylor expansion with the remainder in integral form gives

$$\begin{aligned}
&\mathbb{E}\{\hat{f}(\boldsymbol{x};\mathbf{H})\} - f(\boldsymbol{x}) \\
&\quad = \int_{\mathbb{R}^d}\int_0^1 (1-t)\mathsf{D}^{\otimes 2} f(\boldsymbol{x} - t\mathbf{H}^{1/2}\boldsymbol{z})^\top (\mathbf{H}^{1/2})^{\otimes 2}\boldsymbol{z}^{\otimes 2} K(\boldsymbol{z})\, dt d\boldsymbol{z} \\
&\quad = \boldsymbol{u}(\boldsymbol{x};\mathbf{H})^\top \operatorname{vec}(\mathbf{H}^{1/2}\otimes\mathbf{H}^{1/2}),
\end{aligned}$$

where $\boldsymbol{u}(\boldsymbol{x};\mathbf{H}) = \int_{\mathbb{R}^d}\int_0^1 (1-t)K(\boldsymbol{z})\{\boldsymbol{z}^{\otimes 2}\otimes \mathsf{D}^{\otimes 2} f(\boldsymbol{x} - t\mathbf{H}^{1/2}\boldsymbol{z})\}\, dt d\boldsymbol{z} \in \mathbb{R}^{d^4}$. We have used $\operatorname{vec}(\mathbf{ABC}) = (\mathbf{C}^\top\otimes\mathbf{A})\operatorname{vec}\mathbf{B}$ in the previous display.

For every fixed $\boldsymbol{x}\in\mathbb{R}^d$, using (A1) we have

$$\left\|(1-t)K(\boldsymbol{z})\{\boldsymbol{z}^{\otimes 2}\otimes\mathsf{D}^{\otimes 2} f(\boldsymbol{x} - t\mathbf{H}^{1/2}\boldsymbol{z})\}\right\| \le (1-t)\|\boldsymbol{z}\|^2|K(\boldsymbol{z})| \sup_{\boldsymbol{x}\in\mathbb{R}^d}\|\mathsf{D}^{\otimes 2} f(\boldsymbol{x})\|$$

with $\int_{\mathbb{R}^d}\int_0^1 (1-t)\|\boldsymbol{z}\|^2|K(\boldsymbol{z})| dt d\boldsymbol{z} < \infty$ by (A2). By the Dominated Convergence Theorem it follows that $\boldsymbol{u}(\boldsymbol{x};\mathbf{H}) \to \frac{1}{2}m_2(K)\operatorname{vec}\mathbf{I}_d\otimes\mathsf{D}^{\otimes 2} f(\boldsymbol{x})$ as $\operatorname{vec}\mathbf{H}\to 0$.

Moreover, we can write $\mathrm{ISB}\{\hat{f}(\cdot;\mathbf{H})\} = \boldsymbol{w}(\mathbf{H})^\top\left\{\operatorname{vec}\left(\mathbf{H}^{1/2}\otimes\mathbf{H}^{1/2}\right)\right\}^{\otimes 2}$ with $\boldsymbol{w}(\mathbf{H}) = \int_{\mathbb{R}^d}\boldsymbol{u}(\boldsymbol{x};\mathbf{H})^{\otimes 2} d\boldsymbol{x} \in \mathbb{R}^{d^8}$. Now, the Cauchy-Schwarz inequality yields, for every $\boldsymbol{x}\in\mathbb{R}^d$,

$$\begin{aligned}
\|\boldsymbol{u}(\boldsymbol{x};\mathbf{H})^{\otimes 2}\| &= \|\boldsymbol{u}(\boldsymbol{x};\mathbf{H})\|^2 \\
&\le \left\{\int_{\mathbb{R}^d}\int_0^1 (1-t)\|\boldsymbol{z}\|^2|K(\boldsymbol{z})|\, dt d\boldsymbol{z}\right\} \\
&\quad\times\left\{\int_{\mathbb{R}^d}\int_0^1 (1-t)\|\boldsymbol{z}\|^2|K(\boldsymbol{z})|\left\|\mathsf{D}^{\otimes 2} f(\boldsymbol{x} - t\mathbf{H}^{1/2}\boldsymbol{z})\right\|^2 dt d\boldsymbol{z}\right\}.
\end{aligned}$$

The function on the right-hand side of the previous display, when considered as a function of $\boldsymbol{x}$, has the integral

$$\frac{1}{4}\left\{\int_{\mathbb{R}^d}\|\boldsymbol{z}\|^2|K(\boldsymbol{z})| d\boldsymbol{z}\right\}^2 \int_{\mathbb{R}^d}\left\|\mathsf{D}^{\otimes 2} f(\boldsymbol{x})\right\|^2 d\boldsymbol{x},$$

which is finite by (A1)–(A2). Again by an application of the Dominated Convergence Theorem, we obtain that $\boldsymbol{w}(\mathbf{H}) \to \boldsymbol{w}_0 = \frac{1}{4}m_2(K)^2\int_{\mathbb{R}^d}\left\{\operatorname{vec}\mathbf{I}_d\otimes\mathsf{D}^{\otimes 2} f(\boldsymbol{x})\right\}^{\otimes 2} d\boldsymbol{x}$ as $\operatorname{vec}\mathbf{H}\to 0$. Since

$$\begin{aligned}
&\left\{\operatorname{vec}^\top\mathbf{I}_d\otimes\mathsf{D}^{\otimes 2} f(\boldsymbol{x})^\top\right\}^{\otimes 2}\left\{\operatorname{vec}\left(\mathbf{H}^{1/2}\otimes\mathbf{H}^{1/2}\right)\right\}^{\otimes 2} \\
&\quad = \left[\left\{\operatorname{vec}^\top\mathbf{I}_d\otimes\mathsf{D}^{\otimes 2} f(\boldsymbol{x})^\top\right\}\operatorname{vec}\left(\mathbf{H}^{1/2}\otimes\mathbf{H}^{1/2}\right)\right]^{\otimes 2} \\
&\quad = \left[\operatorname{vec}\left\{\mathsf{D}^{\otimes 2} f(\boldsymbol{x})^\top\left(\mathbf{H}^{1/2}\otimes\mathbf{H}^{1/2}\right)\operatorname{vec}\mathbf{I}_d\right\}\right]^{\otimes 2} \\
&\quad = \left\{\mathsf{D}^{\otimes 2} f(\boldsymbol{x})^\top\operatorname{vec}\mathbf{H}\right\}^{\otimes 2} = \left\{\mathsf{D}^{\otimes 2} f(\boldsymbol{x})^\top\right\}^{\otimes 2}(\operatorname{vec}\mathbf{H})^{\otimes 2}
\end{aligned}$$

we have $\boldsymbol{w}_0^\top\{\text{vec}\,(\mathbf{H}^{1/2}\otimes\mathbf{H}^{1/2})\}^{\otimes 2} = \frac{1}{4}m_2(K)^2\{\text{vec}^\top\mathbf{R}(\mathsf{D}^{\otimes 2}f)\}(\text{vec}\,\mathbf{H})^{\otimes 2}$, so to complete the proof of part $(i)$ we must show that $\text{ISB}\{\hat{f}(\cdot;\mathbf{H})\} - \boldsymbol{w}_0^\top\{\text{vec}\,(\mathbf{H}^{1/2}\otimes\mathbf{H}^{1/2})\}^{\otimes 2} = \{\boldsymbol{w}(\mathbf{H})-\boldsymbol{w}_0\}^\top\{\text{vec}\,(\mathbf{H}^{1/2}\otimes\mathbf{H}^{1/2})\}^{\otimes 2}$ is $o(\|\text{vec}\,\mathbf{H}\|^2)$. Furthermore, noting from the definition of the scalar product that

$$\begin{aligned}\left|\{\boldsymbol{w}(\mathbf{H})-\boldsymbol{w}_0\}^\top\{\text{vec}\,(\mathbf{H}^{1/2}\otimes\mathbf{H}^{1/2})\}^{\otimes 2}\right|\\ \leq \|\boldsymbol{w}(\mathbf{H})-\boldsymbol{w}_0\|\cdot\left\|\text{vec}\,(\mathbf{H}^{1/2}\otimes\mathbf{H}^{1/2})\right\|^2\end{aligned}$$

and taking into account that we have already shown that $\|\boldsymbol{w}(\mathbf{H})-\boldsymbol{w}_0\|\to 0$ as $\text{vec}\,\mathbf{H}\to 0$, it suffices to show that $\|\text{vec}(\mathbf{H}^{1/2}\otimes\mathbf{H}^{1/2})\|^2/\|\text{vec}\,\mathbf{H}\|^2$ is bounded.

We have $\|\text{vec}(\mathbf{H}^{1/2}\otimes\mathbf{H}^{1/2})\|^2 = \text{vec}^\top(\mathbf{H}^{1/2}\otimes\mathbf{H}^{1/2})\,\text{vec}(\mathbf{H}^{1/2}\otimes\mathbf{H}^{1/2}) = \text{tr}(\mathbf{H}\otimes\mathbf{H}) = \text{tr}^2\,\mathbf{H}$ and, similarly, $\|\text{vec}\,\mathbf{H}\|^2 = \text{tr}(\mathbf{H}^2)$. Since from the Schur decomposition we can express $\text{tr}(\mathbf{H}^k) = \sum_{i=1}^d \lambda_i^k$, where $\lambda_1,\ldots,\lambda_d > 0$ denote the eigenvalues of $\mathbf{H}$, we arrive at $1 \leq \|\text{vec}(\mathbf{H}^{1/2}\otimes\mathbf{H}^{1/2})\|^2/\|\text{vec}\,\mathbf{H}\|^2 \leq d$ (the second inequality is due to the Cauchy-Schwarz inequality), which finishes the proof of part $(i)$.

To show $(ii)$, recall that $\text{IV}\{\hat{f}(\cdot;\mathbf{H})\} = n^{-1}|\mathbf{H}|^{-1/2}R(K) - n^{-1}R(K_\mathbf{H}*f)$ from Equation (2.4). Hence it is enough to show that the sequence $R(K_\mathbf{H}*f)$ is bounded. A change of variables, together with Cauchy-Schwarz inequality and Fubini's theorem, implies that

$$\begin{aligned}R(K_\mathbf{H}*f) &= \int_{\mathbb{R}^d}(K_\mathbf{H}*f)(\boldsymbol{x})^2 d\boldsymbol{x} = \int_{\mathbb{R}^d}\left\{\int_{\mathbb{R}^d}K(\boldsymbol{z})f(\boldsymbol{x}-\mathbf{H}^{1/2}\boldsymbol{z})d\boldsymbol{z}\right\}^2 d\boldsymbol{x}\\ &\leq \left\{\int_{\mathbb{R}^d}|K(\boldsymbol{z})|d\boldsymbol{z}\right\}\cdot\left\{\int_{\mathbb{R}^d}\int_{\mathbb{R}^d}|K(\boldsymbol{z})|f(\boldsymbol{x}-\mathbf{H}^{1/2}\boldsymbol{z})^2\,d\boldsymbol{z}d\boldsymbol{x}\right\}\\ &= \left\{\int_{\mathbb{R}^d}|K(\boldsymbol{z})|\,d\boldsymbol{z}\right\}^2 R(f),\end{aligned}$$

which completes the proof. ∎

### 2.9.2 *Asymptotically optimal bandwidth*

Although there is no explicit formula for the asymptotically optimal bandwidth $\mathbf{H}_{\text{AMISE}}$, it is possible to give a more detailed description of its form, as shown next.

Begin by writing $\mathbf{H} = \lambda\mathbf{A}$ for a symmetric, positive definite matrix $\mathbf{A}$ with $|\mathbf{A}| = 1$ and $\lambda > 0$. This is achieved by taking $\lambda = |\mathbf{H}|^{1/d}$ and $\mathbf{A} = |\mathbf{H}|^{-1/d}\mathbf{H}$. In this form, $\lambda$ represents the size of $\mathbf{H}$ and $\mathbf{A}$ accounts for its orientation.

Then, from Equation (2.10),

$$\text{AMISE}(\lambda \mathbf{A}) = n^{-1}\lambda^{-d/2}R(K) + \tfrac{1}{4}m_2(K)^2 Q(\mathbf{A})\lambda^2, \tag{2.13}$$

where $Q(\mathbf{A}) = (\text{vec}^\top \mathbf{A})\mathbf{R}(\mathsf{D}^{\otimes 2}f)\,\text{vec}\,\mathbf{A} > 0$, since it is a quadratic form in $\text{vec}\,\mathbf{A}$ and $\mathbf{R}(\mathsf{D}^{\otimes 2}f)$ is positive definite. For every fixed $\mathbf{A}$, the value of $\lambda = \lambda_0(\mathbf{A})$ that minimises Equation (2.13) is

$$\lambda_0(\mathbf{A}) = \left[dR(K)/\{m_2(K)^2 Q(\mathbf{A})\}\right]^{2/(d+4)} n^{-2/(d+4)}. \tag{2.14}$$

Substituting the value of $\lambda_0$ in Equation (2.14) into Equation (2.13) gives

$$\min_{\lambda>0} \text{AMISE}(\lambda \mathbf{A}) = \frac{d+4}{4}\left\{d^{-1}m_2(K)^2 R(K)^{4/d} Q(\mathbf{A})\right\}^{d/(d+4)} n^{-4/(d+4)}.$$

From this, it follows that the optimal choice of the orientation $\mathbf{A}$ is $\mathbf{A}_0 = \text{argmin}_{\mathbf{A}\in\mathcal{F},|\mathbf{A}|=1} Q(\mathbf{A})$. Note that $\mathbf{A}_0$ does not depend on the sample size $n$, which yields the $O(n^{-4/(d+4)})$ minimal MISE rate announced in Section 2.8. Moreover, this also indicates that the orientation of the optimal bandwidth only depends on the ISB, whilst its magnitude should be taken to balance the IV and the ISB.

Finally the asymptotically optimal bandwidth can be written as $\mathbf{H}_{\text{AMISE}} = \lambda_0(\mathbf{A}_0)\mathbf{A}_0 = \mathbf{C}_0 n^{-2/(d+4)}$ for some symmetric, positive definite matrix $\mathbf{C}_0$ which does not depend on $n$, which reveals that the optimal bandwidth is of order $n^{-2/(d+4)}$.

### 2.9.3 *Vector versus vector half parametrisations*

Initially in the literature the squared error formulas were expressed in terms of the vector half operator $\text{vech}\,\mathbf{H}$, which stacks the lower triangular half of the $d \times d$ matrix $\mathbf{H}$ into a $\frac{1}{2}d(d+1)$-vector rather than $\text{vec}\,\mathbf{H}$ as we have done. Compare, for example,

$$\text{vech}\begin{bmatrix} h_1^2 & h_{12} \\ h_{12} & h_2^2 \end{bmatrix} = \begin{bmatrix} h_1^2 & h_{12} & h_2^2 \end{bmatrix}^\top$$

$$\text{vec}\begin{bmatrix} h_1^2 & h_{12} \\ h_{12} & h_2^2 \end{bmatrix} = \begin{bmatrix} h_1^2 & h_{12} & h_{12} & h_2^2 \end{bmatrix}^\top.$$

The integrated squared bias from Equation (2.7) can be rewritten as

$$\begin{aligned}&\tfrac{1}{4}m_2(K)^2(\text{vec}^\top \mathbf{H})\mathbf{R}(\mathsf{D}^{\otimes 2}f)\,\text{vec}\,\mathbf{H}\\ &\quad= \tfrac{1}{4}m_2(K)^2(\text{vech}^\top \mathbf{H})\mathbf{D}_d^\top\left\{\int_{\mathbb{R}}[\text{vech}\,\mathsf{H}f(\boldsymbol{x})][\text{vech}^\top \mathsf{H}f(\boldsymbol{x})]\,d\boldsymbol{x}\right\}\mathbf{D}_d(\text{vech}\,\mathbf{H})\end{aligned}$$

where $\mathbf{D}_d$ is the $d$-th order duplication matrix such that $\mathbf{D}_d \operatorname{vech} \mathbf{H} = \operatorname{vec} \mathbf{H}$ for a symmetric matrix $\mathbf{H}$, see Magnus & Neudecker (1999, Chapter 3.8, pp. 48–52). This second form has the advantage of being the minimal quadratic form, and $\int_{\mathbb{R}} [\operatorname{vech} \mathsf{H} f(\boldsymbol{x})][\operatorname{vech}^\top \mathsf{H} f(\boldsymbol{x})] \, d\boldsymbol{x}$ does not appear to be unduly more complicated than the more concise $\mathbf{R}(\mathsf{D}^{\otimes 2} f)$. Indeed Wand (1992); Wand & Jones (1994); Duong & Hazelton (2005*a*,*b*) made advances in bandwidth selectors using this vector half form.

The limitations of these vector half expressions become apparent when attempting to develop the algorithms for data-based unconstrained selectors. To progress, a more tractable method of denoting higher order derivatives of $f$ is required. By drawing on the vectorised derivatives, the required density derivative functionals can be expressed as $\boldsymbol{\psi}_{2r} = \mathbb{E}\{\mathsf{D}^{\otimes 2r} f(\boldsymbol{X})\}$. As $\boldsymbol{\psi}_{2r} = (-1)^r \operatorname{vec} \mathbf{R}(\mathsf{D}^{\otimes r} f)$, then they are naturally paired with $\operatorname{vec} \mathbf{H}$ rather than $\operatorname{vech} \mathbf{H}$.

Furthermore, derivatives with respect to $\operatorname{vec} \mathbf{H}$ are easier to calculate from the identification table of Magnus & Neudecker (1999, Table 9.6.2, p. 176) as its entries are expressed in terms of differentials $d \operatorname{vec} \mathbf{H}$. On the other hand, derivatives with respect to $\operatorname{vech} \mathbf{H}$ must always include the duplication matrices $\mathbf{D}_d$ to transform the $d \operatorname{vec} \mathbf{H}$ to $d \operatorname{vech} \mathbf{H}$ differentials.

Chapter 3

# Bandwidth selectors for density estimation

The selection of the bandwidth (or smoothing parameter) is the single most crucial factor in determining the performance of a kernel estimator. So much so that this single question is the generator of a significant proportion of the literature. Well-written reviews of the main methodologies available for the univariate case can be found in Berlinet & Devroye (1994), Cao et al. (1994) and Jones et al. (1996*a*,*b*) or, more recently, in Heidenreich et al. (2013). In the previous chapter, we gave some preliminary advice for practical bandwidth selection. In this chapter, we examine the different bandwidth selectors in more detail and elucidate the reasons for this advice.

The phrase 'bandwidth selection' may seem unusual, and the reader may ask why it is not called bandwidth estimation. Population parameters, e.g., the density function $f$, depend only on the population distribution, and not on the data. An estimator aims to determine the value of a parameter from a data sample, e.g., $\hat{f}(\cdot;\mathbf{H})$ estimates $f$. As the bandwidth is an inherent quantity of $\hat{f}(\cdot;\mathbf{H})$, and although we also refer to it as the smoothing parameter, it is not a population parameter in the same way as the density function. To distinguish between these two cases, we select the bandwidth whilst we estimate the density.

In Chapter 2 we introduced the oracle bandwidths $\mathbf{H}_{\text{MISE}}, \mathbf{H}_{\text{AMISE}}$ which depend on the unknown target density $f$. This dependence implies that they cannot be computed using only a data sample drawn from $f$ and so they are not available for practical data analysis. The usual approach to automatic, data-based bandwidth selection is based on first estimating either the MISE or AMISE, which is then subsequently minimised to yield a bandwidth $\hat{\mathbf{H}} = \text{argmin}_{\mathbf{H}\in\mathcal{F}} \widehat{\text{(A)MISE}}\{\hat{f}(\cdot;\mathbf{H})\}$ that is computed solely from the data. Different bandwidth selectors arise from different estimators of the (A)MISE, and especially of the integrated squared bias.

In this chapter, Sections 3.1–3.3 introduce the normal scale, maximal smoothing and normal mixture selectors, based on replacing the target density with a given parametric density in the error formulas. Sections 3.4–3.7 investigate the unbiased cross validation, biased cross validation, plug-in, and smoothed cross validation selectors. Section 3.8 carries out an empirical comparison among all these selectors. Section 3.9 sets up a key mathematical result to quantify the convergence rate of each data-based bandwidth. Section 3.10 fills in the previously omitted mathematical details of the considered bandwidth selectors.

## 3.1 Normal scale bandwidths

Normal scale bandwidths are often the first selectors to be developed for a kernel estimator as they are the simplest in terms of mathematical and computational complexity, as intuited by Silverman (1986, pp. 45–47). To obtain a normal scale selector, the unknown target density $f$ is replaced by a normal density $\phi_{\mathbf{\Sigma}}(\cdot - \boldsymbol{\mu})$ with mean $\boldsymbol{\mu}$ and variance $\mathbf{\Sigma}$ in the squared error formulas, and the kernel $K$ is taken to be the normal kernel.

Due to the neat mathematical properties of normal densities, in this case the MISE and AMISE of $\hat{f}$ become

$$\begin{aligned}
\mathrm{MISE}_{\mathrm{NS}}\{\hat{f}(\cdot;\mathbf{H})\} &= n^{-1}|\mathbf{H}|^{-1/2}(4\pi)^{-d/2} \\
&\quad + (2\pi)^{-d/2}\{(1-n^{-1})|2\mathbf{H}+2\mathbf{\Sigma}|^{-1/2} - 2|\mathbf{H}+2\mathbf{\Sigma}|^{-1/2} + |2\mathbf{\Sigma}|^{-1/2}\} \\
\mathrm{AMISE}_{\mathrm{NS}}\{\hat{f}(\cdot;\mathbf{H})\} &= n^{-1}|\mathbf{H}|^{-1/2}(4\pi)^{-d/2} \\
&\quad + \tfrac{1}{16}(4\pi)^{-d/2}|\mathbf{\Sigma}|^{-1/2}\{2\,\mathrm{tr}(\mathbf{H}\mathbf{\Sigma}^{-1}\mathbf{H}\mathbf{\Sigma}^{-1}) + \mathrm{tr}^2(\mathbf{H}\mathbf{\Sigma}^{-1})\}
\end{aligned}$$

as shown in Wand & Jones (1993) and Wand (1992), respectively.

There is no explicit formula for the minimiser of $\mathrm{MISE}_{\mathrm{NS}}$, not even in the univariate case. However, Wand (1992) showed that the minimiser of $\mathrm{AMISE}_{\mathrm{NS}}$ within the class $\mathcal{F}$ of unconstrained bandwidths is given by

$$\mathbf{H}_{\mathrm{NS}} = \{4/(d+2)\}^{2/(d+4)} n^{-2/(d+4)} \mathbf{\Sigma} \tag{3.1}$$

(see also Section 5.8 below for an alternative derivation of this result). If we replace the population variance $\mathbf{\Sigma}$ with an estimator, usually the sample variance $\mathbf{S}$, then we obtain a data-based, normal scale bandwidth

$$\hat{\mathbf{H}}_{\mathrm{NS}} = \{4/(d+2)\}^{2/(d+4)} n^{-2/(d+4)} \mathbf{S}. \tag{3.2}$$

This is also commonly referred to as a ‘rule of thumb’ selector. As the normal density is amongst the smoothest densities available, the normal scale selector tends to yield bandwidths which lead to oversmoothing for non-normal data.

Even if the unconstrained optimal bandwidth has a relatively simple explicit formulation in the normal case, as given by Equation (3.1), when kernel density estimation is used as an intermediate tool for other more involved estimation problems it is not uncommon to use a constrained bandwidth $\mathbf{H} = \operatorname{diag}(h_1^2, \ldots, h_d^2) \in \mathcal{D}$ or $\mathbf{H} = h^2\mathbf{I}_d \in \mathcal{A}$ (see Azzalini & Torelli, 2007, Chen et al., 2016, and also Chapters 6–7). In these cases, it is also useful to have normal scale bandwidth selectors prior to developing more sophisticated bandwidth selection methods.

Silverman (1986, p. 86–87) provided some heuristic advice, but it can be verified that $\mathrm{AMISE}_{\mathrm{NS}}\{\hat{f}(\cdot;\mathbf{H})\}$ is minimized for $\mathbf{H} = h^2\mathbf{I}_d$ by taking

$$h_{\mathrm{NS}} = \left[4d|\mathbf{\Sigma}|^{1/2}/\{2\operatorname{tr}(\mathbf{\Sigma}^{-2}) + \operatorname{tr}^2(\mathbf{\Sigma}^{-1})\}\right]^{1/(d+4)} n^{-1/(d+4)}. \tag{3.3}$$

Analogously, the $i$-th diagonal entry of the optimal diagonal bandwidth in the normal case can be shown to be $h_{\mathrm{NS},i}^2$, where

$$h_{\mathrm{NS},i} = \left[4d|\mathbf{\Delta}|^{1/2}/\{2\operatorname{tr}(\mathbf{\Delta}^{-2}) + \operatorname{tr}^2(\mathbf{\Delta}^{-1})\}\right]^{1/(d+4)} \sigma_i n^{-1/(d+4)}. \tag{3.4}$$

Here, $\sigma_i^2$ denotes the $i$-th marginal variance, and $\mathbf{\Delta} = (\operatorname{diag}\mathbf{\Sigma})^{-1}\mathbf{\Sigma}$ with $\operatorname{diag}\mathbf{\Sigma} = \operatorname{diag}(\sigma_1^2, \ldots, \sigma_d^2) \in \mathcal{M}_{d\times d}$ (see Section 5.8). This generalizes the bivariate formula given in Scott (2015, Equation (6.42), p. 163) to an arbitrary dimension. As in Equation (3.2), data-based constrained normal-scale bandwidths are obtained by replacing $\mathbf{\Sigma}$ in Equations (3.3)–(3.4) with the sample variance $\mathbf{S}$.

It is worth noting that, despite the fact that optimal matrices obtained from Equations (3.3) and (3.4) are constrained, their entries depend on the unconstrained variance matrix $\mathbf{\Sigma}$, i.e., on the marginal variances as well as on the covariances.

## 3.2 Maximal smoothing bandwidths

The maximal smoothing selector is similar in spirit to the normal scale selector. As noted in Section 2.9.2, the size of the asymptotic oracle bandwidth $\mathbf{H}_{\mathrm{AMISE}}$ is inversely proportional to $C(f) = \min_{\mathbf{A}\in\mathcal{F},|\mathbf{A}|=1}(\operatorname{vec}^T\mathbf{A})\mathbf{R}(\mathsf{D}^{\otimes 2}f)\operatorname{vec}\mathbf{A}$. The maximal smoothing bandwidth is the one that corresponds to the density $f$ which minimises $C(f)$ amongst all those densities with a given variability.

Terrell (1990) showed that, amongst those densities with identity covariance matrix, the density that minimises $C(f)$ is the triweight density

$$f(\boldsymbol{x}) = \tfrac{1}{6}\{(d+8)\pi\}^{-d/2}\Gamma(d/2+4)\{1-\boldsymbol{x}^\top\boldsymbol{x}/(d+8)\}^3\mathbf{1}\{\boldsymbol{x}^\top\boldsymbol{x} \le d+8\}.$$

The resulting maximal smoothing bandwidth, obtained by replacing the unknown density in the AMISE formula with the triweight density, is

$$\hat{\mathbf{H}}_{\text{MS}} = \left\{ \frac{(d+8)^{(d+6)/2}\pi^{d/2}R(K)}{16(d+2)\Gamma(d/2+4)} \right\}^{2/(d+4)} n^{-2/(d+4)}\mathbf{S}. \tag{3.5}$$

This bandwidth can be interpreted as the largest bandwidth that should be used for any data whose sample variance is $\mathbf{S}$. Thus it deliberately oversmooths the data, even more so than the normal scale bandwidth. Equations (3.2) and (3.5) have the same form, with only different coefficients.

## 3.3 Normal mixture bandwidths

The normal scale selector can be extended to a more general case of a normal mixture $\sum_{\ell=1}^{q} w_\ell \phi_{\boldsymbol{\Sigma}_\ell}(\cdot - \boldsymbol{\mu}_\ell)$ where the $\ell$-th component has mean $\boldsymbol{\mu}_\ell$ and variance $\boldsymbol{\Sigma}_\ell$, and the weights $w_\ell$ are non-negative and sum to 1. The corresponding MISE and AMISE formulas are

$$\begin{aligned} \text{MISE}_{\text{NM}}\{\hat{f}(\cdot;\mathbf{H})\} &= n^{-1}|\mathbf{H}|^{-1/2}(4\pi)^{-d/2} \\ &\quad + \boldsymbol{w}^\top[(1-n^{-1})\boldsymbol{\Omega}_2 - 2\boldsymbol{\Omega}_1 + \boldsymbol{\Omega}_0]\boldsymbol{w} \\ \text{AMISE}_{\text{NM}}\{\hat{f}(\cdot;\mathbf{H})\} &= n^{-1}|\mathbf{H}|^{-1/2}(4\pi)^{-d/2} + \tfrac{1}{4}\boldsymbol{w}^\top\boldsymbol{\Xi}\boldsymbol{w} \end{aligned}$$

where $\boldsymbol{w} = (w_1,\ldots,w_q)$, $\boldsymbol{\Omega}_a$ and $\boldsymbol{\Xi}$ are $q\times q$ matrices whose $(\ell,\ell')$ elements are $[\boldsymbol{\Omega}_a]_{\ell,\ell'} = \phi_{a\mathbf{H}+\boldsymbol{\Sigma}_\ell+\boldsymbol{\Sigma}_{\ell'}}(\boldsymbol{\mu}_\ell - \boldsymbol{\mu}_{\ell'})$ from Wand & Jones (1993, Theorem 1), and $[\boldsymbol{\Xi}]_{\ell,\ell'} = \phi_{\boldsymbol{\Sigma}_\ell+\boldsymbol{\Sigma}_{\ell'}}(\boldsymbol{\mu}_\ell - \boldsymbol{\mu}_{\ell'})[2\operatorname{tr}(\mathbf{H}\mathbf{A}_{\ell,\ell'}\mathbf{H}\mathbf{B}_{\ell,\ell'}) + \operatorname{tr}^2(\mathbf{H}\mathbf{C}_{\ell,\ell'})]$, with $\mathbf{A}_{\ell,\ell'} = (\boldsymbol{\Sigma}_\ell + \boldsymbol{\Sigma}_{\ell'})^{-1}$, $\mathbf{B}_{\ell,\ell'} = \mathbf{A}_{\ell,\ell'} - 2\mathbf{A}_{\ell,\ell'}(\boldsymbol{\mu}_\ell - \boldsymbol{\mu}_{\ell'})(\boldsymbol{\mu}_\ell - \boldsymbol{\mu}_{\ell'})^\top\mathbf{A}_{\ell,\ell'}$, and $\mathbf{C}_{\ell,\ell'} = \mathbf{A}_{\ell,\ell'} - \mathbf{A}_{\ell,\ell'}(\boldsymbol{\mu}_\ell - \boldsymbol{\mu}_{\ell'})(\boldsymbol{\mu}_\ell - \boldsymbol{\mu}_{\ell'})^\top\mathbf{A}_{\ell,\ell'}$ from Wand (1992, Theorem 1).

These exact formulas are of a limited utility for non-normal mixture data. So Cwik & Koronacki (1997) proposed to initially fit a normal mixture $\sum_{\ell=1}^{\hat{q}} \hat{w}_\ell \phi_{\hat{\boldsymbol{\Sigma}}_\ell}(\cdot - \hat{\boldsymbol{\mu}}_\ell)$ to the data, where $\hat{q}, \hat{w}_\ell, \hat{\boldsymbol{\mu}}_\ell, \hat{\boldsymbol{\Sigma}}_\ell$ are the estimated parameters, and then to substitute them into the above AMISE formula to obtain

$$\text{NM}(\mathbf{H}) = n^{-1}(4\pi)^{-d/2}|\mathbf{H}|^{-1/2} + \tfrac{1}{4}\hat{\boldsymbol{w}}^\top\hat{\boldsymbol{\Xi}}\hat{\boldsymbol{w}}$$

where $\hat{\boldsymbol{w}} = (\hat{w}_1,\ldots,\hat{w}_{\hat{q}})$ and $\hat{\boldsymbol{\Xi}}$ is a $\hat{q}\times\hat{q}$ matrix equivalent of $\boldsymbol{\Xi}$. Therefore $\hat{\mathbf{H}}_{\text{NM}} = \operatorname{argmin}_{\mathbf{H}\in\mathcal{F}} \text{NM}(\mathbf{H})$. The NM selector has a straightforward implementation in Algorithm 1.

## 3.4 Unbiased cross validation bandwidths

The previous selectors were highly dependent on a reference parametric distribution (i.e., normal, triweight or normal mixture). In keeping with the non-

**Algorithm 1** Normal mixture bandwidth selector for density estimation

**Input:** $\{\boldsymbol{X}_1, \ldots, \boldsymbol{X}_n\}$
**Output:** $\hat{\mathbf{H}}_{\text{NM}}$
1: Fit an optimal normal mixture model with $\hat{q}$ components
2: $\hat{\mathbf{H}}_{\text{NM}} :=$ minimiser of $\text{NM}(\mathbf{H})$

parametric spirit of kernel estimators, we explore bandwidth selectors which reduce this dependence on parametric reference distributions.

Unbiased cross validation (UCV), also known as least squares cross validation (LSCV), is a leave-one-out cross validation method for selecting the bandwidth, introduced by Rudemo (1982) and Bowman (1984). This method admits motivations from the ISE and MISE points of view. To begin, note that the ISE can be expanded as

$$\text{ISE}\{\hat{f}(\cdot;\mathbf{H})\} = \int_{\mathbb{R}^d} \hat{f}(\boldsymbol{x};\mathbf{H})^2\, d\boldsymbol{x} - 2\int_{\mathbb{R}^d} \hat{f}(\boldsymbol{x};\mathbf{H}) f(\boldsymbol{x})\, d\boldsymbol{x} + R(f).$$

In this expansion, the first term is fully known since it only involves the kernel estimator, the third term does not affect the choice of the minimiser, and the integral in the second one can be expressed as the conditional expectation $\mathbb{E}\{\hat{f}(\boldsymbol{X};\mathbf{H})|\boldsymbol{X}_1,\ldots,\boldsymbol{X}_n\}$ for a random variable $\boldsymbol{X} \sim f$ which is independent of $\boldsymbol{X}_1,\ldots,\boldsymbol{X}_n$. So if we denote the leave-one-out density estimator as

$$\hat{f}_{-i}(\boldsymbol{x};\mathbf{H}) = (n-1)^{-1} \sum_{j=1, j\neq i}^{n} K_{\mathbf{H}}(\boldsymbol{x} - \boldsymbol{X}_j),$$

then a sensible estimator of this conditional expectation is $n^{-1}\sum_{i=1}^n \hat{f}_{-i}(\boldsymbol{X}_i;\mathbf{H})$. This leads to the UCV criterion

$$\text{UCV}(\mathbf{H}) = \int_{\mathbb{R}^d} \hat{f}(\boldsymbol{x};\mathbf{H})^2\, d\boldsymbol{x} - 2n^{-1}\sum_{i=1}^{n} \hat{f}_{-i}(\boldsymbol{X}_i;\mathbf{H})$$

from which follows $\hat{\mathbf{H}}_{\text{UCV}} = \text{argmin}_{\mathbf{H}\in\mathcal{F}} \text{UCV}(\mathbf{H})$.

After expanding the first term, this criterion can also be expressed as

$$\begin{aligned}\text{UCV}(\mathbf{H}) &= n^{-1}|\mathbf{H}|^{-1/2}R(K) \\ &\quad + \{n(n-1)\}^{-1} \sum_{\substack{i,j=1 \\ j\neq i}}^{n} \{(1-n^{-1})K_{\mathbf{H}} * K_{\mathbf{H}} - 2K_{\mathbf{H}}\}(\boldsymbol{X}_i - \boldsymbol{X}_j).\end{aligned} \tag{3.6}$$

The expected value of the UCV criterion is

$$\begin{aligned}\mathbb{E}\{\mathrm{UCV}(\mathbf{H})\} &= n^{-1}|\mathbf{H}|^{-1/2}R(K)\\ &\quad + (1-n^{-1})R(K_{\mathbf{H}} * f) - 2\int_{\mathbb{R}^d}(K_{\mathbf{H}} * f)(\boldsymbol{x})f(\boldsymbol{x})\,d\boldsymbol{x}\\ &= \mathrm{MISE}\{\hat{f}(\cdot;\mathbf{H})\} - R(f)\end{aligned}$$

from Equation (2.4). The UCV is an unbiased estimator of the MISE, ignoring the $R(f)$ constant which does not involve the bandwidth, giving the method its name.

It is usual to make the identification $1-n^{-1} \approx 1$ inside the double sum in Equation (3.6) since, although this yields the loss of exact unbiasedness, the resulting bandwidth is asymptotically equivalent (Chacón & Duong, 2011, Theorem 1) and its formulation is slightly simpler, leading to

$$\mathrm{UCV}(\mathbf{H}) = n^{-1}|\mathbf{H}|^{-1/2}R(K) + \{n(n-1)\}^{-1}\sum_{\substack{i,j=1\\ j\neq i}}^{n}(K_{\mathbf{H}} * K_{\mathbf{H}} - 2K_{\mathbf{H}})(\boldsymbol{X}_i - \boldsymbol{X}_j). \tag{3.7}$$

If the normal kernel $K = \phi$ is used, then the UCV criterion has an even simpler form in Equation (3.8) due to the convolution properties of normal densities, for example $\phi_{\boldsymbol{\Sigma}_1} * \phi_{\boldsymbol{\Sigma}_2} = \phi_{\boldsymbol{\Sigma}_1+\boldsymbol{\Sigma}_2}$ (Wand & Jones, 1993), which gives

$$\mathrm{UCV}(\mathbf{H}) = n^{-1}|\mathbf{H}|^{-1/2}(4\pi)^{-d/2} + \{n(n-1)\}^{-1}\sum_{\substack{i,j=1\\ j\neq i}}^{n}(\phi_{2\mathbf{H}} - 2\phi_{\mathbf{H}})(\boldsymbol{X}_i - \boldsymbol{X}_j). \tag{3.8}$$

The UCV selector has a straightforward implementation in Algorithm 2, which contains only one step for the numerical minimisation of the UCV criterion.

**Algorithm 2** UCV bandwidth selector for density estimation

**Input:** $\{\boldsymbol{X}_1,\ldots,\boldsymbol{X}_n\}$
**Output:** $\hat{\mathbf{H}}_{\mathrm{UCV}}$
1: $\hat{\mathbf{H}}_{\mathrm{UCV}} :=$ minimiser of $\mathrm{UCV}(\mathbf{H})$

The existence of multiple local minima of the UCV curve for 1-dimensional data is a well-documented phenomenon, as well as a tendency for undersmoothing (Hall & Marron, 1991). These two issues are usually addressed by restraining the numerical optimisation within a fixed range, bounded away from very small bandwidths, and with an initial large bandwidth (e.g., a normal scale or maximal smoothing).

## 3.5 Biased cross validation bandwidths

A different flavour of cross validation arises if the error estimation step focuses on the AMISE rather than the MISE. Recall from Section 2.6 that

$$\mathrm{AMISE}\{\hat{f}(\cdot;\mathbf{H})\} = n^{-1}|\mathbf{H}|^{-1/2}R(K) + \tfrac{1}{4}m_2(K)^2\{\mathrm{vec}^\top \mathbf{R}(\mathsf{D}^{\otimes 2}f)\}(\mathrm{vec}\,\mathbf{H})^{\otimes 2}.$$

The biased cross validation criterion is based on replacing the unknown $f$ in such AMISE expression for the kernel density estimator $\hat{f}$, resulting in

$$\mathrm{BCV}(\mathbf{H}) = n^{-1}|\mathbf{H}|^{-1/2}R(K) + \tfrac{1}{4}m_2(K)^2\{\mathrm{vec}^\top \mathbf{R}(\mathsf{D}^{\otimes 2}\hat{f}(\cdot;\mathbf{H}))\}(\mathrm{vec}\,\mathbf{H})^{\otimes 2} \tag{3.9}$$

and $\hat{\mathbf{H}}_{\mathrm{BCV}} = \mathrm{argmin}_{\mathbf{H}\in\mathcal{F}}\mathrm{BCV}(\mathbf{H})$. Scott & Terrell (1987) called it biased cross validation (BCV) due to the fact that it targets the AMISE instead of the MISE, but for this reason it is alternatively referred to as asymptotic cross validation by Jones & Kappenman (1992).

The initial hope was that the biasedness would lead to a decrease in the wide variability of the UCV, and thus an overall decrease in the MISE of $\hat{f}$. Subsequent research has highlighted the importance of using a different, pilot bandwidth to estimate the curvature matrix $\mathbf{R}(\mathsf{D}^{\otimes 2}f)$, despite the extra computational burden, and so the BCV has not attracted sufficient interest since its introduction.

Algorithm 3 for the BCV selector is straightforward.

**Algorithm 3** BCV bandwidth selector for density estimation

**Input:** $\{\boldsymbol{X}_1,\ldots,\boldsymbol{X}_n\}$
**Output:** $\hat{\mathbf{H}}_{\mathrm{BCV}}$
1: $\hat{\mathbf{H}}_{\mathrm{BCV}} :=$ minimiser of $\mathrm{BCV}(\mathbf{H})$

## 3.6 Plug-in bandwidths

Plug-in bandwidth selectors were first introduced for multivariate data for constrained matrices by Wand & Jones (1994), who extended the univariate methodology of Sheather & Jones (1991). The current approach for unconstrained bandwidth matrices was first introduced by Duong & Hazelton (2003) and later refined by Chacón & Duong (2010).

For the plug-in class of bandwidth selectors, we begin with an alternative form of the AMISE to those previously exhibited in Equation (2.10), namely

$$\mathrm{AMISE}\{\hat{f}(\cdot;\mathbf{H})\} = n^{-1}|\mathbf{H}|^{-1/2}R(K) + \tfrac{1}{4}m_2(K)^2\boldsymbol{\psi}_4^\top(\mathrm{vec}\,\mathbf{H})^{\otimes 2}. \tag{3.10}$$

The vector $\boldsymbol{\psi}_4 = \operatorname{vec} \mathbf{R}(\mathsf{D}^{\otimes 2} f) \in \mathbb{R}^{d^4}$ corresponds to the case $s = 2$ of the more general functional $\boldsymbol{\psi}_{2s} \in \mathbb{R}^{d^{2s}}$, defined as

$$\boldsymbol{\psi}_{2s} = (-1)^s \operatorname{vec} \mathbf{R}(\mathsf{D}^{\otimes s} f) = \int_{\mathbb{R}^d} \mathsf{D}^{\otimes 2s} f(\boldsymbol{x}) f(\boldsymbol{x})\, d\boldsymbol{x},$$

where the last equality follows by element-wise integration by parts, subject to some regularity conditions.

As $\boldsymbol{\psi}_{2s} = \mathbb{E}\{\mathsf{D}^{\otimes 2s} f(\boldsymbol{X})\}$ for $\boldsymbol{X} \sim f$, a natural estimator is

$$\hat{\boldsymbol{\psi}}_{2s} \equiv \hat{\boldsymbol{\psi}}_{2s}(\mathbf{G}) = n^{-1} \sum_{i=1}^{n} \mathsf{D}^{\otimes 2s} \tilde{f}(\boldsymbol{X}_i; \mathbf{G}) = n^{-2} \sum_{i,j=1}^{n} \mathsf{D}^{\otimes 2s} L_{\mathbf{G}}(\boldsymbol{X}_i - \boldsymbol{X}_j), \quad (3.11)$$

where $\tilde{f}$ is a kernel density estimator based on a pilot kernel $L$ and a pilot bandwidth $\mathbf{G}$, which may be potentially different from the $K$ and $\mathbf{H}$ used in $\hat{f}$. The special case of $s = 2$ yields the plug-in estimator of the AMISE,

$$\operatorname{PI}(\mathbf{H}; \mathbf{G}) = n^{-1} |\mathbf{H}|^{-1/2} R(K) + \tfrac{1}{4} m_2(K)^2 \hat{\boldsymbol{\psi}}_4(\mathbf{G})^\top (\operatorname{vec} \mathbf{H})^{\otimes 2}, \quad (3.12)$$

and the bandwidth selector $\hat{\mathbf{H}}_{\mathrm{PI}} = \operatorname{argmin}_{\mathbf{H} \in \mathcal{F}} \operatorname{PI}(\mathbf{H}; \mathbf{G})$. Note that $\hat{\mathbf{H}}_{\mathrm{PI}}$ depends on the choice of the pilot bandwidth $\mathbf{G}$.

From Equation (3.12), it appears that we have not made any progress towards the goal of selecting $\mathbf{H}$ as we have introduced the problem of selecting the pilot bandwidth $\mathbf{G}$. Whilst it is crucial to the performance of the density estimator $\hat{f}$ that $\mathbf{H}$ is selected carefully, the selection of the pilot bandwidth $\mathbf{G}$ should affect the final estimate less markedly as it has a lesser, secondary effect on $\hat{f}$ via the calculation of $\hat{\boldsymbol{\psi}}_4(\mathbf{G})$, see Sheather & Jones (1991).

In any case, the analogous optimality criterion for selecting $\mathbf{G}$ is

$$\operatorname{MSE}(\hat{\mathbf{H}}_{\mathrm{PI}}) = \mathbb{E}\left\{ \|\operatorname{vec}(\hat{\mathbf{H}}_{\mathrm{PI}} - \mathbf{H}_{\mathrm{AMISE}})\|^2 \right\}$$

and it can be shown that

$$\operatorname{MSE}(\hat{\mathbf{H}}_{\mathrm{PI}}) = \mathrm{const} \cdot \mathbb{E}\left\{ \|\hat{\boldsymbol{\psi}}_4(\mathbf{G}) - \boldsymbol{\psi}_4\|^2 \right\} \{1 + o(1)\} \quad (3.13)$$

where the constant does not involve $\mathbf{G}$ or the data (Duong & Hazelton, 2005*a*, Lemma 1), so that it can be ignored when optimising the $\operatorname{MSE}(\hat{\mathbf{H}}_{\mathrm{PI}})$ with respect to $\mathbf{G}$. The performance of $\hat{\mathbf{H}}_{\mathrm{PI}}$ is directly related to the mean squared error $\operatorname{MSE}\{\hat{\boldsymbol{\psi}}_4(\mathbf{G})\} = \mathbb{E}\left\{ \|\hat{\boldsymbol{\psi}}_4(\mathbf{G}) - \boldsymbol{\psi}_4\|^2 \right\}$.

The leading asymptotic term of this MSE is

$$\begin{aligned} \operatorname{AMSE}\{\hat{\boldsymbol{\psi}}_4(\mathbf{G})\} = \| & n^{-1} |\mathbf{G}|^{-1/2} (\mathbf{G}^{-1/2})^{\otimes 4} \mathsf{D}^{\otimes 4} L(\mathbf{0}) \\ & + \tfrac{1}{2} m_2(L) (\operatorname{vec}^\top \mathbf{G} \otimes \mathbf{I}_{d^4}) \boldsymbol{\psi}_6 \|^2 \end{aligned} \quad (3.14)$$

as developed by Chacón & Duong (2010, Theorem 1). The asymptotically optimal $\mathbf{G}$ is the minimiser of this AMSE in Equation (3.14). As it contains the unknown quantity $\boldsymbol{\psi}_6$, in practice it is necessary to replace $\boldsymbol{\psi}_6$ with another kernel estimator $\hat{\boldsymbol{\psi}}_6$, whose bandwidth is chosen as a normal scale pilot bandwidth $\hat{\mathbf{G}}_{\text{NS},6} = \{2/(d+6)\}^{2/(d+8)} 2\mathbf{S} n^{-2/(d+8)}$, which is obtained as the minimiser of $\text{AMSE}\{\hat{\boldsymbol{\psi}}_6(\mathbf{G})\}$ in the normal case, see Chacón & Duong (2010, Equation (8)). This is known as a 2-stage selector as the number of stages enumerates the number of kernel functional estimation, in this case $\hat{\boldsymbol{\psi}}_4, \hat{\boldsymbol{\psi}}_6$. Two stages of kernel functional estimation has been theoretically and empirically recommended by Aldershof (1991); Park & Marron (1992); Wand & Jones (1995); Tenreiro (2003).

A normal scale selector $\hat{\mathbf{G}}_{\text{NS},6}$ for the sixth order integrated functional leads to a minor loss of efficiency in the presence of departures from non-normality as compared to a normal scale selector $\hat{\mathbf{H}}_{\text{NS}}$ for the density estimate $\hat{f}$. This insensitivity of the pilot bandwidth to an imprecise estimation can be extended to afford a further reduction in complexity without overly compromising the accuracy of $\hat{f}$ by using the scalar bandwidth class $\mathbf{G} \in \mathcal{A}$, as there is an analytic expression for the minimiser of $\text{AMSE}\{\hat{\boldsymbol{\psi}}_4(g^2\mathbf{I}_d)\}$, namely

$$\hat{g}_4 = [2A_1/\{-A_2 + (A_2^2 + 4A_1A_3)^{1/2}\}]^{1/(d+6)} n^{-1/(d+6)} \tag{3.15}$$

where the constants are $A_1 = (2d+8)\mathsf{D}^{\otimes 4}L(\mathbf{0})^\top \mathsf{D}^{\otimes 4}L(\mathbf{0}), A_2 = (d+2)m_2(L)\mathsf{D}^{\otimes 4}L(\mathbf{0})^\top(\operatorname{vec}\mathbf{I}_d \otimes \mathbf{I}_d)\hat{\boldsymbol{\psi}}_6(\hat{g}^2_{\text{NS},6}\mathbf{I}_d)$, and $A_3 = m_2(L)^2 \hat{\boldsymbol{\psi}}_6(\hat{g}^2_{\text{NS},6}\mathbf{I}_d)^\top \times (\operatorname{vec}\mathbf{I}_d \operatorname{vec}^\top \mathbf{I}_{d^4} \otimes \mathbf{I}_d)\hat{\boldsymbol{\psi}}_6(\hat{g}^2_{\text{NS},6}\mathbf{I}_d)$. The normal scale scalar pilot is

$$\hat{g}_{\text{NS},6} = [2B_1/\{-B_2 + (B_2^2 + 4B_1B_3)^{1/2}\}]^{1/(d+8)} n^{-1/(d+8)}$$

where $B_1 = 30(2\pi)^{-d} d(d+2)(d+4)(d+6), B_2 = -\frac{15}{16} 2^{-d/2}(2\pi)^{-d}(d+4)|\mathbf{S}|^{-1/2}\nu_4(\mathbf{S}^{-1})$, and $B_3 = \hat{\boldsymbol{\psi}}_{\text{NS},8}^\top(\operatorname{vec}\mathbf{I}_d \operatorname{vec}^\top \mathbf{I}_d \otimes \mathbf{I}_{d^6})\hat{\boldsymbol{\psi}}_{\text{NS},8}$. For the definition of $\nu_4(\mathbf{S}^{-1})$ see Section 5.1.3, and $\hat{\boldsymbol{\psi}}_{\text{NS},8}$ see Section 5.8.2. Whilst scalar bandwidths usually result in simple explicit expressions, in this case it is the unconstrained pilot bandwidth $\hat{\mathbf{G}}_{\text{NS},6}$ which exhibits a more concise form than $\hat{g}_{\text{NS},6}$. A pilot bandwidth of class $\mathcal{A}$ is sufficient for most cases if the data are pre-scaled to have the same marginal variance, though for cases where unconstrained pilots of class $\mathcal{F}$ are more appropriate, see Chacón & Duong (2010).

The 2-stage plug-in selector is given in Algorithm 4, with the option in steps 1 and 3 to use an unconstrained or scalar pilot, as provided by Duong & Hazelton (2003). The result is $\hat{\mathbf{H}}_{\text{PI}}$ which is a generalisation of the diagonal bandwidths from Wand & Jones (1994).

**Algorithm 4** Two-stage plug-in bandwidth selector for density estimation

**Input:** $\{\boldsymbol{X}_1, \ldots, \boldsymbol{X}_n\}$
**Output:** $\hat{\mathbf{H}}_{\text{PI}}$

1: Compute 6th order normal scale pilot bandwidth
   (a) Unconstrained pilot $\hat{\mathbf{G}}_6 := \hat{\mathbf{G}}_{\text{NS},6}$
   (b) Scalar pilot $\hat{\mathbf{G}}_6 := \hat{g}^2_{\text{NS},6}\mathbf{I}_d$
2: Compute 6th order kernel functional estimate $\hat{\boldsymbol{\psi}}_6(\hat{\mathbf{G}}_6)$ /* Stage 1 */
3: Plug $\hat{\boldsymbol{\psi}}_6(\hat{\mathbf{G}}_6)$ into formula for pilot bandwidth $\hat{\mathbf{G}}_4$
   (a) Unconstrained pilot $\hat{\mathbf{G}}_4 :=$ minimiser of $\widehat{\text{AMSE}}\{\hat{\boldsymbol{\psi}}_4(\mathbf{G})\}$
   (b) Scalar pilot $\hat{\mathbf{G}}_4 := \hat{g}^2_4\mathbf{I}_d$
4: Compute 4th order kernel functional estimate $\hat{\boldsymbol{\psi}}_4(\hat{\mathbf{G}}_4)$ /* Stage 2 */
5: $\hat{\mathbf{H}}_{\text{PI}} :=$ minimiser of $\text{PI}(\mathbf{H};\hat{\mathbf{G}}_4)$

### 3.7 Smoothed cross validation bandwidths

A third main flavour of cross validation, known as smoothed cross validation (SCV), has shown more promise than biased cross validation in the goal to reduce the large variability of unbiased cross validation. This is achieved by an improved estimator of the integrated squared bias, posited by Hall et al. (1992). Instead of estimating the asymptotic ISB, the SCV criterion aims to estimate the exact ISB by replacing the true density $f$ by a pilot kernel density estimator $\tilde{f}(\boldsymbol{x};\mathbf{G}) = n^{-1}\sum_{i=1}^n L_{\mathbf{G}}(\boldsymbol{x} - \boldsymbol{X}_i)$ to obtain

$$\begin{aligned}\widehat{\text{ISB}}(\mathbf{H};\mathbf{G}) &= \int_{\mathbb{R}^d} [\{K_{\mathbf{H}} * \tilde{f}(\cdot;\mathbf{G})\}(\boldsymbol{x}) - \tilde{f}(\boldsymbol{x};\mathbf{G})]^2\, d\boldsymbol{x} \\ &= n^{-2}\sum_{i,j=1}^n (\bar{K}_{\mathbf{H}} * \bar{L}_{\mathbf{G}} - 2K_{\mathbf{H}} * \bar{L}_{\mathbf{G}} + \bar{L}_{\mathbf{G}})(\boldsymbol{X}_i - \boldsymbol{X}_j)\end{aligned}$$

where $\bar{K} = K * K$ and $\bar{L} = L * L$. Adding the dominant term of the IV yields

$$\begin{aligned}\text{SCV}(\mathbf{H};\mathbf{G}) &= n^{-1}|\mathbf{H}|^{-1/2}R(K) \\ &\quad + n^{-2}\sum_{i,j=1}^n (\bar{K}_{\mathbf{H}} * \bar{L}_{\mathbf{G}} - 2K_{\mathbf{H}} * \bar{L}_{\mathbf{G}} + \bar{L}_{\mathbf{G}})(\boldsymbol{X}_i - \boldsymbol{X}_j) \qquad (3.16)\end{aligned}$$

as the SCV estimator of the (A)MISE. The SCV selector is then $\hat{\mathbf{H}}_{\text{SCV}} = \text{argmin}_{\mathbf{H}\in\mathcal{F}}\text{SCV}(\mathbf{H};\mathbf{G})$. In contrast to the - estimator of the AMISE in Equation (3.12), no estimator of the integrated density functional $\boldsymbol{\psi}_4$ is required in Equation (3.16), with the trade-off of the more computationally intensive double sums.

If there are no duplicates in the data $\boldsymbol{X}_1, \ldots, \boldsymbol{X}_n$ and if we consider $L_0$

as a Dirac delta function, then the unbiased cross validation is a special case of the smoothed cross validation as $\mathrm{SCV}(\mathbf{H};\mathbf{0}) \equiv \mathrm{UCV}(\mathbf{H})$, as observed by Hall et al. (1992). So SCV differs from the UCV by pre-smoothing the data differences $\boldsymbol{X}_i - \boldsymbol{X}_j$ by $\bar{L}_{\mathbf{G}}$.

If, instead of using the dominant term of the IV, the exact IV is employed, the resulting estimate of the MISE is

$$\begin{aligned}\mathrm{BMISE}(\mathbf{H};\mathbf{G}) &= n^{-1}|\mathbf{H}|^{-1/2}R(K)\\ &\quad + n^{-2}\sum_{i,j=1}^{n}\{(1-n^{-1})\bar{K}_{\mathbf{H}} * \bar{L}_{\mathbf{G}} - 2K_{\mathbf{H}} * \bar{L}_{\mathbf{G}} + \bar{L}_{\mathbf{G}}\}(\boldsymbol{X}_i - \boldsymbol{X}_j).\end{aligned} \tag{3.17}$$

The only difference with Equation (3.16) is the $1-n^{-1}$ factor in the first term of the double sum, hence the difference between the two criteria is asymptotically negligible. Equation (3.17) shows that there is little effect on bandwidth selection in substituting $n^{-1}|\mathbf{H}|^{-1/2}R(K)$ for the exact variance, at least in an asymptotic sense, and this can be extended from the SCV to any bandwidth selector, as shown in Chacón & Duong (2011, Theorem 1). This echoes the asymptotically negligible difference between the exact and asymptotic variance in the MISE and AMISE, respectively, in Equations (2.4) and (2.10).

In this form, it can be shown that BMISE is the smoothed bootstrap estimate of the MISE, obtained by replacing the target $f$ by a pilot estimator $\tilde{f}(\cdot;\mathbf{G})$ everywhere in the MISE formula, including in the expectation operator which is taken with respect to $\tilde{f}(\cdot;\mathbf{G})$ instead of $f$. This bootstrap approach to bandwidth selection was introduced in the univariate case by Taylor (1989) with $\mathbf{G} = \mathbf{H}$, and more recently Horová et al. (2013) developed a closely related multivariate analogue. Faraway & Jhun (1990) noticed the advantages of using a different pilot bandwidth $\mathbf{G}$, and its theoretical properties were studied in Marron (1992) and Cao (1993). Due to its equivalence to the SCV criterion, only the latter will be treated here in detail.

Similar to the UCV, if normal kernels $K = L = \phi$ are used, then the SCV has a simpler form in Equation (3.18):

$$\begin{aligned}\mathrm{SCV}(\mathbf{H};\mathbf{G}) &= n^{-1}|\mathbf{H}|^{-1/2}(4\pi)^{-d/2}\\ &\quad + n^{-2}\sum_{i,j=1}^{n}(\phi_{2\mathbf{H}+2\mathbf{G}} - 2\phi_{\mathbf{H}+2\mathbf{G}} + \phi_{2\mathbf{G}})(\boldsymbol{X}_i - \boldsymbol{X}_j).\end{aligned} \tag{3.18}$$

The analogous pilot bandwidth selection problem for the SCV is slightly different from the PI selector. The MSE has the same form $\mathrm{MSE}(\hat{\mathbf{H}}_{\mathrm{SCV}}) = \mathbb{E}\{\|\mathrm{vec}(\hat{\mathbf{H}}_{\mathrm{SCV}} - \mathbf{H}_{\mathrm{MISE}})\|^2\}$ which has leading asymptotic term $\text{AMSE}(\hat{\mathbf{H}}_{\text{SCV}}) = \text{const} \cdot \text{AMSE}^*\{\hat{\boldsymbol{\psi}}_4(\mathbf{G})\}$ where the constant does not involve $\mathbf{G}$, see Chacón & Duong (2011, Theorem 2), and

$$\begin{aligned}\text{AMSE}^*\{\hat{\boldsymbol{\psi}}_4(\mathbf{G})\} = \|&n^{-1}|\mathbf{G}|^{-1/2}(\mathbf{G}^{-1/2})^{\otimes 4}\mathsf{D}^{\otimes 4}\bar{L}(\mathbf{0}) \\ &+ \tfrac{1}{2}m_2(\bar{L})(\text{vec}^\top \mathbf{G} \otimes \mathbf{I}_{d^4})\boldsymbol{\psi}_6\|^2\end{aligned} \quad (3.19)$$

which is the $\text{AMSE}\{\hat{\boldsymbol{\psi}}_4(\mathbf{G})\}$ in Equation (3.14) except that the convolved kernel $\bar{L}$ rather than $L$ is used. Whilst $\hat{\boldsymbol{\psi}}_4$ itself is not needed to evaluate the SCV criterion, we are still required to compute the corresponding $\hat{\mathbf{G}}_4$. For a 2-stage SCV selector, we are also required to compute $\hat{\mathbf{G}}_6$ and $\hat{\boldsymbol{\psi}}_6$, with the former selected as the minimiser of an analogous $\text{AMSE}^*\{\hat{\boldsymbol{\psi}}_6(\mathbf{G})\}$. To begin the procedure, the normal scale pilot is $\hat{\mathbf{G}}^*_{\text{NS},6} = \{2/(d+6)\}^{2/(d+8)}\mathbf{S}n^{-2/(d+8)}$.

If scalar pilot selectors are preferred, then the scalar minimiser of $\widehat{\text{AMSE}}^*\{\hat{\boldsymbol{\psi}}_{2s}(g^2\mathbf{I}_d)\}$ is

$$\hat{g}^*_4 = [2A_1/\{-A_2 + (A_2^2 + 4A_1A_3)^{1/2}\}]^{1/(d+6)}n^{-1/(d+6)} \quad (3.20)$$

where the constants are $A_1 = (2d+8)\mathsf{D}^{\otimes 4}\bar{L}(\mathbf{0})^\top \mathsf{D}^{\otimes 4}\bar{L}(\mathbf{0}), A_2 = (d+2)m_2(\bar{L})[\text{vec}\,\mathbf{I}_d \otimes \mathsf{D}^{\otimes 4}\bar{L}(\mathbf{0})]^\top \hat{\boldsymbol{\psi}}_6(\hat{g}^{*2}_{\text{NS},6}\mathbf{I}_d)$, and $A_3 = m_2(\bar{L})^2 \hat{\boldsymbol{\psi}}_6(\hat{g}^{*2}_{\text{NS},6}\mathbf{I}_d)^\top \times (\text{vec}\,\mathbf{I}_d\,\text{vec}^\top \mathbf{I}_d \otimes \mathbf{I}_{d^4})\hat{\boldsymbol{\psi}}_6(\hat{g}^{*2}_{\text{NS},6}\mathbf{I}_d)$. The normal scale scalar pilot is

$$\hat{g}^*_{\text{NS},6} = [2B_1/\{-B_2 + (B_2^2 + 4B_1B_3)^{1/2}\}]^{1/(d+8)}n^{-1/(d+8)}$$

where $B_1 = \frac{15}{32}(4\pi)^{-d}d(d+2)(d+4)(d+6), B_2 = -\frac{15}{64}(4\pi)^{-d}(d+4) \times |\mathbf{S}|^{-1/2}\nu_4(\mathbf{S}^{-1})$ and $B_3 = 4\hat{\boldsymbol{\psi}}^\top_{\text{NS},8}(\text{vec}\,\mathbf{I}_d\,\text{vec}^\top \mathbf{I}_d \otimes \mathbf{I}_{d^6})\hat{\boldsymbol{\psi}}_{\text{NS},8}$.

The 2-stage SCV selector is given in Algorithm 5, with the option in Steps 3–4, provided by Chacón & Duong (2011), to use an unconstrained or a scalar pilot. The result is $\hat{\mathbf{H}}_{\text{SCV}}$, which is a generalisation of the univariate selector of Hall et al. (1992).

Whilst we do not have in general explicit expressions for the unconstrained data-based selectors, or even the oracle (A)MISE selectors, we are able to establish that all these selectors are of order $n^{-2/(d+4)}$. Analogously we can show that optimal pilot selector $\hat{\mathbf{G}}_4$ for both the plug-in and SCV is of order $n^{-2/(d+6)}$.

## 3.8 Empirical comparison of bandwidth selectors

To explore the performance of all the previously exposed bandwidth selection methods in practice we focus on one target density, the Trimodal III normal mixture density introduced in Wand & Jones (1993) as density (K),

**Algorithm 5** Two-stage SCV bandwidth selector for density estimation

**Input:** $\{\boldsymbol{X}_1,\ldots,\boldsymbol{X}_n\}$
**Output:** $\hat{\mathbf{H}}_{\text{SCV}}$
1: Compute 6th order normal scale pilot bandwidth
 (a) Unconstrained pilot $\hat{\mathbf{G}}_6 := \hat{\mathbf{G}}^*_{\text{NS},6}$
 (b) Scalar pilot $\hat{\mathbf{G}}_6 := \hat{g}^{*2}_{\text{NS},6}\mathbf{I}_d$
2: Compute 6th order kernel functional estimate $\hat{\boldsymbol{\psi}}_6(\hat{\mathbf{G}}_6)$ /* Stage 1 */
3: Plug $\hat{\boldsymbol{\psi}}_6(\hat{\mathbf{G}}_6)$ into formula for pilot bandwidth $\hat{\mathbf{G}}_4$
 (a) Unconstrained pilot $\hat{\mathbf{G}}_4 :=$ minimiser of $\widehat{\text{AMSE}}^*\{\hat{\boldsymbol{\psi}}_4(\mathbf{G})\}$
 (b) Scalar pilot $\hat{\mathbf{G}}_4 := \hat{g}^{*2}_4\mathbf{I}_d$
4: $\mathbf{H}_{\text{SCV}} :=$ minimiser of $\text{SCV}(\mathbf{H};\hat{\mathbf{G}}_4)$ /* Stage 2 */

and also included in Chacón (2009) as density #9. This density can be regarded as representative of a medium level of estimation difficulty, since it consists of three well-separated modal regions, albeit with different sizes and covariance structures. Its distribution is $\frac{3}{7}N((-1,0),\frac{1}{25}[9,\frac{63}{10};\frac{63}{10},\frac{49}{4}]) + \frac{3}{7}N((1,2/\sqrt{3}),\frac{1}{25}[9,0;0,\frac{49}{4}]) + \frac{1}{7}N((1,-2/\sqrt{3}),\frac{1}{25}[9,0;0,\frac{49}{4}])$ and its quartile contour plots are displayed in Figure 3.1. While it is inadvisable to draw general conclusions from the study of the performance of the selectors for this particular density, our broader experience suggests that the main patterns obtained for this moderately difficult example are closely followed in many other similar situations.

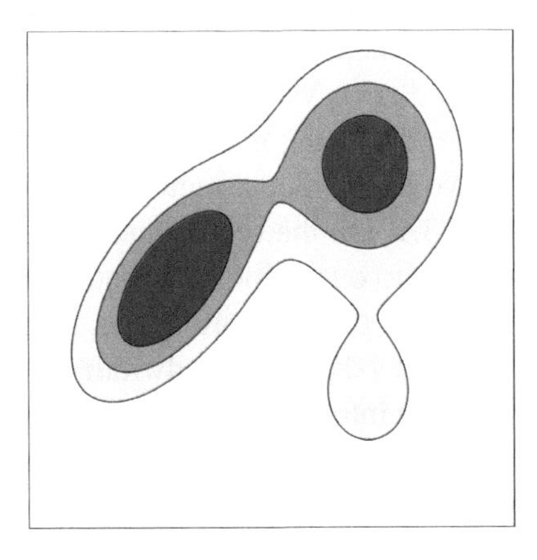

Figure 3.1 *Contour plot of a moderately difficult-to-estimate normal mixture density, the Trimodal III density, having three modal regions with different sizes and covariance structures.*

We drew $N = 500$ samples of size $n = 1000$ from this target density.

For each sample, the normal scale, normal mixture, UCV, BCV, PI and SCV bandwidth selectors were computed, and the ISEs of the kernel density estimator using such bandwidths were obtained. The advantage of normal mixture densities is that, in addition to being easy to simulate from to produce a wide variety of density shapes, the ISE has a closed form. When the target normal mixture density is $f = \sum_{\ell=1}^{q} w_\ell \phi_{\mathbf{\Sigma}_\ell}(\cdot - \boldsymbol{\mu}_\ell)$ and the kernel is the standard normal density, Duong (2004) showed that the ISE can be written as $\mathrm{ISE}\{\hat{f}(\cdot;\mathbf{H})\} = n^{-2}\sum_{i,i'=1}^{n} \phi_{2\mathbf{H}}(\boldsymbol{X}_i - \boldsymbol{X}_{i'}) - 2n^{-1}\sum_{i=1}^{n}\sum_{\ell=1}^{q} w_\ell \phi_{\mathbf{H}+\mathbf{\Sigma}_\ell}(\boldsymbol{X}_i - \boldsymbol{\mu}_\ell) + \sum_{\ell,\ell'=1}^{q} w_\ell \phi_{\mathbf{\Sigma}_\ell+\mathbf{\Sigma}_{\ell'}}(\boldsymbol{\mu}_\ell - \boldsymbol{\mu}_{\ell'})$.

To obtain an impression of the typical performance of each bandwidth selection method we recorded the sample number (out of the $N = 500$ samples considered) that produced the median ISE for each of the bandwidth selectors considered, and computed the corresponding density estimate using that sample and bandwidth selector, as shown in Figure 3.2.

From these median-performance displays it can be concluded that most of the bandwidth selectors considered have a reasonably good median performance. Here, the normal mixture bandwidth $\hat{\mathbf{H}}_{\mathrm{NM}}$ should be close to asymptotic oracle bandwidth, since the target density is indeed a normal mixture density. In Figure 3.2(b) it can be seen that the median performance of this bandwidth is able to recover the trimodal structure, although perhaps with too irregular contours. The other selector that typically discovers the trimodal structure, with slightly smoother contours, is the plug-in bandwidth $\hat{\mathbf{H}}_{\mathrm{PI}}$ in Figure 3.2(e). The median performance estimate of the unbiased cross validation bandwidth $\hat{\mathbf{H}}_{\mathrm{UCV}}$ in Figure 3.2(c) also shows the lower rightmost modal region, though there is an indication of undersmoothing in the upper modal region. Whilst the smoothed cross validation selector $\hat{\mathbf{H}}_{\mathrm{SCV}}$ does not reveal the trimodal structure as clearly as the normal mixture, UCV, and PI selectors, the orientation of the two larger modal regions are the closest to those of the target density in Figure 3.1. On the other hand, the normal scale bandwidth $\hat{\mathbf{H}}_{\mathrm{NS}}$ in Figure 3.2(a) leads to oversmoothed contours as it does not reveal the lower mode. Even smoother contours are shown in the median performance estimate of the biased cross validation bandwidth $\hat{\mathbf{H}}_{\mathrm{BCV}}$ in Figure 3.2(d).

To gain a deeper insight into the behaviour of these bandwidth selectors it is useful to explore further graphical features of the scaled kernels employed to construct the estimates. The individual scaled bivariate normal kernels, which comprise the density estimate, exhibit elliptical contours whose size and orientation are determined by the bandwidth matrix $\mathbf{H}$. The entries of $\mathbf{H}$ have a clear statistical meaning: the two diagonal entries provide the variance of each of the two coordinates in this scaled distribution and the off-diagonal entry the covariance between them.

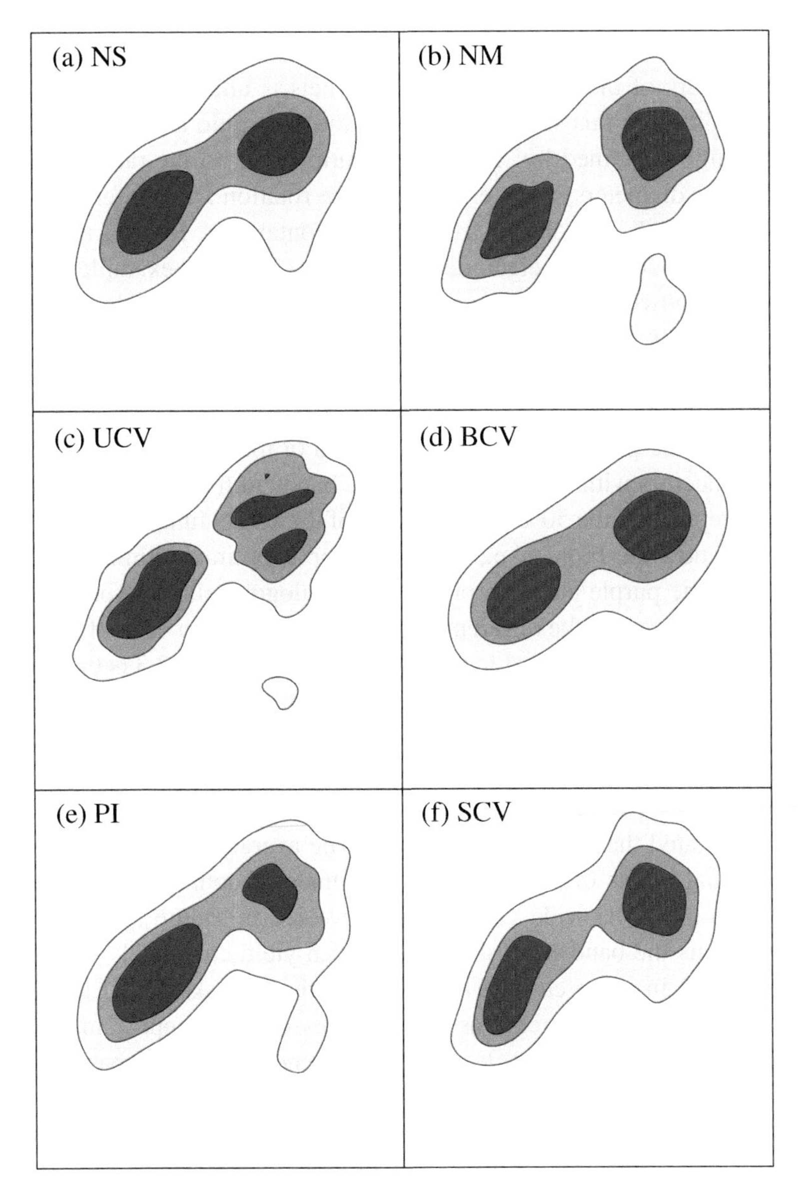


Figure 3.2 *Different bandwidth selectors for the density estimates for the* $n = 1000$ *data from the Trimodal III normal mixture density. (a) Normal scale* $\hat{\mathbf{H}}_{\mathrm{NS}}$. *(b) Normal mixture* $\hat{\mathbf{H}}_{\mathrm{NM}}$. *(c) Unbiased cross validation* $\hat{\mathbf{H}}_{\mathrm{UCV}}$. *(d) Biased cross validation* $\hat{\mathbf{H}}_{\mathrm{BCV}}$. *(e) Plug-in* $\hat{\mathbf{H}}_{\mathrm{PI}}$. *(f) Smoothed cross validation* $\hat{\mathbf{H}}_{\mathrm{SCV}}$.

From a graphical point of view, it is perhaps more instructive to utilise the size-rotation parametrisation from Section 2.9.2, i.e., let $\mathbf{H} = \lambda \mathbf{A}$ with $|\mathbf{A}| = 1$ so that $\lambda = |\mathbf{H}|^{1/d}$ can be regarded as the size of $\mathbf{H}$. The ellipse that forms the contours of the individual scaled kernels is unambiguously determined if, in addition to this size, we are given the axes ratio and the rotation angle of the ellipse defined by $\mathbf{A}$. The axes ratio refers to the ratio of the major and minor diameters of the ellipse, and the rotation angle refers to the angle subtended of the major axis from the horizontal axis. Both of the quantities are by-products of the eigen-decomposition of $\mathbf{H}$. For example, the MISE-optimal bandwidth for this normal mixture density and sample size $n = 1000$ is numerically computed to be $\mathbf{H}_{\text{MISE}} = [0.053, 0.027; 0.027, 0.072]$. Its size is $|\mathbf{H}_{\text{MISE}}|^{1/2} = 0.056$ and, for the associated ellipse, the axes ratio is 2.666 and the rotation angle is 55.12 degrees.

Figure 3.3 provides a visual impression of the performance of the bandwidth selectors with respect to the oracle bandwidth $\mathbf{H}_{\text{MISE}}$.. The thick black ellipse represents the 50% contour curve of the kernel function scaled with the oracle bandwidth $\mathbf{H}_{\text{MISE}}$, i.e., the scaling for an optimal estimate of the target density. The purple curves represent the analogous ellipses for the different scalings induced by the different bandwidth selectors for each of the 500 simulated samples. The easiest feature to compare in Figure 3.3 is the bandwidth size. The fact that $\hat{\mathbf{H}}_{\text{NS}}$ and $\hat{\mathbf{H}}_{\text{BCV}}$ both oversmooth is clearly shown, as their ellipses present little variability but their sizes are certainly larger than the size of the optimal ellipse. A slight tendency for under-rotation is also observed for $\hat{\mathbf{H}}_{\text{NS}}$, and $\hat{\mathbf{H}}_{\text{BCV}}$ tends to produce nearly circular ellipses which are markedly less elongated than the optimal ellipse. The average size of $\hat{\mathbf{H}}_{\text{UCV}}$ seems to follow that of the oracle, by sometimes undersmoothing while other times oversmoothing. Its wide variability is revealed in Figure 3.3(c), which not only affects the bandwidth size but also can yield completely wrong ellipse orientations in some cases. These characteristics make $\hat{\mathbf{H}}_{\text{UCV}}$ insufficiently reliable in practice. In contrast, the ellipses of $\hat{\mathbf{H}}_{\text{NM}}$ appear markedly less variable. This bandwidth selector shows a tendency to undersmooth, probably because the size of its target, the asymptotic oracle $\mathbf{H}_{\text{AMISE}}$, is smaller than the size of the non-asymptotic counterpart $\mathbf{H}_{\text{MISE}}$ (the univariate case, Marron & Wand (1992) conjectured that this relationship is true for all densities). The ellipses of both $\hat{\mathbf{H}}_{\text{PI}}$ and $\hat{\mathbf{H}}_{\text{SCV}}$ are tightly concentrated around the ellipse of the oracle bandwidth, with only a slight bias towards oversmoothing. The performance of these two selectors is similar to that of $\hat{\mathbf{H}}_{\text{NM}}$, even if the former were not constructed on the basis of a normal mixture assumption.

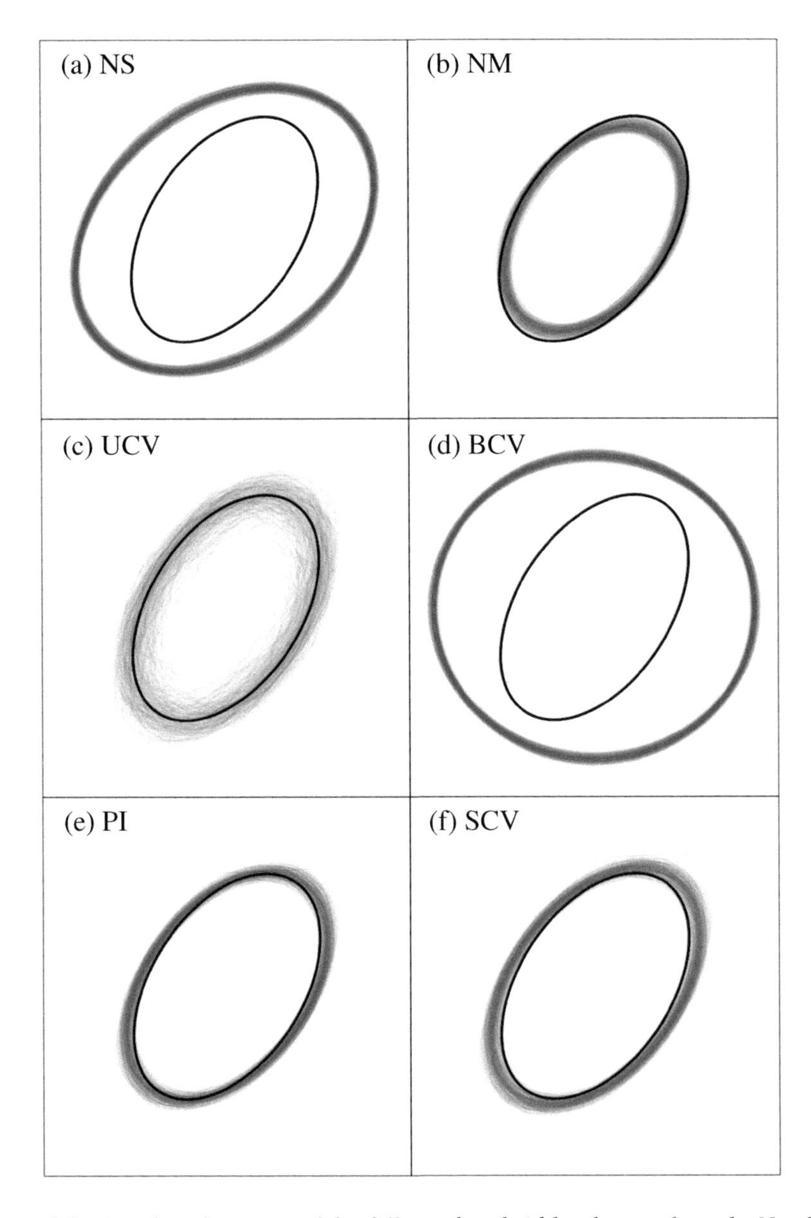


Figure 3.3 *Visual performance of the different bandwidth selectors from the* $N = 500$ *samples of size* $n = 1000$ *from the Trimodal III normal mixture density. The target 50% contour ellipse for* $\mathbf{H}_{\text{MISE}}$ *is the solid black curve, the 50% contour ellipses for the bandwidth selectors are the purple curves. (a) Normal scale* $\hat{\mathbf{H}}_{\text{NS}}$. *(b) Normal mixture* $\hat{\mathbf{H}}_{\text{NM}}$. *(c) Unbiased cross validation* $\hat{\mathbf{H}}_{\text{UCV}}$. *(d) Biased cross validation* $\hat{\mathbf{H}}_{\text{BCV}}$. *(e) Plug-in* $\hat{\mathbf{H}}_{\text{PI}}$. *(f) Smoothed cross validation* $\hat{\mathbf{H}}_{\text{SCV}}$.

## 3.9 Theoretical comparison of bandwidth selectors

The different selectors $\hat{\mathbf{H}}$ exhibited in the previous sections are all the same asymptotic order $n^{-2/(d+4)}$, because it can be shown that they are all consistent for $\mathbf{H}_{\text{AMISE}}$, in the sense that $\mathbf{H}_{\text{AMISE}}^{-1}\hat{\mathbf{H}}$ converges in probability to the identity matrix. However, it would be incorrect then to infer that their relative rate of convergence to the oracle bandwidth $\mathbf{H}_{\text{AMISE}}$ is also of the same order. The different convergence rates reflect the differences in their estimation performance.

A common measure used to quantify the theoretical asymptotic performance between the different selectors is the relative rate of convergence . We say that a bandwidth selector $\hat{\mathbf{H}}$ converges to the oracle bandwidth $\mathbf{H}_{\text{AMISE}}$ at relative rate $n^{-\alpha}$, for $\alpha > 0$, when

$$\operatorname{vec}(\hat{\mathbf{H}} - \mathbf{H}_{\text{AMISE}}) = O_P(\mathbf{J}_{d^2} n^{-\alpha}) \operatorname{vec} \mathbf{H}_{\text{AMISE}} \tag{3.21}$$

where $\mathbf{J}_{d^2}$ is the $d^2 \times d^2$ matrix of all ones, as posited by Duong & Hazelton (2005*a*). The $\mathbf{J}_{d^2}$ matrix is used here instead of the identity $\mathbf{I}_{d^2}$ to cover the cases whenever $\mathbf{H}_{\text{AMISE}}$ contains zero off-diagonal elements, e.g., when the target density is unimodal with a diagonal variance matrix, but as the data-based selector $\hat{\mathbf{H}}$ belongs to the unconstrained class $\mathcal{F}$, the $\mathbf{J}_{d^2}$ ensures that Equation (3.21) remains well-defined by utilising the overall rate of $\mathbf{H}_{\text{AMISE}}$ as the relative rate of convergence to a zero element is undefined.

Observe that the squared norm of the left-hand side of Equation (3.21), i.e., $\|\operatorname{vec}(\hat{\mathbf{H}} - \mathbf{H}_{\text{AMISE}})\|^2$, is indeed the $\text{MSE}(\hat{\mathbf{H}})$ in Equation (3.13). This close relationship is the basis for the computation of these relative convergence rates.

The rates for all the unconstrained selectors are summarised in Table 3.1, extending Jones (1992, Table 3) for the selectors in class $\mathcal{A}$. The performance decreases with increasing dimension. The rates for the BCV and PI are established by Duong & Hazelton (2005*a*) and Chacón & Duong (2010), and the rates for UCV and SCV by Duong & Hazelton (2005*b*) and Chacón & Duong (2011). For $d = 1$ for the PI and SCV selectors, bias annihilation is possible yielding a rate of $n^{-5/14}$ (Sheather & Jones, 1991; Jones et al., 1991) whereas for $d > 1$ only bias minimisation is possible so the rate is $n^{-2/(d+6)}$. For $d = 2, 3$ the BCV and UCV selectors rate of $n^{-\min\{d,4\}/(2d+8)}$ is slower than the SCV and PI selectors rate, but this swaps over for $d \geq 4$.

The convergence rates remain the same for constrained matrices for the cross validation selectors (UCV, BCV, SCV), though the constrained plug-in selectors $\hat{\mathbf{H}}_{\text{PI},\mathcal{D}}$ or $\hat{\mathbf{H}}_{\text{PI},\mathcal{A}}$ have a faster rate of convergence $n^{-\min\{8,d+4\}/(2d+12)}$, again due to bias annihilation, indicating that selecting the off-diagonal elements of the bandwidth matrix, which determine the orientation of the kernel, is the most difficult aspect of unconstrained plug-in selection (Wand & Jones, 1994).

| Selector | Class | Convergence rate to $\mathbf{H}_{\text{AMISE}}$ |
|---|---|---|
| $\hat{\mathbf{H}}_{\text{UCV}}$ | $\mathcal{A}, \mathcal{D}, \mathcal{F}$ | $n^{-\min\{d,4\}/(2d+8)}$ |
| $\hat{\mathbf{H}}_{\text{BCV}}$ | $\mathcal{A}, \mathcal{D}, \mathcal{F}$ | $n^{-\min\{d,4\}/(2d+8)}$ |
| $\hat{\mathbf{H}}_{\text{SCV}}$ | $d = 1$ | $n^{-5/14}$ |
| $\hat{\mathbf{H}}_{\text{SCV}}$ | $d > 1, \mathcal{A}, \mathcal{D}, \mathcal{F}$ | $n^{-2/(d+6)}$ |
| $\hat{\mathbf{H}}_{\text{PI}}$ | $d = 1$ or $d > 1, \mathcal{A}, \mathcal{D}$ | $n^{-\min\{8,d+4\}/(2d+12)}$ |
| $\hat{\mathbf{H}}_{\text{PI}}$ | $d > 1, \mathcal{F}$ | $n^{-2/(d+6)}$ |

Table 3.1 *Convergence rates to* $\mathbf{H}_{\text{AMISE}}$ *for the unbiased cross validation* $\hat{\mathbf{H}}_{\text{UCV}}$, *biased cross validation* $\hat{\mathbf{H}}_{\text{BCV}}$, *smoothed cross validation* $\hat{\mathbf{H}}_{\text{SCV}}$ *and plug-in* $\hat{\mathbf{H}}_{\text{PI}}$ *selectors.*

Similar to Equation (3.21), we can establish that $\text{vec}(\mathbf{H}_{\text{AMISE}} - \mathbf{H}_{\text{MISE}}) = O(\mathbf{J}_{d^2} n^{-2/(d+4)}) \text{vec}\,\mathbf{H}_{\text{MISE}}$, i.e., the convergence rate of $\mathbf{H}_{\text{AMISE}}$ to $\mathbf{H}_{\text{MISE}}$ is $O(n^{-2/(d+4)})$, which is faster than all the rates in Table 3.1, implying that the rates of convergence of $\hat{\mathbf{H}}$ to $\mathbf{H}_{\text{MISE}}$ remain the same.

## 3.10 Further mathematical analysis of bandwidth selectors

We provide more details about the behaviour of kernel estimators: whilst they are not required for practical data analysis, they contribute to a deeper understanding and appreciation of their important properties. It is these properties which underlie their widespread use in practise. The literature for kernel density estimators is vast, and it is not possible to cover all the mathematical results which have been developed since their introduction, so we provide an outline of the most prominent ones.

### *3.10.1 Relative convergence rates of bandwidth selectors*

The $n^{-\alpha}$ order in probability in the relative convergence rate of a bandwidth selector $\hat{\mathbf{H}}$, i.e., $\text{vec}(\hat{\mathbf{H}} - \mathbf{H}_{\text{AMISE}}) = O_P(\mathbf{J}_{d^2} n^{-\alpha}) \text{vec}\,\mathbf{H}_{\text{AMISE}}$ from Equation (3.21), is difficult to establish directly. Fortunately, if we include some stronger regularity conditions so that

$$\text{MSE}(\hat{\mathbf{H}}) = \mathbb{E}\{\|\text{vec}(\hat{\mathbf{H}} - \mathbf{H}_{\text{AMISE}})\|^2\}$$

exists, then Duong & Hazelton (2005*a*) demonstrate that this MSE has a tractable asymptotic form in terms of the derivative of the difference

$\widehat{\text{AMISE}} - \text{AMISE}$, which we reproduce as Theorem 2. This theorem elucidates that the convergence rate of a data-based selector depends on the quality of its corresponding estimator of the (A)MISE.

For brevity, denote $\text{AMISE}(\mathbf{H}) \equiv \text{AMISE}\{\hat{f}(\cdot;\mathbf{H})\}$ and $\hat{\boldsymbol{\alpha}}(\mathbf{H}) = \mathsf{D}_{\mathbf{H}}(\widehat{\text{AMISE}} - \text{AMISE})(\mathbf{H})$, with $\mathsf{D}_{\mathbf{H}}$ standing for the gradient operator with respect to $\operatorname{vec}\mathbf{H}$. Similarly, denote by $\mathsf{H}_{\mathbf{H}}$ the Hessian operator with respect to $\operatorname{vec}\mathbf{H}$.

**Theorem 2** *Suppose that (A1)–(A3) in Conditions A hold and that $\hat{\mathbf{H}}$ is consistent for $\mathbf{H}_{\text{AMISE}}$. Then the mean squared error of a bandwidth selector can be expanded as* $\text{MSE}(\hat{\mathbf{H}}) = \text{AMSE}(\hat{\mathbf{H}})\{1+o(1)\}$ *where*

$$\begin{aligned}\text{AMSE}(\hat{\mathbf{H}}) = \operatorname{tr}\big[&\{\mathsf{H}_{\mathbf{H}}\text{AMISE}(\mathbf{H}_{\text{AMISE}})\}^{-2}\big(\text{Var}\{\hat{\boldsymbol{\alpha}}(\mathbf{H}_{\text{AMISE}})\}\\ &+\{\mathbb{E}\,\hat{\boldsymbol{\alpha}}(\mathbf{H}_{\text{AMISE}})\}\{\mathbb{E}\,\hat{\boldsymbol{\alpha}}^\top(\mathbf{H}_{\text{AMISE}})\}\big)\big]\end{aligned}$$

*and* $\mathsf{H}_{\mathbf{H}}\text{AMISE}(\mathbf{H}_{\text{AMISE}})$ *is a constant symmetric positive definite matrix which does not depend on the data. Thus if* $\text{MSE}(\hat{\mathbf{H}}) = O(n^{-2\alpha})\|\operatorname{vec}\mathbf{H}^2_{\text{AMISE}}\|$, *then the relative rate of convergence of $\hat{\mathbf{H}}$ to $\mathbf{H}_{\text{AMISE}}$ is $O_P(n^{-\alpha})$.*

**Proof (Proof of Theorem 2)** We expand $\mathsf{D}_{\mathbf{H}}\widehat{\text{AMISE}}$ as follows:

$$\begin{aligned}\mathsf{D}_{\mathbf{H}}\widehat{\text{AMISE}}(\hat{\mathbf{H}}) &= \mathsf{D}_{\mathbf{H}}(\widehat{\text{AMISE}} - \text{AMISE})(\hat{\mathbf{H}}) + \mathsf{D}_{\mathbf{H}}\text{AMISE}(\hat{\mathbf{H}})\\ &= \mathsf{D}_{\mathbf{H}}(\widehat{\text{AMISE}} - \text{AMISE})(\hat{\mathbf{H}}) + \big\{\mathsf{D}_{\mathbf{H}}\text{AMISE}(\mathbf{H}_{\text{AMISE}})\\ &\quad + \mathsf{H}_{\mathbf{H}}\text{AMISE}(\mathbf{H}_{\text{AMISE}})\operatorname{vec}(\hat{\mathbf{H}} - \mathbf{H}_{\text{AMISE}})\big\}\{1+o_p(1)\}.\end{aligned}$$

Now we have $\mathsf{D}_{\mathbf{H}}\widehat{\text{AMISE}}(\hat{\mathbf{H}}) = \mathsf{D}_{\mathbf{H}}\text{AMISE}(\mathbf{H}_{\text{AMISE}}) = \mathbf{0}$ so that

$$\begin{aligned}&\operatorname{vec}(\hat{\mathbf{H}} - \mathbf{H}_{\text{AMISE}})\\ &\quad = -\{\mathsf{H}_{\mathbf{H}}\text{AMISE}(\mathbf{H}_{\text{AMISE}})\}^{-1}\mathsf{D}_{\mathbf{H}}(\widehat{\text{AMISE}} - \text{AMISE})(\hat{\mathbf{H}})\{1+o_p(1)\}\\ &\quad = -\{\mathsf{H}_{\mathbf{H}}\text{AMISE}(\mathbf{H}_{\text{AMISE}})\}^{-1}\mathsf{D}_{\mathbf{H}}(\widehat{\text{AMISE}} - \text{AMISE})(\mathbf{H}_{\text{AMISE}})\\ &\qquad \times\{1+o_p(1)\},\end{aligned}$$

the last equality due to the fact that $\hat{\mathbf{H}}$ is consistent for $\mathbf{H}_{\text{AMISE}}$. Taking expectations and variances respectively completes the result. This is essentially a modified version of the proof of Duong & Hazelton (2005*a*, Lemma 1) which has been updated to the current notation.

To compute the derivatives of $\text{AMISE}(\mathbf{H})$, the required differentials are $d|\mathbf{H}|^{-1/2}$ and $d\{(\operatorname{vec}^\top\mathbf{H})\mathbf{R}(\mathsf{D}^{\otimes 2}f)\operatorname{vec}\mathbf{H}\}$. We have $d|\mathbf{H}|^{-1/2} = -\frac{1}{2}|\mathbf{H}|^{-1/2}\operatorname{vec}^\top\mathbf{H}^{-1}d\operatorname{vec}\mathbf{H}$, as $d|\mathbf{H}|^{-1/2} = -|\mathbf{H}|^{-1}(d|\mathbf{H}|^{1/2})$ and $d|\mathbf{H}|^{1/2} = \frac{1}{2}|\mathbf{H}|^{-1}d|\mathbf{H}| = \frac{1}{2}|\mathbf{H}|^{1/2}(\operatorname{vec}^\top\mathbf{H}^{-1})d\operatorname{vec}\mathbf{H}$, as $d|\mathbf{H}| = |\mathbf{H}|(\operatorname{vec}^\top\mathbf{H}^{-1})d\operatorname{vec}\mathbf{H}$, from Magnus & Neudecker (1999, Chapter 9.10, p. 178). We also have $d\{(\text{vec}^\top \mathbf{H})\mathbf{R}(\mathsf{D}^{\otimes 2}f)\,\text{vec}\,\mathbf{H}\} = 2(\text{vec}^\top \mathbf{H})\mathbf{R}(\mathsf{D}^{\otimes 2}f)d\,\text{vec}\,\mathbf{H}$ from Magnus & Neudecker (1999, Chapter 9.8), since $\mathbf{R}(\mathsf{D}^{\otimes 2}f)$ is symmetric.

So the first derivative of $\text{AMISE}(\mathbf{H})$ with respect to $\text{vec}\,\mathbf{H}$ is thus

$$\mathsf{D}_{\mathbf{H}}\text{AMISE}(\mathbf{H}) = -\tfrac{1}{2}n^{-1}R(K)|\mathbf{H}|^{-1/2}\,\text{vec}\,\mathbf{H}^{-1} + \tfrac{1}{2}m_2(K)^2\mathbf{R}(\mathsf{D}^{\otimes 2}f)\,\text{vec}\,\mathbf{H}.$$

Moreover, since $d(\text{vec}\,\mathbf{H}^{-1}) = -(\mathbf{H}^{-1}\otimes\mathbf{H}^{-1})d\,\text{vec}\,\mathbf{H}$ from Magnus & Neudecker (1999, Chapter 9.13, p. 183), the Hessian matrix is

$$\begin{aligned}\mathsf{H}_{\mathbf{H}}\text{AMISE}(\mathbf{H}) &= \tfrac{1}{2}n^{-1}R(K)|\mathbf{H}|^{-1/2}(\text{vec}\,\mathbf{H}^{-1}\,\text{vec}^\top\mathbf{H}^{-1} + \mathbf{H}^{-1}\otimes\mathbf{H}^{-1})\\ &\quad + \tfrac{1}{2}m_2(K)^2\mathbf{R}(\mathsf{D}^{\otimes 2}f).\end{aligned}$$

For any $\mathbf{H} = O(n^{-2/(d+4)})$, this implies that $\mathsf{H}_{\mathbf{H}}\text{AMISE}(\mathbf{H})$ is a constant matrix, i.e., does not depend on $n$ or the data. Applying the result for the inverse of the sum of a non-singular matrix and the outer product of the two vectors (Bartlett, 1951) ensures that $\mathsf{H}_{\mathbf{H}}\text{AMISE}(\mathbf{H})$ is non-singular. ■

With the general statement in Theorem 2, we can compute the rates contained in Table 3.1. We illustrate it with the special case plug-in selector in Theorem 3, which was first developed by Chacón & Duong (2010, Theorem 1).

**Theorem 3** *Suppose that (A1)–A(3) in Conditions A hold. Then the* $\text{AMSE}(\hat{\mathbf{H}}_{\text{PI}}) = O(n^{-4/(d+6)})\|\text{vec}\,\mathbf{H}^2_{\text{AMISE}}\|$, *and the relative rate of convergence of* $\hat{\mathbf{H}}_{\text{PI}}$ *to* $\mathbf{H}_{\text{AMISE}}$ *is* $n^{-2/(d+6)}$.

**Proof (Proof of Theorem 3)** Recall that $\text{PI}(\mathbf{H};\mathbf{G}) = n^{-1}|\mathbf{H}|^{-1/2}R(K) + \frac{1}{4}m_2(K)^2\hat{\boldsymbol{\psi}}_4(\mathbf{G})^\top(\text{vec}\,\mathbf{H})^{\otimes 2}$ so the key quantity of the discrepancy $\boldsymbol{\alpha}(\mathbf{H}) = \mathsf{D}_{\mathbf{H}}(\text{PI}-\text{AMISE})(\mathbf{H})$ is obtained as

$$\hat{\boldsymbol{\alpha}}(\mathbf{H}) = \tfrac{1}{2}m_2(K)^2\{(\text{vec}^\top\mathbf{H})\otimes\mathbf{I}_{d^2}\}\{\hat{\boldsymbol{\psi}}_4(\mathbf{G}) - \boldsymbol{\psi}_4\}$$

since $\mathbf{R}(\mathsf{D}^{\otimes 2}f)\,\text{vec}\,\mathbf{H} = \text{vec}\{\mathbf{I}_{d^2}\mathbf{R}(\mathsf{D}^{\otimes 2}f)\,\text{vec}\,\mathbf{H}\} = \{(\text{vec}^\top\mathbf{H})\otimes\mathbf{I}_{d^2}\}\boldsymbol{\psi}_4$. It follows that $\mathbb{E}\{\hat{\boldsymbol{\alpha}}(\mathbf{H})\} = \frac{1}{2}m_2(K)^2\{(\text{vec}^\top\mathbf{H})\otimes\mathbf{I}_{d^2}\}[\mathbb{E}\{\hat{\boldsymbol{\psi}}_4(\mathbf{G})\} - \boldsymbol{\psi}_4]$ and $\text{Var}\{\hat{\boldsymbol{\alpha}}(\mathbf{H})\} = \frac{1}{4}m_2(K)^4\{(\text{vec}^\top\mathbf{H})\otimes\mathbf{I}_{d^2}\}\,\text{Var}\{\hat{\boldsymbol{\psi}}_4(\mathbf{G})\}(\text{vec}\,\mathbf{H}\otimes\mathbf{I}_{d^2})$. As $\mathsf{H}_{\mathbf{H}}\text{AMISE}(\mathbf{H}_{\text{AMISE}}) = O(\mathbf{J}_{d^2})$ from Theorem 2, then $\text{AMSE}(\hat{\mathbf{H}}_{\text{PI}})$ is the same order as the squared bias and variance of $\hat{\boldsymbol{\psi}}_4(\mathbf{G})$, multiplied by $\|\text{vec}\,\mathbf{H}^2_{\text{AMISE}}\|$. Their leading terms are given in Chacón & Duong (2010, Theorem 1) as

$$\begin{aligned}\text{Bias}\{\hat{\boldsymbol{\psi}}_4(\mathbf{G})\} &= \{n^{-1}|\mathbf{G}|^{-1/2}(\mathbf{G}^{-1/2})^{\otimes 4}\mathsf{D}^{\otimes 4}L(\mathbf{0})\\ &\quad + \tfrac{1}{2}m_2(L)(\text{vec}^\top\mathbf{G}\otimes\mathbf{I}_{d^4})\boldsymbol{\psi}_6\}\{1+o(1)\}\\ \text{Var}\{\hat{\boldsymbol{\psi}}_4(\mathbf{G})\} &= \{4n^{-1}\,\text{Var}\{\mathsf{D}^{\otimes 4}f(\boldsymbol{X})\}\\ &\quad + 2n^{-2}\psi_0(\mathbf{G}^{-1/2})^{\otimes 4}\mathbf{R}(\mathsf{D}^{\otimes 4}L)(\mathbf{G}^{-1/2})^{\otimes 4}\}\{1+o(1)\}.\end{aligned}$$

Theorem 2 in this same paper asserts that the pilot bandwidth which minimises the $\mathrm{MSE}\{\hat{\boldsymbol{\psi}}_4(\mathbf{G})\}$ is order $n^{-2/(d+6)}$. Using this bandwidth order, then $\mathrm{AMISE}(\hat{\mathbf{H}}_{\mathrm{PI}}) = O(n^{-4/(d+6)})\|\mathrm{vec}\,\mathbf{H}^2_{\mathrm{AMISE}}\|$. ■

For smoothed cross validation, the calculations are similar to the plug-in case but are more complicated as a higher order approximation of the AMISE is required in $(\mathrm{SCV}-\mathrm{AMISE})(\mathbf{H})$. The resulting $\mathrm{MSE}(\hat{\mathbf{H}}_{\mathrm{SCV}})$ is the same order as $\mathrm{AMSE}^*\{\hat{\boldsymbol{\psi}}_4(\mathbf{G})\}$ which is the same as $\mathrm{AMSE}\{\hat{\boldsymbol{\psi}}_4(\mathbf{G})\}$ except that the convolved kernel $\bar{L}$ replaces the kernel $L$, see Chacón & Duong (2011). With these similarities between the plug-in and SCV analysis, it is not surprising that the relative convergence rate of $\hat{\mathbf{H}}_{\mathrm{SCV}}$ is the same as for $\hat{\mathbf{H}}_{\mathrm{PI}}$.

For unbiased cross validation, whilst $(\mathrm{UCV}-\mathrm{MISE})(\mathbf{H})$ involves a double sum, there is no pilot bandwidth to consider. The convergence rate of $\hat{\mathbf{H}}_{\mathrm{UCV}}$ is established in Duong & Hazelton (2005*b*, Theorem 1), and updated to our current notation by Chacón & Duong (2013, Theorem 1).

For biased cross validation, for the required discrepancy $(\mathrm{BCV}-\mathrm{AMISE})(\mathbf{H}) = \frac{1}{4}m_2(K)^2(\hat{\boldsymbol{\psi}}_4(\mathbf{H}) - \boldsymbol{\psi}_4)^\top(\mathrm{vec}\,\mathbf{H})^{\otimes 2}$, it is more involved to compute the derivative with respect to $\mathrm{vec}\,\mathbf{H}$ than for the PI selector, since for the latter the term $\hat{\boldsymbol{\psi}}_4(\mathbf{G})$ does not depend on $\mathbf{H}$. The convergence rate of $\hat{\mathbf{H}}_{\mathrm{BCV}}$ is obtained in Duong & Hazelton (2005*a*, Theorem 2). Observe that the UCV and BCV selectors have the same convergence rate as they both do not use pilot selectors.

### 3.10.2 *Optimal pilot bandwidth selectors*

An equivalent definition of the relative convergence rate of a data-based pilot bandwidth $\hat{\mathbf{G}}$ to $\mathbf{G}_{\mathrm{AMSE}}$ follows directly from Equation (3.21). Results treating the unconstrained case have not been established, though e.g., Duong (2004, Lemmas 6 and 14) establish that this relative rate is $O_p(n^{-2/d+8)})$ for the scalar class $\mathcal{A}$, respectively, for the plug-in (Equation 3.15) and SCV (Equation 3.20) pilot selectors. It is conjectured that this rate will remain unchanged for the unconstrained class $\mathcal{F}$.

Separate pilot bandwidths for each of the elements of $\mathbf{R}(\mathsf{D}^{\otimes 2}f)$ is espoused in Wand & Jones (1994), i.e., considering different pilot bandwidths $\mathrm{argmin}_{g>0}\,\mathrm{AMSE}\{\hat{\psi}_{r_1,\ldots,r_d}(g)\}$ for each combination of $r_1,\ldots,r_d = 0,\ldots,4, r_1+\ldots r_d = 4$, which are the multi-indices that enumerate all the fourth order partial derivatives comprising $\mathbf{R}(\mathsf{D}^{\otimes 2}f)$. If $\hat{\mathbf{R}}(\mathsf{D}^{\otimes 2}f)$ is obtained by the element-wise replacement of the $\psi_{r_1,\ldots,r_d}$ by $\hat{\psi}_{r_1,\ldots,r_d}(g)$, then an alternative form of the plug-in criterion in Equation (3.12) is $\mathrm{PI}^*(\mathbf{H}) = n^{-1}|\mathbf{H}|^{-1/2}R(K) + \frac{1}{4}m_2(K)^2(\mathrm{vec}^\top\mathbf{H})\hat{\mathbf{R}}(\mathsf{D}^{\otimes 2}f)\,\mathrm{vec}\,\mathbf{H}$. With a diagonal bandwidth matrix $\mathbf{H}\in\mathcal{D}$, then $\mathrm{argmin}_{\mathbf{H}\in\mathcal{D}}\,\mathrm{PI}^*(\mathbf{H})$ is well-defined. However this is not the case for $\mathbf{H} \in \mathcal{F}$ as $\hat{\mathbf{R}}(\mathsf{D}^{\otimes 2}f)$ is no longer guaranteed to be positive definite. Duong & Hazelton (2003) observe that positive definiteness is reinstated by using a single pilot bandwidth for all the elements, and derived their ad hoc pilot as the minimiser of the sum of all AMSEs of the individual scalar estimators:

$$\mathrm{SAMSE}(g) = \sum_{\substack{r_1,\ldots,r_d=0 \\ r_1+\cdots+r_d=4}}^{4} \mathrm{AMSE}\{\psi_{r_1,\ldots,r_d}(g)\}.$$

This has been subsequently refined for the more rigorous optimality criterion $\mathrm{MSE}(\hat{\mathbf{H}}_{\mathrm{PI}}) = \|\mathbb{E}\{(\mathrm{vec}\,\hat{\mathbf{H}}_{\mathrm{PI}} - \mathbf{H}_{\mathrm{AMISE}}\}\|^2$ and for unconstrained pilot matrices by Chacón & Duong (2010, Theorems 1–2). As stated previously in Equation (3.13), $\mathrm{MSE}(\hat{\mathbf{H}}_{\mathrm{PI}}) = \mathrm{const} \cdot \mathrm{AMSE}\{\hat{\boldsymbol{\psi}}_4(\mathbf{G})\}\{1+o(1)\}$ where the constant does not affect the optimisation with respect to $\mathbf{G}$. Furthermore the squared bias dominates the corresponding variance in the $\mathrm{AMSE}\{\hat{\boldsymbol{\psi}}_4(\mathbf{G})\}$, so optimal selection can be based on a bias minimisation of the type $\mathbf{G}_4 = \mathrm{argmin}_{\mathbf{G}\in\mathcal{F}}\|\mathrm{Bias}\{\hat{\boldsymbol{\psi}}_4(\mathbf{G})\}\|^2$, which is Equation (3.14).

For smoothed cross validation, the calculations are similar but more complicated (Chacón & Duong, 2011, Theorems 1–3) with the result that $\mathrm{MSE}(\hat{\mathbf{H}}_{\mathrm{SCV}}) = \|\mathbb{E}\{(\mathrm{vec}\,\hat{\mathbf{H}}_{\mathrm{SCV}} - \mathbf{H}_{\mathrm{MISE}}\}\|^2 = \mathrm{const} \cdot \mathrm{AMSE}^*\{\hat{\boldsymbol{\psi}}_4(\mathbf{G})\}$.

It should be noted also that, through the use of Fourier analysis and a careful choice of the truncation frequency (which is equivalent to using the infinite-order sinc kernel and a pilot bandwidth), recently Wu et al. (2014) were able to obtain an estimator $\tilde{\boldsymbol{\psi}}_4$ such that the difference $\tilde{\boldsymbol{\psi}}_4 - \boldsymbol{\psi}_4$ is of order $n^{-1/2}$ in probability for sufficiently smooth densities, resulting in a relative rate of convergence of the corresponding $\tilde{\mathbf{H}}_{\mathrm{PI}}$ to $\mathbf{H}_{\mathrm{AMISE}}$ of order $n^{-1/2}$ as well. Nonetheless, the relative rate of convergence of $\tilde{\mathbf{H}}_{\mathrm{PI}}$ to $\mathbf{H}_{\mathrm{MISE}}$ remains $n^{-2/(d+4)}$, since this is the order of the approximation of $\mathbf{H}_{\mathrm{AMISE}}$ to $\mathbf{H}_{\mathrm{MISE}}$.

### *3.10.3 Convergence rates with data-based bandwidths*

The computation of a data-based bandwidth $\hat{\mathbf{H}}$ is an intermediate step to the computation of the density estimate $\hat{f}(\boldsymbol{x};\hat{\mathbf{H}})$. So an entirely rigorous mathematical analysis requires the examination of the MISE of this 'nested' estimator $\hat{f}(\boldsymbol{x};\hat{\mathbf{H}})$. Despite this, the majority of results for kernel estimators are usually stated, as we have done, in terms of the $\mathrm{MISE}\{\hat{f}(\cdot;\mathbf{H})\}$ where the bandwidth is non-random and does not depend on the data. This appears to be a startling oversight as there are few reported results for $\hat{f}(\boldsymbol{x};\hat{\mathbf{H}})$. However this is much less serious than it first appears, as using the standard results for the convergence of random variables, then $\mathrm{MISE}\{\hat{f}(\cdot;\hat{\mathbf{H}})\}$ is automatically assured once the convergence of the $\text{MISE}\{\hat{f}(\cdot;\mathbf{H})\}$ to zero and the $\hat{\mathbf{H}}$ to $\mathbf{H}_{\text{AMISE}}$ are established, as the MISE is an smooth integral operator. Tenreiro (2001) establishes that, for all $d$, the difference between $\text{MISE}\{\hat{f}(\cdot;\hat{\mathbf{H}})\}$ and $\text{MISE}\{\hat{f}(\cdot;\mathbf{H})\}$ is $O(n^{-(d+8)/(2d+8)})$ for all the bandwidth selectors considered in Table 3.1, which is dominated by the minimal $\text{MISE}\{\hat{f}(\cdot;\mathbf{H})\}$ rate of $O(n^{-4/(d+4)})$. This implies, from a practical data analysis point of view, that we can use $\hat{f}(\boldsymbol{x};\hat{\mathbf{H}})$ with confidence even if we only establish the convergence of $\hat{f}(\boldsymbol{x};\mathbf{H})$.

Likewise, the relative convergence rates for $\hat{\mathbf{H}}_{\text{PI}}, \hat{\mathbf{H}}_{\text{SCV}}$ in Table 3.1 are established assuming that the optimal pilot bandwidths $\mathbf{G}_{\text{AMSE}}$ are non-random, whereas for practical data analysis, the $\hat{\mathbf{G}}$ are computed from the data as in Algorithms 4 and 5. Since $\hat{\mathbf{G}}$ converges to $\mathbf{G}_{\text{AMSE}}$, then the regularity imposed by (A1)–(A3) in Conditions A implies that the $\hat{\mathbf{H}}_{\text{PI}}, \hat{\mathbf{H}}_{\text{SCV}}$ remain convergent. Tenreiro (2003) establishes the convergence rate for a univariate plug-in selector with data-based pilot bandwidths, though with slightly different conditions to ours. The unconstrained multivariate case remains unestablished.

Chapter 4

# Modified density estimation

The previous chapters treated the standard setup for density estimation, in which the normal kernel, with a fixed, global bandwidth is employed to analyse data with an infinite support (or a support without a rigid boundary) and without heavy tails. When these classical conditions do not hold, various modifications have been introduced. Section 4.1 develops variable bandwidth estimators which apply varying amounts of smoothing to tackle heavy-tailed data. Sections 4.2–4.3 present transformation and boundary kernel estimators to tackle bounded data. Sections 4.4–4.5 examine the role of the kernel function and the alternatives to the normal kernel. Section 4.6 fills in the previously omitted mathematical details of the considered modified density estimators.

## 4.1 Variable bandwidth density estimators

The standard kernel density estimator $\hat{f}$ in Equation (2.2) assigns a probability mass to the neighbourhood of each data point $\boldsymbol{X}_i$ according to the action of the scaled kernel function $K_{\mathbf{H}}(\boldsymbol{x}-\boldsymbol{X}_i)$, $i = 1,\ldots,n$. As $\mathbf{H}$ is a constant matrix, this applies a constant amount of smoothing throughout the data space. Variable bandwidth estimators extend $\hat{f}$ by varying the bandwidth. These variable estimators are also known as adaptive estimators. The underlying idea is that, for data-sparse regions, a large bandwidth applies smoothing over a larger region to compensate for the few nearby data points, and conversely, for data-dense regions, a small bandwidth applies smoothing in a smaller region due to the presence of many nearby data points. Variable density estimators are divided into two main classes depending on the method of varying the bandwidth: there exist several competing nomenclatures, and we adopt that of Sain & Scott (1996).

### 4.1.1 *Balloon density estimators*

The first class is the balloon density estimators where the bandwidth varies with the estimation point $\boldsymbol{x}$

$$\hat{f}_{\text{ball}}(\boldsymbol{x};\mathbf{H}(\boldsymbol{x})) = n^{-1}\sum_{i=1}^{n} K_{\mathbf{H}(\boldsymbol{x})}(\boldsymbol{x}-\boldsymbol{X}_i).$$

To account for the idea of using a larger bandwidth in data-sparse regions, in their original paper Loftsgaarden & Quesenberry (1965) propose class $\mathcal{A}$ nearest neighbour selectors $\mathbf{H}(\boldsymbol{x}) = \delta_{(k)}(\boldsymbol{x})^2\mathbf{I}_d$ where $\delta_{(k)}(\boldsymbol{x})$ is the distance from $\boldsymbol{x}$ to the $k$-th nearest data sample point ($k$-th nearest neighbour). General bandwidth functions in the above equation were not introduced until Jones (1990).

The pointwise AMSE of $\hat{f}_{\text{ball}}$ is the almost exactly the same as that for the fixed bandwidth estimator $\hat{f}$, namely

$$\begin{aligned}\text{AMSE}\{\hat{f}_{\text{ball}}(\boldsymbol{x};\mathbf{H}(\boldsymbol{x}))\} &= n^{-1}|\mathbf{H}(\boldsymbol{x})|^{-1/2}R(K)f(\boldsymbol{x}) \\ &\quad + \tfrac{1}{4}m_2(K)^2\{\mathsf{D}^{\otimes 2}f(\boldsymbol{x})^\top \operatorname{vec}\mathbf{H}(\boldsymbol{x})\}^2.\end{aligned}$$

Analogously, $\mathbf{H}_{\text{AMSE}}(\boldsymbol{x}) = \operatorname{argmin}_{\mathbf{H}\in\mathcal{F}} \text{AMSE}\{\hat{f}_{\text{ball}}(\boldsymbol{x};\mathbf{H})\}$ is a suitable oracle target bandwidth. As noted in Terrell & Scott (1992, Section 5), the existence of an explicit formula for this oracle bandwidth depends on the definiteness properties of the Hessian matrix $\mathsf{H}f(\boldsymbol{x})$. Observe that $\{\mathsf{D}^{\otimes 2}f(\boldsymbol{x})^\top \operatorname{vec}\mathbf{A}\}^2 = \operatorname{tr}^2\{\mathsf{H}f(\boldsymbol{x})\mathbf{A}\}$ and that the minimum of this functional over the matrices $\mathbf{A}$ with $|\mathbf{A}| = 1$ is attained at $|\mathsf{H}f(\boldsymbol{x})|^{1/d}\mathsf{H}f(\boldsymbol{x})^{-1}$, with a minimum value of $d^2|\mathsf{H}f(\boldsymbol{x})|^{2/d}$, see Celeux & Govaert (1995, Theorem A.1). Therefore, writing $\mathbf{H} = \lambda\mathbf{A}$ with $|\mathbf{A}| = 1$ and reasoning as in Section 2.9.2, if $\mathsf{H}f(\boldsymbol{x})$ is positive definite, then

$$\mathbf{H}_{\text{AMSE}}(\boldsymbol{x}) = \left\{\frac{R(K)f(\boldsymbol{x})}{m_2(K)^2 d}\right\}^{2/(d+4)} |\mathsf{H}f(\boldsymbol{x})|^{1/(d+4)}\mathsf{H}f(\boldsymbol{x})^{-1}n^{-2/(d+4)}.$$

The disadvantage of this formula for $\mathbf{H}_{\text{AMSE}}(\boldsymbol{x})$ is that is not applicable for those estimation points whose Hessian is not positive definite. So to progress with its analysis, we let $\mathbf{H}(\boldsymbol{x}) = h(\boldsymbol{x})^2\mathbf{H}_0$, for a fixed matrix $\mathbf{H}_0$. While this is not maximally general, it remains in the $\mathcal{F}$ class and the explicit optimiser is $\mathbf{H}_{\text{AMSE}}(\boldsymbol{x}) = h_{\text{AMSE}}(\boldsymbol{x})^2\mathbf{H}_0$, where

$$h_{\text{AMSE}}(\boldsymbol{x}) = \left[\frac{dR(K)f(\boldsymbol{x})}{m_2(K)^2|\mathbf{H}_0|^{1/2}\{\mathsf{D}^{\otimes 2}f(\boldsymbol{x})^\top \operatorname{vec}\mathbf{H}_0\}^2}\right]^{1/(d+4)} n^{-1/(d+4)}$$

which is almost identical to Terrell & Scott (1992, Proposition 6). The sub-optimality of $h_{\text{AMSE}}(\boldsymbol{x})^2\mathbf{H}_0$ is compensated by the varying nature of the bandwidths $h_{\text{AMSE}}(\boldsymbol{x})$ which adapt the smoothing according to the local data density. In high density regions the curvature is high, so $h_{\text{AMSE}}$ is smaller, and in low density regions the curvature is flatter, so $h_{\text{AMSE}}$ is larger. The minimal AMSE is $O(n^{-1}h_{\text{AMSE}}(\boldsymbol{x})^{-d} + h_{\text{AMSE}}(\boldsymbol{x})^2) = O(n^{-4/(d+4)})$ which is the same rate as for fixed bandwidth estimator $\hat{f}$, though it has been established that the coefficient for the former is smaller than the latter (Cao, 2001).

A disadvantage of $\hat{f}_{\text{ball}}$ is that it is not guaranteed to be a proper density function: whilst it is non-negative, its integral is not one and is not even guaranteed to be finite for general $f$. Another undesirable property is that the potentially unbounded domain of $\mathbf{H}_{\text{AMSE}}(\boldsymbol{x})$ could lead to very distant points to $\boldsymbol{x}$ contributing to the density estimate at $\boldsymbol{x}$, disrupting the local averaging nature of kernel estimators. Moreover, despite the existence of some proposals for data-based selection of the local bandwidth $\mathbf{H}(\boldsymbol{x})$ (Hall & Schucany, 1989; Hall, 1993; Hazelton, 1996; Sain, 2001), for some classes of densities it has been shown that it is impossible to construct consistent data-based local bandwidth selectors (Devroye & Lugosi, 2001*b*). All this has not facilitated the widespread use of balloon density estimators.

### 4.1.2 *Sample point density estimators*

Another type of variable kernel estimator is the sample point estimator, as devised by Breiman et al. (1977), where a different bandwidth is employed to rescale the kernel around each data point

$$\hat{f}_{\text{SP}}(\boldsymbol{x};\mathbf{H}) \equiv \hat{f}_{\text{SP}}(\boldsymbol{x};\mathbf{H}_1,\ldots,\mathbf{H}_n) = n^{-1}\sum_{i=1}^{n} K_{\mathbf{H}_i}(\boldsymbol{x}-\boldsymbol{X}_i).$$

The estimator $\hat{f}_{\text{SP}}$ resolves some of the disadvantages of the balloon estimator $\hat{f}_{\text{ball}}$. The former remains a proper density function if $K$ is a second order kernel. There are as many bandwidths as data points $n$ which, for modest sample sizes, is less likely to lead to overparametrisation than the bandwidth function for $\hat{f}_{\text{ball}}$, which requires a different bandwidth at each estimation point. Due to the dependence on the data points, it is also common to write $\mathbf{H}_i = \mathbf{H}(\boldsymbol{X}_i)$.

Similar to Loftsgaarden & Quesenberry (1965), Breiman et al. (1977) proposed the selectors $\mathbf{H}_i = \delta_{(k)}(\boldsymbol{X}_i)^2\mathbf{I}_d$ where $\delta_{(k)}(\boldsymbol{X}_i)$ is the $k$-th nearest neighbour distance to $\boldsymbol{X}_i$. Sain (2002) established that this nearest neighbour-based choice is asymptotically equivalent to $\mathbf{H}_i = f(\boldsymbol{X}_i)^{-2/d}h^2\mathbf{I}_d$, and also reported that Breiman et al. (1977) themselves admitted that the numerical performance of their selectors did not perform as well for the univariate case as in the bivariate case, though the latter were not able to explain this apparent inconsistency. This was resolved by Abramson (1982) who proposed the eponymous selector $\mathbf{H}_{\text{Ab},i} = f(\boldsymbol{X}_i)^{-1}h^2\mathbf{I}_d$ for all dimensions $d$. Observe that for $d = 1$, the Breiman et al. bandwidths are proportional to $f(\boldsymbol{X}_i)^{-2}$ which is a different order of magnitude compared to the $f(\boldsymbol{X}_i)^{-1}$ Abramsom bandwidths, which explains the former's unsatisfactory performance. For $d = 2$, these bandwidths coincide so their bivariate performances are identical.

The calculation of the AMSE with the Abramsom selector involves some complex algebra. The resulting expressions in Abramson (1982), Hall & Marron (1988), Terrell & Scott (1992), or Bowman & Foster (1993*a*) are not in a suitable form for data-based bandwidth selection. The most intuitive derivation is the multivariate generalisation of Jones et al. (1994) which yields

$$\text{AMSE}\{\hat{f}_{\text{SP}}(\boldsymbol{x};\mathbf{H}_{\text{Ab}})\} = n^{-1}h^{-d}R(K)f(\boldsymbol{x})^{1+d/2} + \tfrac{1}{576}h^8\{\boldsymbol{m}_4(K)^\top\boldsymbol{\lambda}(\boldsymbol{x})\}^2 \tag{4.1}$$

where $\boldsymbol{m}_4(K) = \int_{\mathbb{R}^d}\boldsymbol{x}^{\otimes 4}K(\boldsymbol{x})\,d\boldsymbol{x}$ and

$$\begin{aligned}\boldsymbol{\lambda}(\boldsymbol{x}) &= \mathsf{D}^{\otimes 4}\{f(\boldsymbol{x})^{-1}\}\\ &= 24\{\mathsf{D}f(\boldsymbol{x})\}^{\otimes 4}/f(\boldsymbol{x})^5 - 36[\mathsf{D}^{\otimes 2}f(\boldsymbol{x})\otimes\{\mathsf{D}f(\boldsymbol{x})\}^{\otimes 2}]/f(\boldsymbol{x})^4\\ &\quad + [6\{\mathsf{D}^{\otimes 2}f(\boldsymbol{x})\}^{\otimes 2} + 8\mathsf{D}^{\otimes 3}f(\boldsymbol{x})\otimes\mathsf{D}f(\boldsymbol{x})]/f(\boldsymbol{x})^3 - \mathsf{D}^{\otimes 4}f(\boldsymbol{x})/f(\boldsymbol{x})^2.\end{aligned}$$

This has an optimiser at

$$h_{\text{Ab}}(\boldsymbol{x}) = \left[\frac{72dR(K)f(\boldsymbol{x})^{1+d/2}}{\{\boldsymbol{m}_4(K)^\top\boldsymbol{\lambda}(\boldsymbol{x})\}^2}\right]^{1/(d+8)} n^{-1/(d+8)}.$$

These bandwidths induce in $\hat{f}_{\text{SP}}$ a bias of order $h^4$, which improves on the order $h^2$ bias of a fixed, global bandwidth. This bias order is the same as that achieved using a fourth order kernel for a fixed bandwidth estimator, although without the side effect of possible negativity, see Section 4.4. The minimal MSE rate of $\hat{f}_{\text{SP}}$ is thus of order $n^{-8/(d+8)}$, which is faster than the $n^{-4/(d+4)}$ rate for $\hat{f}_{\text{ball}}$. This has contributed to the widespread use of the Abramson bandwidths. A more recent proposal posited by Sain (2002) is a bandwidth function which is constant within the bins of an estimation grid, but it appears not to have challenged the prevalence of the older Abramson bandwidths.

### *4.1.3 Bandwidth selectors for variable kernel estimation*

The formulas for $h_{\text{AMSE}}(\boldsymbol{x})$ and $h_{\text{Ab}}(\boldsymbol{x})$ involve the unknown density $f$ and its derivatives. As usual, they can be estimated using pilot estimators, although the important practical issues of these pilot estimators need to be resolved before variable kernel estimators reach the same level of maturity as fixed bandwidth estimators. Nonetheless, we still wish to illustrate the potential gains of the former for finite samples so, for simplicity, we take the normal scale bandwidth $\hat{\mathbf{H}}_{\mathrm{NS},r} = \{4/(d+2r)\}^{2/(d+2r+2)} n^{-2/(d+2r+2)} \mathbf{S}$ for the $r$-th order density derivative estimator $\mathsf{D}^{\otimes r} \hat{f}_{\mathrm{pilot}}$, which is developed in Equation (5.18) in Chapter 5. Then we take $\hat{\mathbf{H}}_{\mathrm{ball}}(\boldsymbol{x}) = \hat{h}_{\mathrm{ball}}(\boldsymbol{x})^2 \mathbf{I}_d$ where

$$\hat{h}_{\mathrm{ball}}(\boldsymbol{x}) = \left[ \frac{d \hat{f}_{\mathrm{pilot}}(\boldsymbol{x}; \hat{\mathbf{H}}_{\mathrm{NS},2})}{(4\pi)^{d/2} \{\mathsf{D}^{\otimes 2} \hat{f}_{\mathrm{pilot}}(\boldsymbol{x}; \hat{\mathbf{H}}_{\mathrm{NS},2})^\top \operatorname{vec} \mathbf{I}_d\}^2} \right]^{1/(d+4)} n^{-1/(d+4)}$$

for the balloon estimator, and $\hat{\mathbf{H}}_{\mathrm{Ab},i} = \hat{f}_{\mathrm{pilot}}(\boldsymbol{X}_i; \hat{\mathbf{H}}_{\mathrm{NS},4})^{-1} \hat{h}^2_{\mathrm{Ab},i} \mathbf{I}_d$ where

$$\hat{h}_{\mathrm{Ab},i} = \hat{h}_{\mathrm{Ab}}(\boldsymbol{X}_i) = \left[ \frac{8d \hat{f}_{\mathrm{pilot}}(\boldsymbol{X}_i; \hat{\mathbf{H}}_{\mathrm{NS},4})^{1+d/2}}{(4\pi)^{d/2} \{(\operatorname{vec}^\top \mathbf{I}_d)^{\otimes 2} \hat{\boldsymbol{\lambda}}(\boldsymbol{X}_i; \hat{\mathbf{H}}_{\mathrm{NS},4})\}^2} \right]^{1/(d+8)} n^{-1/(d+8)}$$

for the sample point estimator. Whenever the numerator and/or denominator is zero, this will lead to undefined density estimates: for any $\boldsymbol{x}$ where this is the case for $\hat{h}_{\mathrm{ball}}(\boldsymbol{x})$, then $\hat{f}(\boldsymbol{x})$ is set to 0, and for any $\boldsymbol{X}_i$ where this is the case for $\hat{h}_{\mathrm{Ab},i}$, then $\boldsymbol{X}_i$ is excluded from the computation. An optimal strategy to deal with these cases is one of the outstanding issues in variable kernel estimation.

We note that these two classes of balloon and sample point estimators are not mutually exclusive and could even be combined (Jones, 1990).

**Example 4.1** The fixed and variable density estimates on a subset of the World Bank development indicators data from Section 1.1, the annual GDP growth rate (%) and the annual inflation rate (%), are displayed in Figure 4.1. Figure 4.1(a) is the scatter plot of the $n = 177$ nations with complete measurements. Most of these are concentrated around (5%, 5%), though there are a few extreme outlying pairs. All of these data, including the outliers were utilised in calculations as no pre-processing was applied to remove them, indicating that variable kernel methods robustly handle outliers. In the lower left of this panel is the normal kernel with fixed bandwidth $\hat{\mathbf{H}}_{\mathrm{PI}} = [2.80, 0.63; 0.63, 2.65]$. Figure 4.1(b) is the fixed density estimate with this $\hat{\mathbf{H}}_{\mathrm{PI}}$. The constant smoothing applied by a fixed bandwidth results in the bumps centred at the data points in the tails. Since these GDP growth and inflation rates have similar marginal spreads, the variable bandwidth functions from class $\mathcal{A}$ are suitable without any re-scaling. Figure 4.1(c) is the bandwidth function $\hat{\mathbf{H}}_{\mathrm{ball}}(\boldsymbol{x}) = \hat{h}_{\mathrm{ball}}(\boldsymbol{x})^2 \mathbf{I}_2$ for $\boldsymbol{x}$ over a grid: the grid points with no contours imply that $\hat{h}_{\mathrm{ball}}$ is zero or undefined. The contours are

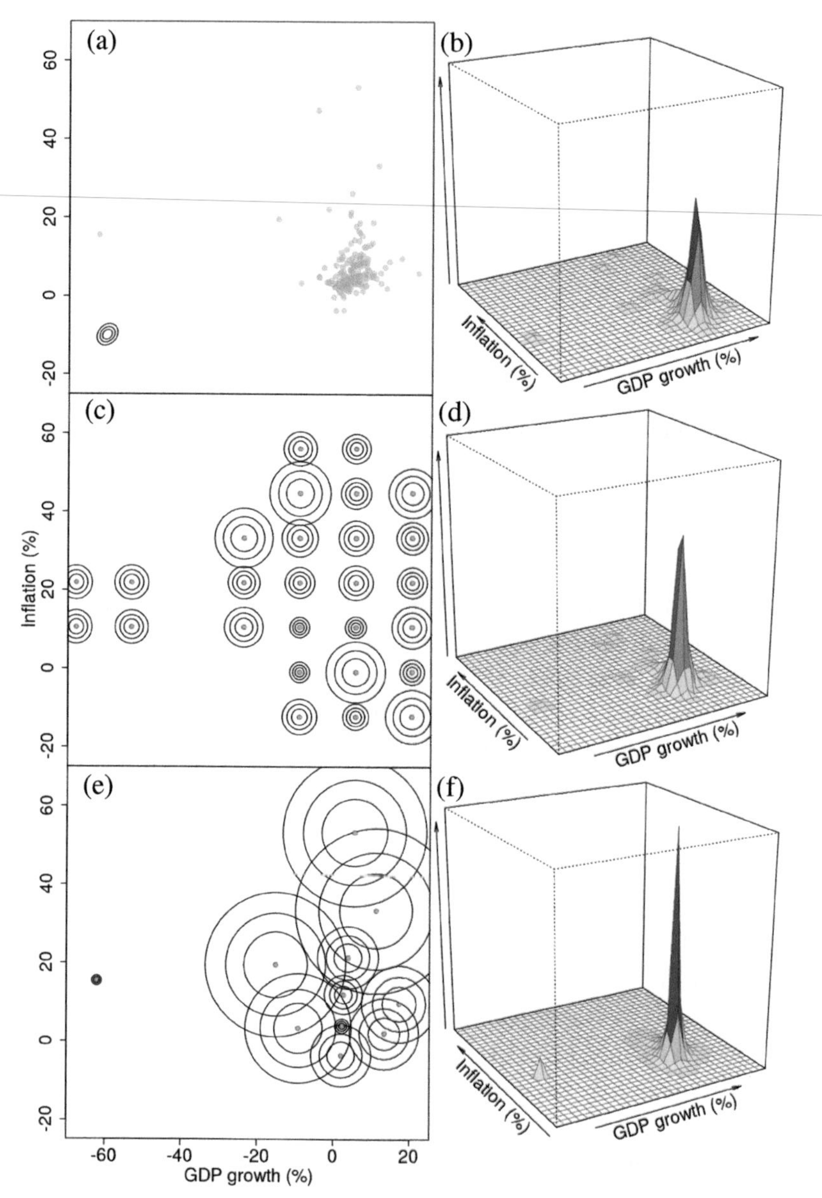


Figure 4.1 *Variable bandwidth density estimates for the World Bank GDP-inflation data. The x-axis is the annual GDP growth rate (%) and the y-axis is the inflation rate (%). (a) Scatter plot of the $n = 177$ data pairs, with a normal kernel with bandwidth* $\hat{\mathbf{H}}_{\text{PI}} = [2.80, 0.63; 0.63, 2.65]$. *(b) Fixed bandwidth density estimate $\hat{f}$ with $\hat{\mathbf{H}}_{\text{PI}}$. (c) Balloon variable bandwidth function $\hat{\mathbf{H}}_{\text{ball}}$. (d) Balloon variable density estimate $\hat{f}_{\text{ball}}$ with $\hat{\mathbf{H}}_{\text{ball}}$. (e) Abramson bandwidth function $\hat{\mathbf{H}}_{\text{Ab},i}$. (f) Sample point variable density estimate $\hat{f}_{\text{SP}}$ with $\hat{\mathbf{H}}_{\text{Ab},i}$.*

the most peaked near the densest data region, and tend to increase in size as we move away from this region. The balloon density estimate with this $\hat{\mathbf{H}}_{\text{ball}}(\boldsymbol{x})$ is shown in Figure 4.1(d), where we observe that the bumps for the fixed bandwidth estimate in (b) are attenuated and the height of the main modal region is increased. Figure 4.1(e) shows the Abramsom bandwidths $\hat{\mathbf{H}}_{\text{Ab},i} = \hat{f}_{\text{pilot}}(\boldsymbol{X}_i; \hat{\mathbf{H}}_{\text{NS},4})^{-1}\hat{h}^2_{\text{Ab},i}\mathbf{I}_2$ centred on (a subset of) the data points, and the induced kernel contours are more clearly inversely proportional to the data density. The sample point estimator is shown in Figure 4.1(f). In comparison to the balloon estimator, the main modal height is higher, and the tail bumps more attenuated, though the minor mode at the isolated data point (–61.8%, 13.3%) is more pronounced. □

## 4.2 Transformation density estimators

The standard kernel density estimator $\hat{f}$ in Equation (2.2) assigns a probability mass to the neighbourhood of each data point $\boldsymbol{X}_i$ according the action of the scaled kernel function $K_{\mathbf{H}}(\boldsymbol{x} - \boldsymbol{X}_i)$, $i = 1, \ldots, n$. In the case where the data support contains a rigid boundary, for a data point $\boldsymbol{X}_i$ sufficiently close to this boundary, the scaled kernel assigns a positive probability mass outside of the support, which leads to a local under-estimation of the density and hence an increased bias of the estimator.

A major class of modified density estimators for bias reduction are the transformation density estimators, introduced by Devroye & Györfi (1985); Silverman (1986). Let $\boldsymbol{t} : \mathbb{R}^d \to \mathbb{R}^d$ be a known function which is component-wise monotone, i.e., the component functions $t_j, j = 1, \ldots, d$ are invertible. The usual relation between the density $\hat{f}_{\boldsymbol{Y}}$ of $\boldsymbol{Y} = \boldsymbol{t}(\boldsymbol{X})$ and the density $f$ of $\boldsymbol{X}$ is $f_{\boldsymbol{Y}}(\boldsymbol{y}) = J_{\boldsymbol{t}}(\boldsymbol{x})^{-1} f(\boldsymbol{x})$ where $J_{\boldsymbol{t}}(\boldsymbol{x})$ is the Jacobian of $\boldsymbol{t}$. Inverting this equation, we have the transformation density estimator

$$\hat{f}_{\text{trans}}(\boldsymbol{x}; \mathbf{H}) = J_{\boldsymbol{t}}(\boldsymbol{t}^{-1}(\boldsymbol{y}))\hat{f}_{\boldsymbol{Y}}(\boldsymbol{y}; \mathbf{H}) = J_{\boldsymbol{t}}(\boldsymbol{x})\hat{f}_{\boldsymbol{Y}}(\boldsymbol{t}(\boldsymbol{x}); \mathbf{H}). \tag{4.2}$$

When the data support is $(0, \infty)^d$, a suitable transformation is $t(x_j) = \log(x_j)$. Each inverse function is $t_j^{-1}(x_j) = \exp(x_j)$, and Jacobian $J_{\boldsymbol{t}}(\boldsymbol{x}) = 1/(x_1 \cdots x_d)$, which leads to the logarithm transformation density estimator

$$\hat{f}_{\text{trans}}(\boldsymbol{x}; \mathbf{H}) = \exp[-(y_1 + \cdots + y_d)]\hat{f}_{\boldsymbol{Y}}(\boldsymbol{y}; \mathbf{H}) = 1/(x_1 \cdots x_d)\hat{f}_{\boldsymbol{Y}}(\log(\boldsymbol{x}); \mathbf{H}). \tag{4.3}$$

Since the transformed $\boldsymbol{Y}_i$ have infinite support, we can apply any of the bandwidth selection and density estimation methods in Chapters 2–3, and then back transform to the bounded support of $\boldsymbol{X}_i$ in Equation (4.3).

**Example 4.2** We illustrate the logarithm transformation density estimator on a subset of the World Bank development indicators data: carbon dioxide ($CO_2$) emissions per capita (thousands Kg) and the gross domestic product (GDP) per capita (thousands of current USD). The scatter plot in Figure 4.2(a) shows that most observations are located close to the origin and the $CO_2$ emissions and the GDP tend to follow a positive correlation trend.

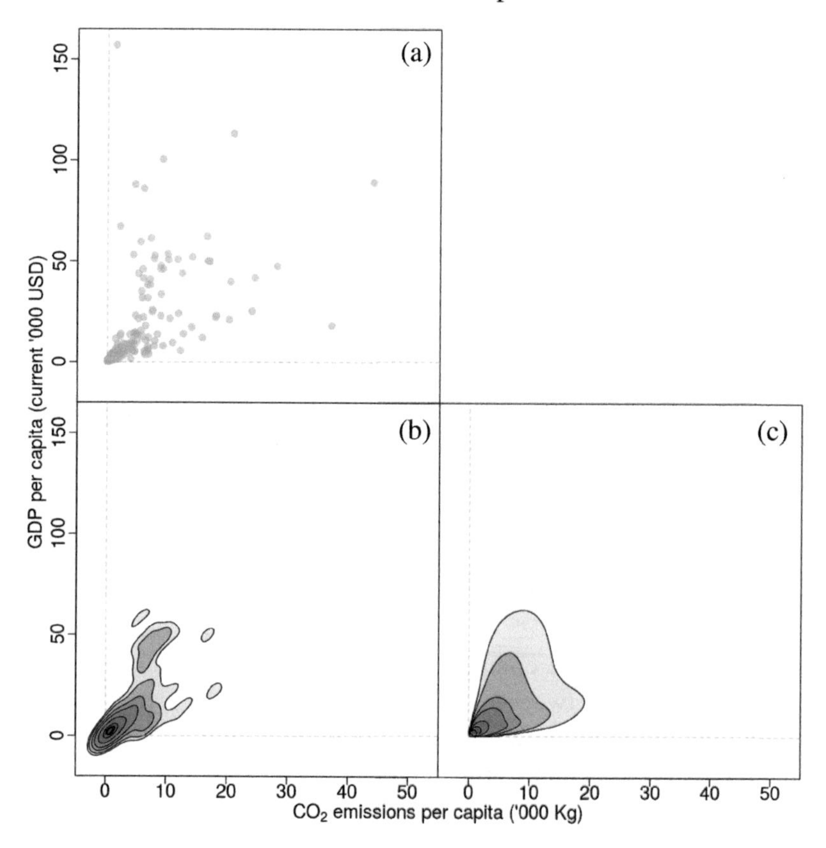


Figure 4.2 *Transformation density estimate for the World Bank $CO_2$-GDP data. The horizontal axis is the $CO_2$ emissions per capita (thousands Kg) and the vertical axis is the GDP per capita (thousands current USD). (a) Scatter plot of the $n = 245$ data pairs. (b) Standard density estimate, with bandwidth* $[1.44, 2.83; 2.83, 15.37]$. *(c) Logarithm transformation density estimate with bandwidth* $[0.32, 0.26; 0.26, 0.29]$.

If we apply a standard density estimate with a plug-in bandwidth $\hat{\mathbf{H}}_{\text{PI}} = [1.44, 2.83; 2.83, 15.37]$ without taking into account of the boundary near the origin, we observe that it assigns a significant probability mass to negative values for the $CO_2$ emissions and GDP in Figure 4.2(b). This is rectified by the log transformation estimator in Figure 4.2(c) with a plug-in bandwidth

$[0.32, 0.26; 0.26, 0.29]$ computed from the log-transformed data pairs. For the log transformation density estimate in Figure 4.2(c), we observe that it also removes the undersmoothed contours in Figure 4.2(b). In addition to correcting for the bounded data support, it mitigates the appearance of spurious bumps in the upper tail from utilising a fixed, global bandwidth with highly spatially inhomogeneous data. □

From Equation (4.3), the expected value of $\hat{f}_{\rm trans}$ at an estimation point $\boldsymbol{x}$ is

$$\begin{aligned}
&\mathbb{E}\{\hat{f}_{\rm trans}(\boldsymbol{x};\mathbf{H})\} \\
&\quad = 1/(x_1\cdots x_d)\{f_{\boldsymbol{Y}}(\log(\boldsymbol{x})) + \tfrac{1}{2}m_2(K)(\operatorname{vec}^\top\mathbf{H})\mathsf{D}^{\otimes 2}f_{\boldsymbol{Y}}(\log(\boldsymbol{x}))\}\{1+o(1)\} \\
&\quad = f(\boldsymbol{x}) + \tfrac{1}{2}m_2(K)(\operatorname{vec}^\top\mathbf{H})\mathsf{D}^{\otimes 2}f_{\boldsymbol{Y}}(\log(\boldsymbol{x}))/(x_1\cdots x_d)\{1+o(1)\}.
\end{aligned}$$

For a fixed $\mathbf{H}$, the bias of $\hat{f}_{\rm trans}$ depends on the ratio $\mathsf{D}^{\otimes 2}f_{\boldsymbol{Y}}(\log(\boldsymbol{x}))/(x_1\cdots x_d)$. For the World Bank data in Figure 4.2, this ratio does appear to tend to zero as $\boldsymbol{x}\to\mathbf{0}$, so there is no evidence of a systematic bias at the origin, which is also confirmed visually.

For the case of non-zero bias at the boundary, other approaches can be pursued to improve on the logarithm transformation density estimator. Possibilities include using a transformation from the richer shifted power family: $t(x_j) = x_j + \lambda_1)^{\lambda_2}\operatorname{sign}(\lambda_2)$ for $\lambda_2 \neq 0$, and $t(x_j) = \log(x_j+\lambda_1)$ for $\lambda_2 = 0$, see Wand et al. (1991). The logarithm transformation is the special case $\lambda_1 = \lambda_2 = 0$, and so to utilise the richness of this family, these authors propose data-based estimators for $\lambda_1, \lambda_2$. Ruppert & Cline (1994) introduced $t(x_j) = G^{-1}(\hat{F}_{X_j}(x_j; g_j))$, where $\hat{F}_{X_j}(\cdot; g_j)$ is the kernel estimator of the $j$-th marginal cumulative distribution function of $\boldsymbol{X}$ with bandwidth $g_j$, and $G$ is any known cumulative distribution which can be chosen to control the boundary bias, though estimating these extra parameters and/or functions have not won widespread favour over the simpler logarithm transformation.

When the data are supported on a hyper-rectangle, rather than $(0,\infty)^d$, then, without loss of generality, we can consider the unit hyper-rectangle as the data support. In this case, the transformation density estimator is applicable if we use the logit transformation $t_j(x_j) = \log(x_j/(1-x_j))$ or the probit transformation $t_j(x_j) = \Phi^{-1}(x_j)$ where $\Phi$ is the probit function (quantile function of standard normal random variable). Their properties are similar to those already explored, though we note that as the boundary region occupies more volume, more care is required to control the boundary bias. Geenens (2014) posited local polynomial adjustments of univariate probit density estimators in the boundary region to accomplish this control, accompanied by their bandwidth selection strategies.

Transformation density estimators were not initially introduced as boundary correction methods, but to improve density estimation under conditions such as spatial inhomogeneity and heavy tails, as reviewed in Geenens (2014), and as illustrated in Figure 4.2.

## 4.3 Boundary kernel density estimators

Whereas transformation estimators apply a transformation function which implicitly modifies the kernel function in order to reduce the bias in the boundary region, an alternative approach involves explicitly modifying the kernel functions in the boundary region from the usual symmetric to asymmetric functions to account for the rigid boundary.

### *4.3.1 Beta boundary kernels*

The simplest bounded support region is a (unit) hyper-rectangle. A suitable starting point here is the beta boundary kernel of Chen (1999). The univariate scaled beta boundary kernel is $K_{\text{beta}(1)}(y;x,h) = \text{Beta}(y;\alpha_{1,h}(x),\alpha_{2,h}(x))$ for $0 \le x,y \le 1$, where $\text{Beta}(y;\alpha_1,\alpha_2) = y^{\alpha_1-1}(1-y)^{\alpha_2-1}/B(\alpha_1,\alpha_2)$ is the density function for a beta random variable with shape parameters $\alpha_1,\alpha_2$,

$$\begin{cases} \alpha_{1,h}(x) = \rho_h(x), \alpha_{2,h}(x) = (1-x)/h^2 & x \in [0,2h^2) \\ \alpha_{1,h}(x) = x/h^2, \alpha_{2,h}(x) = (1-x)/h^2 & x \in [2h^2,1-2h^2] \\ \alpha_{1,h}(x) = x/h^2, \alpha_{2,h}(x) = \rho_h(1-x) & x \in (1-2h^2,1] \end{cases}$$

and $\rho_h(u) = 2h^4 + 5/2 - (4h^4 + 6h^2 + 9/4 - u^2 - u/h^2)^{1/2}$, following the notation of Jones & Henderson (2007). At a point $x$, we compute the mean of $X_i{}^{\alpha_{1,h}(x)-1}(1-X_i)^{\alpha_{2,h}(x)-1}/B(\alpha_{1,h}(x)+\alpha_{2,h}(x))$, $i=1,\dots,d$. Compare this to the balloon variable estimator $\hat{f}_{\text{ball}}(x;h(x))$ with symmetric $\text{Beta}(3/2,3/2)$ kernels where we compute the mean of $8(x-X_i)^{1/2}(h(x)+X_i-x)^{1/2}/[\pi h(x)^2]$.

A multivariate beta density was introduced by Lee (1996) and Olkin & Liu (2003), though it is not apparent how it can be modified to mimic the properties of $K_{\text{beta}(1)}$ in order for it to be a spherically symmetric boundary kernel. So we use the less general product kernel $K_{\text{beta}}(\boldsymbol{y};\boldsymbol{x},\boldsymbol{h}) = \prod_{j=1}^d K_{\text{beta}(1)}(y_j;x_j,h_j)$ for $\boldsymbol{x},\boldsymbol{y} \in [0,1]^d$. This leads to a beta boundary density estimator

$$\hat{f}_{\text{beta}}(\boldsymbol{x};\boldsymbol{h}) = n^{-1}\sum_{i=1}^{n} K_{\text{beta}}(\boldsymbol{X}_i;\boldsymbol{x},\boldsymbol{h}).$$

Chen (1999) shows that the univariate asymptotic expected value is

$$\mathbb{E}\{\hat{f}_{\text{beta}}(x;h)\} = f(x) + \begin{cases} h^2\rho^*(x)f'(x)\{1+o(1)\} & x \in [0,2h^2) \\ \frac{1}{2}h^2x(1-x)f''(x)\{1+o(1)\} & x \in [2h^2, 1-2h^2] \\ h^2\rho^*(1-x)f'(x)\{1+o(1)\} & x \in (1-2h^2, 1] \end{cases}$$

where $\rho^*(x) = (1-x)[\rho(x) - x/h^2]/[1+h^2\rho(x) - x]$, which is $O(h^2)$ in all of $[0,1]$; and that the variance is $O(n^{-1}h^{-1})$ in the interior and $O(n^{-1}h^{-2})$ in the boundary, though without fully elucidating the coefficients. The multivariate bias and variance have not been established.

Without a tractable expression for the $\text{MISE}\{\hat{f}_{\text{beta}}(\cdot;\boldsymbol{h})\}$, for bandwidth selection, Bouezmarni & Rombouts (2010) proposed an unbiased cross validation selector $\hat{\boldsymbol{h}}_{\text{UCV}}$ which does not require it. They established that $\text{MISE}\{\hat{f}_{\text{beta}}(\cdot;\hat{\boldsymbol{h}}_{\text{UCV}})\}$ converges in probability to $\inf_{\boldsymbol{h}>0}\text{MISE}\{\hat{f}_{\text{beta}}(\cdot;\boldsymbol{h})\}$, but did not quantify a relative rate of convergence of $\hat{\boldsymbol{h}}_{\text{UCV}}$. Furthermore, the computational load and the large variability of this UCV selector leads us to search for more efficient and stable selectors.

As a plug-in estimator of $\text{AMISE}\{\hat{f}_{\text{beta}}(\cdot;\boldsymbol{h})\}$ remains elusive, observe that its minimiser is of order $n^{-1/(d+4)}$, if we temporarily ignore the contribution from the boundary region, which is the same order as the optimal bandwidth for $\hat{f}$. Since $K_{\text{beta}(1)}(\cdot, 1/2, h)$ is a $\text{Beta}(1/(2h^2), 1/(2h^2))$ random variable at the midpoint of the unit interval, its associated variance is $h^2/[4(1+h^2)] = h^2/4\{1+o(1)\}$. Noting that the variance associated with a normal kernel $K_{(1)}$ is $h^2$, then $\boldsymbol{h}_{\text{beta}} = 2\,\text{diag}(\mathbf{H}^{1/2}_{\text{AMISE},\mathcal{D}})\{1+o(1)\}$ is a suitable bandwidth for $\hat{f}_{\text{beta}}$, where $\mathbf{H}_{\text{AMISE},\mathcal{D}}$ is the diagonal matrix version of $\mathbf{H}_{\text{AMISE}}$ from Section 2.7.

### 4.3.2 *Linear boundary kernels*

The boundary beta density estimator is restricted to hyper-rectangular data supports. To analyse more general, yet fully known, compact data supports $\Omega \subset \mathbb{R}^d$, we require some further notation. Let $\mathcal{B}(\boldsymbol{x}) = \{h^{-1}(\boldsymbol{x}-\boldsymbol{y}) : \boldsymbol{y} \in \Omega\}$ for a constant scalar $h$. This $h$ defines the extent of the boundary region $\mathcal{B}$ around $\boldsymbol{x}$ where we apply a boundary correction. Let $\Omega(\boldsymbol{x};\mathbf{H}) = \{(1 - h^{-1}\mathbf{H}^{1/2})\boldsymbol{x} - h^{-1}\mathbf{H}^{1/2}\boldsymbol{y} : \boldsymbol{y} \in \Omega\}$. If $\text{vec}\,\mathbf{H} \to 0$ as $n \to \infty$, then the $\{\Omega(\boldsymbol{x};\mathbf{H})\}$ form a sequence of scaled versions of the data support $\Omega$ which are shrinking towards $\boldsymbol{x}$.

The pointwise expected value of the standard estimator $\hat{f}$ is

$$\begin{aligned}\mathbb{E}\{\hat{f}(\boldsymbol{x};\mathbf{H})\} &= \int_{\Omega(\boldsymbol{x};\mathbf{H})} K_{\mathbf{H}}(\boldsymbol{x}-\boldsymbol{y})f(\boldsymbol{y})\,d\boldsymbol{y} = \int_{\mathcal{B}(\boldsymbol{x})} K(\boldsymbol{w})f(\boldsymbol{x}-\mathbf{H}^{1/2}\boldsymbol{w})\,d\boldsymbol{w} \\ &= m_0(\mathcal{B}(\boldsymbol{x});K)f(\boldsymbol{x}) + \boldsymbol{m}_1(\mathcal{B}(\boldsymbol{x});K)^\top \mathbf{H}^{1/2}\mathsf{D}f(\boldsymbol{x}) \\ &\quad + \tfrac{1}{2}\boldsymbol{m}_2(\mathcal{B}(\boldsymbol{x});K)^\top(\mathbf{H}^{1/2})^{\otimes 2}\mathsf{D}^{\otimes 2}f(\boldsymbol{x})\{1+o(1)\}\end{aligned}$$

where $\boldsymbol{m}_r(\mathcal{B}(\boldsymbol{x});K) = \int_{\mathcal{B}(\boldsymbol{x})} \boldsymbol{w}^{\otimes r}K(\boldsymbol{w})\,d\boldsymbol{w} \in \mathbb{R}^{d^r}$ is the $r$-th partial moment of $K$ restricted to $\mathcal{B}(\boldsymbol{x})$. If $\boldsymbol{x}$ is in the interior region, then there is no change to the usual expression in Section 2.6, as $m_0(\mathcal{B}(\boldsymbol{x});K) = 1, \boldsymbol{m}_1(\mathcal{B}(\boldsymbol{x});K) = \mathbf{0}$ and the order $\|\text{vec}\,\mathbf{H}\|$ bias prevails. If $\boldsymbol{x}$ is in the boundary region and $\Omega(\boldsymbol{x};\mathbf{H})$ is not entirely contained within $\Omega$, then $m_0(\mathcal{B}(\boldsymbol{x});K) \neq 1, \boldsymbol{m}_1(\mathcal{B}(\boldsymbol{x});K) \neq \mathbf{0}$. So the bias of $\hat{f}(\boldsymbol{x};\mathbf{H})/m_0(\mathcal{B}(\boldsymbol{x});K)$ is $\boldsymbol{m}_1(\mathcal{B}(\boldsymbol{x});K)^\top \mathbf{H}^{1/2}\mathsf{D}f(\boldsymbol{x})/m_0(\mathcal{B}(\boldsymbol{x});K)$ which is order $\|\text{vec}\,\mathbf{H}^{1/2}\|$. This is the mathematical formulation of the increased boundary bias for the uncorrected density estimator applied to bounded data.

To resolve this boundary bias problem, Gasser & Müller (1979) proposed univariate linear boundary kernels which are the solution of variational problems, e.g., the Epanechnikov kernel has the minimum MISE amongst second order kernels. These were extended to multivariate product kernels in Müller & Stadtmüller (1999) which can correct the boundary bias for an arbitrarily shaped compact support with a piecewise linear boundary by suitably modifying the variational problem. As noted by Hazelton & Marshall (2009), this general approach is somewhat unwieldy for practical data analysis, and they propose a computationally simpler linear boundary kernel as a degree 1 polynomial in $\boldsymbol{x}$ multiplying a given second order kernel $K$, i.e.

$$K_{\text{LB}}(\boldsymbol{x}) = (a_0 + \boldsymbol{a}_1^\top \boldsymbol{x})K(\boldsymbol{x}) \tag{4.4}$$

where

$$\begin{aligned}a_0 &= 1/[m_0(\mathcal{B}(\boldsymbol{x});K) - \boldsymbol{m}_1(\mathcal{B}(\boldsymbol{x});K)^\top \mathbf{M}_2(\mathcal{B}(\boldsymbol{x});K)^{-1}\boldsymbol{m}_1(\mathcal{B}(\boldsymbol{x});K)] \\ \boldsymbol{a}_1 &= -a_0\mathbf{M}_2(\mathcal{B}(\boldsymbol{x});K)^{-1}\boldsymbol{m}_1(\mathcal{B}(\boldsymbol{x});K)\end{aligned}$$

and $\mathbf{M}_2(\mathcal{B}(\boldsymbol{x});K) = \int_{\mathcal{B}(\boldsymbol{x})} \boldsymbol{w}\boldsymbol{w}^\top K(\boldsymbol{w})\,d\boldsymbol{w}$. If $K$ is a product kernel, then $K_{\text{LB}}$ reduces to a Müller & Stadtmüller (1999) product kernel. Moreover, unlike the Müller & Stadtmüller kernel, $K_{\text{LB}}$ is straightforward to be cast as a spherically symmetric kernel.

The scaled linear boundary kernel is $K_{\text{LB},\mathbf{H}}(\boldsymbol{x}) = (a_0 + \boldsymbol{a}_1^\top \mathbf{H}^{-1/2}\boldsymbol{x})K_{\mathbf{H}}(\boldsymbol{x})$ and the corresponding linear boundary kernel estimator is

$$\hat{f}_{\text{LB}}(\boldsymbol{x};\mathbf{H}) = n^{-1}\sum_{i=1}^{n} K_{\text{LB},\mathbf{H}}(\boldsymbol{x}-\boldsymbol{X}_i).$$

Its pointwise AMSE is

$$\begin{aligned}\text{AMSE}\{\hat{f}_{\text{LB}}(\boldsymbol{x};\mathbf{H})\} &= n^{-1}|\mathbf{H}|^{-1/2}\big[a_0^2 m_0(\mathcal{B}(\boldsymbol{x});K^2) + 2a_0\boldsymbol{a}_1^\top \boldsymbol{m}_1(\mathcal{B}(\boldsymbol{x});K^2)\\ &\quad + (\boldsymbol{a}_1^\top)^{\otimes 2}\boldsymbol{m}_2(\mathcal{B}(\boldsymbol{x});K^2)\big] + \tfrac{1}{4}\big[a_0\boldsymbol{m}_2(\mathcal{B}(\boldsymbol{x});K)^\top\\ &\quad + \boldsymbol{m}_3(\mathcal{B}(\boldsymbol{x});K)^\top(\boldsymbol{a}_1 \otimes \mathbf{I}_{d^2})\big]^{\otimes 2}(\mathbf{H}^{1/2})^{\otimes 4}(\mathsf{D}^{\otimes 2}f(\boldsymbol{x}))^{\otimes 2}. \qquad (4.5)\end{aligned}$$

The bias of $\hat{f}_{\text{LB}}$ is $O(\|\text{vec}\,\mathbf{H}\|)$ everywhere in $\Omega$, whilst the variance is controlled to remain $O(n^{-1}|\mathbf{H}|^{-1/2})$. The minimal MSE rate of $\hat{f}_{\text{LB}}$ in $\Omega$ is thus $O(n^{-4/(d+4)})$. For practical data analysis, the $a_0, \boldsymbol{a}_1$ coefficients and the partial moments $\boldsymbol{m}_1(\mathcal{B}(\boldsymbol{x});K)$ etc. can be numerically approximated by Riemann sums for each $\boldsymbol{x}$. This added computational cost is the trade-off of ensuring the standard AMSE convergence behaviour also applies in the boundary regions.

The coefficients in Equation (4.4) require that $\mathbf{M}_2$ is invertible. In the case that $\mathbf{M}_2$ is singular, then $a_0 = [1 - \boldsymbol{a}_1^\top \boldsymbol{m}_1(\mathcal{B}(\boldsymbol{x});K)]/m_0(\mathcal{B}(\boldsymbol{x});K)$ and $\boldsymbol{a}_1$ is the solution to $[\boldsymbol{m}_1(\mathcal{B}(\boldsymbol{x});K)\boldsymbol{m}_1(\mathcal{B}(\boldsymbol{x});K)^\top - m_0(\mathcal{B}(\boldsymbol{x});K)\mathbf{M}_2(\mathcal{B}(\boldsymbol{x});K)]\boldsymbol{a}_1 = \boldsymbol{m}_1(\mathcal{B}(\boldsymbol{x});K)$. The bias and variance properties remain the same.

**Example 4.3** In Figure 4.3 is the comparison of these two boundary kernel estimators with the standard estimator on a subset of the World Bank development indicators data: the number of internet users per 100 inhabitants and the added value of the agricultural production as a ratio of the total GDP (%). Figure 4.3(a) is the scatter plot of the $n = 177$ nations with complete measurements. In the upper right of this panel is the normal kernel with fixed bandwidth $\hat{\mathbf{H}}_{\text{PI}} = [95.6, -21.8; -21.8, 11.9]$. The standard kernel estimator with $\hat{\mathbf{H}}_{\text{PI}}$ is in Figure 4.3(b) and we observe that it exceeds the data support delimited by the dashed grey lines.

The beta boundary kernels based on $2\hat{\mathbf{H}}_{\text{PI},\mathcal{D}}^{1/2}$ are shown n Figure 4.3(c). In the interior regions, the kernel $K_{\text{beta}}$ is symmetric whereas as we approach the boundary regions, it becomes asymmetric with the probability mass being transferred in the direction of the boundary which results in the kernel support becoming correspondingly compressed. The beta boundary kernel estimator is shown in Figure 4.3(d). In addition to not exceeding the data support, the bimodality of the data is more apparent than for the fixed kernel estimator.

The linear boundary kernels $K_{\text{LB}}$ based on the normal kernel with $\hat{\mathbf{H}}_{\text{PI}}$ are shown in Figure 4.3(e). Their supports vary less than those for the beta boundary kernels due to less prominent shifts of the probability mass towards the boundary. The normal linear boundary kernel estimator is shown in Figure 4.3(f). Again, the bounded data support is respected and the bimodality is apparent. It is less noisy than the beta boundary estimator, reflecting our earlier observations for the gains in unconstrained matrices over diagonal

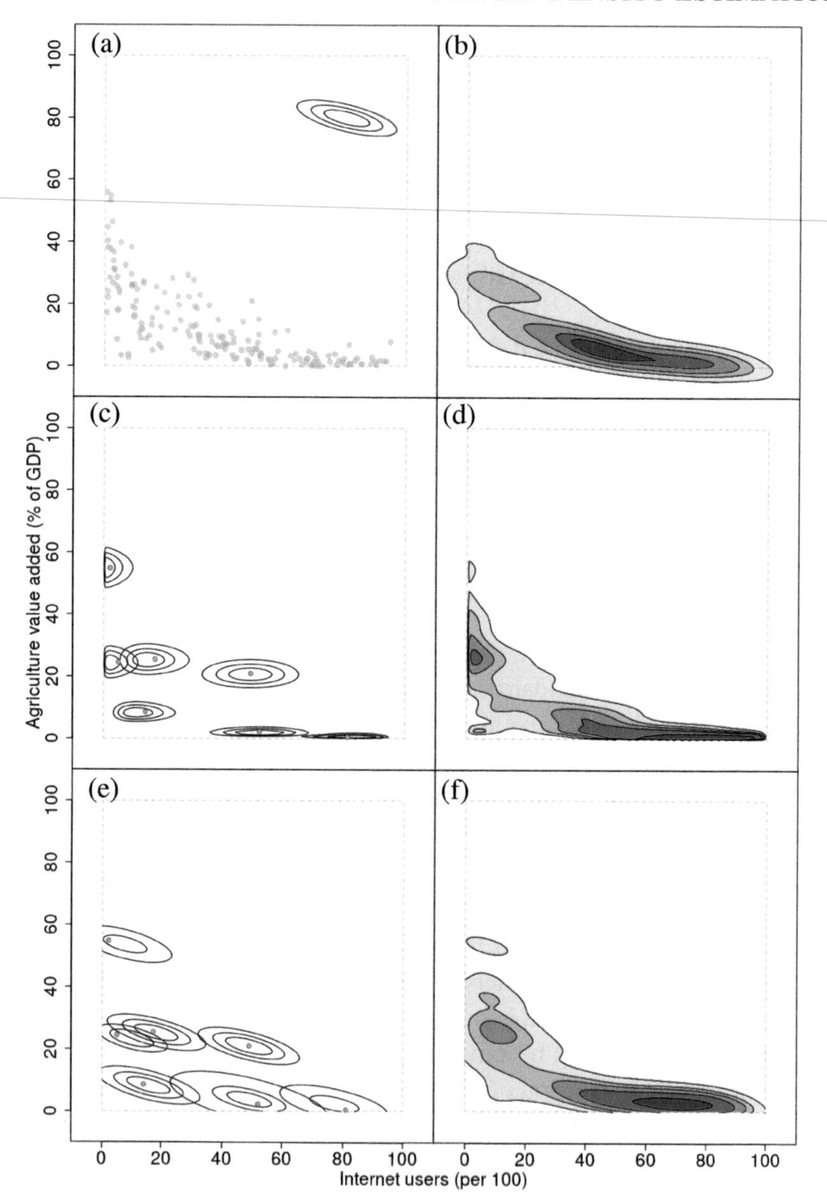


Figure 4.3 *Boundary kernel density estimates for the World Bank internet-agriculture data. The x-axis is the percentage of internet users (%) and the y-axis is the added value of agriculture as a ratio of the GDP (%). (a) Scatter plot of the* $n = 177$ *data pairs, with a normal kernel with* $\hat{\mathbf{H}}_{\text{PI}} = [95.57, -21.78; -21.78, 11.90]$. *(b) Fixed bandwidth density estimate* $\hat{f}$ *with* $\hat{\mathbf{H}}_{\text{PI}}$. *(c) Beta boundary kernels* $K_{\text{beta}}$ *with* $2(\hat{\mathbf{H}}_{\text{PI},\mathcal{D}})^{1/2}$. *(d) Beta boundary density estimate* $\hat{f}_{\text{beta}}$. *(e) Normal linear boundary kernel* $K_{\text{LB}}$ *with* $\hat{\mathbf{H}}_{\text{PI}}$. *(f) Normal linear boundary density estimate* $\hat{f}_{\text{LB}}$.

bandwidth matrices. On the other hand, it appears to be oversmoothed very close to the axes leading to the appearance of locally peaked regions, whereas the beta boundary estimator has monotonically increasing density values. A remedy to this could be combining linear boundary kernels with the variable sample point bandwidths as posited by Marshall & Hazelton (2010). □

## 4.4 Kernel choice

The assumption (A1) in Conditions A defines the general properties of a rich class of kernel functions. In practise, many fewer are utilised, as it has been well-established that the kernel function choice is less important to the performance of density estimators than the bandwidth choice. The most common univariate non-normal kernels belong to the beta family, of which the most well known are the uniform, Epanechnikov, biweight and triweight kernels, all of which are of the form

$$K_{(1)}(x;r) = B(r+1,1/2)^{-1}(1-x^2)^r \mathbf{1}\{x \in [-1,1]\}$$

for some $r \geq 0$, where $B(\alpha_1,\alpha_2) = \Gamma(\alpha_1)\Gamma(\alpha_2)/\Gamma(\alpha_1+\alpha_2)$ is the beta function. The uniform kernel ($r = 0$) can be considered to be the simplest kernel, the Epanechnikov kernel ($r = 1$) yields the minimal MISE (Epanechnikov, 1969), and the biweight ($r = 2$) and triweight ($r = 3$) kernels possess the minimal integrated squared gradient and squared curvature respectively (Terrell, 1990). Moreover, the normal kernel can be considered the limiting case as $r \to \infty$. Silverman (1986, Table 3.1, p. 43) shows that the loss of MISE efficiency is small in using other beta family kernels or the normal kernel instead of the Epanechnikov kernel. See also Wand & Jones (1995, Section 2.7, pp. 28–31) for a concise summary of the choice of an optimal kernel.

For multivariate data, the question of kernel choice is more complicated as there are two main ways of generating a multivariate kernel, given a univariate kernel $K_{(1)}$:

$$K^P(\boldsymbol{x}) = c_P \prod_{i=1}^{d} K_{(1)}(x_i), \ K^S(\boldsymbol{x}) = c_S K_{(1)}((\boldsymbol{x}^\top \boldsymbol{x})^{1/2})$$

where $K^P$ is a product kernel and $K^S$ is a spherically symmetric kernel. The coefficients $c_P, c_S$ are normalisation constants so that the integrals of $K^P, K^S$ remain one. There does not appear to be much difference between these two until we introduce their scaled versions. The product kernel is suitable only with a bandwidth matrix of class $\mathcal{A}$ or $\mathcal{D}$, as its scaled version is $K^P_{\mathbf{H}}(\boldsymbol{x}) = c_P \prod_{i=1}^{d} h_i^{-1} K_{(1)}(x_i/h_i)$, whereas the spherically symmetric kernel can be used with any bandwidth matrix class since $K_{\mathbf{H}}^S(\boldsymbol{x}) = c_S|\mathbf{H}|^{-1/2}K_{(1)}((\boldsymbol{x}^\top\mathbf{H}^{-1}\boldsymbol{x})^{1/2})$. Note that $K_{\mathbf{H}}^S \not\equiv K_{\mathbf{H}}^P$ even for $\mathbf{H} \in \mathcal{D}$ for general $K_{(1)}$. The normal kernel is the only kernel for which this holds, which means it is the only case where computing estimators with constrained or unconstrained matrices does not require a different kernel.

The normalisation constant $c_S$ for a normal kernel is well known as $(2\pi)^{-d/2}$, whereas for a non-normal spherically symmetric kernel, it is non-trivial to compute even if the univariate constant is known. This was only recently solved by Duong (2015) for the beta family kernels.

Consider the family of spherically symmetric kernels $K^S(\cdot;r)$ obtained from the univariate beta kernels $K_{(1)}(\cdot;r)$. Sacks & Ylvisaker (1981) state that the spherically symmetric Epanechnikov kernel $K^S(\cdot;1)$ is optimal in a MISE sense rather than its product kernel counterpart. So the efficiencies of the other kernels in the family can be expressed in the ratio $\text{Eff}(K^S(\cdot;1), K^S(\cdot;r)) = [C(K^S(\cdot;1))/C(K^S(\cdot;r))]^{(d+4)/4}$ where $C(K) = [R(K)^4 m_2(K)^{2d}]^{1/(d+4)}$, see Wand & Jones (1995, p. 103). To achieve the same MISE as the optimal Epanechnikov kernel with a sample size $n$, the kernel $K^S(\cdot;r)$ requires a sample size of $n/\text{Eff}(K^S(\cdot;1), K^S(\cdot;r))$. These are calculated in Duong (2015) and reproduced in Table 4.1 which extends Wand & Jones (1995, Table 2.1, p. 31). For a fixed value of $r$ the efficiency decreases as $d$ increases, although the loss of efficiency is small, except perhaps for the normal kernel for $d = 4$.

| | | **Efficiency** | | | |
|---|---|---|---|---|---|
| **Kernel** | $r$ | $d=1$ | $d=2$ | $d=3$ | $d=4$ |
| Uniform | 0 | 0.930 | 0.889 | 0.862 | 0.844 |
| Epanechnikov | 1 | 1.000 | 1.000 | 1.000 | 1.000 |
| Biweight | 2 | 0.994 | 0.988 | 0.982 | 0.977 |
| Triweight | 3 | 0.987 | 0.972 | 0.958 | 0.945 |
| Normal | $\infty$ | 0.951 | 0.889 | 0.820 | 0.750 |

Table 4.1 *Efficiencies* $\text{Eff}(K^S(\cdot;1), K^S(\cdot;r))$ *for spherically symmetric beta family kernels, for* $d = 1,2,3,4$ *and* $r = 0,1,2,3,\infty$.

Whilst the normal kernel is more complicated than the polynomial beta family kernels, and has a lower efficiency, the former allows for important mathematical and computational simplifications and avoids any possible problems with the non-existence of higher order derivatives of the kernel function when computing data-based bandwidth selectors. These desirable properties translate into an overwhelming current utilisation of normal kernels for multivariate density estimation.

## 4.5 Higher order kernels

The kernels considered in the previous section are second order kernels, which are non-negative functions whose first moment is zero, and whose second moment is non-zero, as required by (A2) in Conditions A. More generally a $k$-th order kernel has the property that the first $(k-1)$ moments are zero and the $k$-th moment is non-zero for some even integer $k$. Let $\hat{f}(\boldsymbol{x};\mathbf{H},k)$ be the density estimator constructed via a $k$-th order kernel, then

$$\begin{aligned}\mathbb{E}\{\hat{f}(\boldsymbol{x};\mathbf{H},k)\} &= \int_{\mathbb{R}^d} K(\boldsymbol{w}) f(\boldsymbol{x}-\mathbf{H}^{1/2}\boldsymbol{w})\,d\boldsymbol{w} \\ &= \int_{\mathbb{R}^d} K(\boldsymbol{w}) f(\boldsymbol{x}) \sum_{i=0}^{k} \frac{1}{j!}(-\boldsymbol{w}^\top \mathbf{H}^{1/2})^{\otimes j} \mathsf{D}^{\otimes j} f(\boldsymbol{x})\,d\boldsymbol{w}\{1+o(1)\} \\ &= f(\boldsymbol{x}) + \tfrac{1}{k!}\boldsymbol{m}_k^\top(K)(\mathbf{H}^{1/2})^{\otimes k}\mathsf{D}^{\otimes k} f(\boldsymbol{x})\{1+o(1)\}\end{aligned}$$

where $\boldsymbol{m}_k(K) = \int_{\mathbb{R}^d} \boldsymbol{x}^{\otimes k} K(\boldsymbol{x})\,d\boldsymbol{x}$. The bias is now of order $\|\mathrm{vec}\,\mathbf{H}^{k/2}\|$, which decreases faster as the order $k$ increases, so higher order kernels offer a method for bias reduction. Moreover, it can be analogously shown that the asymptotic IV remains unaffected, so that the MISE of $\hat{f}(\boldsymbol{x};\mathbf{H},k)$ is of order $O(n^{-2k/(2k+d)})$, provided the density is $k$-times continuously differentiable, with bounded and square integrable $k$-th order partial derivatives. The trade-off is that for $k > 2$, $\hat{f}(\boldsymbol{x};\mathbf{H},k)$ is no longer guaranteed to be a non-negative function. It is also noteworthy that Jones & Foster (1993) asserted that the most gain in practical data analysis performance is most likely to be from the step from second to fourth order kernels, rather than for even higher order kernels.

Whilst second order kernels can be taken as any symmetric unimodal probability density function, higher order kernels are less easily encountered. For univariate kernels, one such construction of a $(k+2)$-th order kernel is a $k$-th polynomial multiplying the univariate normal kernel, as posited by Wand & Schucany (1990). Another is a recursive formula where a $(k+2)$-th order kernel is $\frac{1}{k}[(k+1)K_{(1)}(x;k)+xK'_{(1)}(x;k)]$ given a $k$-th order kernel $K_{(1)}(x;k)$, as introduced by Rustagi et al. (1991) and popularised by Jones & Foster (1993). The extension of these approaches to multivariate product kernels is straightforward, though appears not to be have been published yet, and the case of spherically symmetric kernels remains an open problem.

Due to the lack of convincing gains in bias reduction for finite sample sizes, the increased computation time and difficulty of construction, and the loss of non-negativity of the resulting density estimator, higher order kernels for density estimation have not entered into widespread use; see Marron &

Wand (1992), Jones & Foster (1993), Marron (1994), Wand & Jones (1995, Chapter 2.8, pp. 32–35) or Jones & Signorini (1997).

From a theoretical point of view, the minimax rates of convergence for density estimation in the MISE sense can be shown to depend solely on the degree of smoothness of the target density. If $f$ is $p$-times differentiable, with bounded $p$-th order partial derivatives, then the fastest possible MISE rate for any density estimator (of kernel or any other type) is of order $n^{-2p/(2p+d)}$ (Tsybakov, 2009). Moreover, for extremely smooth densities (e.g., infinitely differentiable densities), the best MISE rate improves to $n^{-1}$ up to logarithmic coefficients (Watson & Leadbetter, 1963).

Since the convergence rate for second-order kernel density estimators is $n^{-4/(d+4)}$ when the density $f$ is twice differentiable with bounded second-order partial derivatives, this means that kernel estimators attain the fastest possible convergence rate for such class of densities. On the other hand, as noted above, the fastest attainable MISE rate for a kernel estimator with a $k$-th order kernel is $n^{-2k/(2k+d)}$, which implies that using a $k$-th order kernel implicitly hinders the performance of the kernel density estimator for densities which more than $k$-times differentiable. An option to avoid such limitations is to employ infinite-order kernels, i.e, kernels $K$ such that $\boldsymbol{m}_k(K) = \mathbf{0}$ for all $k$.

Kernels with null moments of all orders can be constructed by imposing that the Fourier transform of the kernel be identically equal to 1 on a neighbourhood about the origin. These are known as superkernels and their utility in density estimation has been explored in Devroye (1992) and Chacón et al. (2007), where it is shown that superkernel density estimators are rate-adaptive, in the sense that their MISE rate of convergence is the fastest possible as determined only by the degree of smoothness of the true density. They even seem to defy the curse of dimensionality by achieving $n^{-1}$ rates (up to logarithmic coefficients) for extremely smooth densities, regardless the dimension $d$ (Politis & Romano, 1999). In practice, however, the problem of bandwidth selection for superkernel density estimators has been insufficiently addressed (Politis, 2003; Amezziane & McMurry, 2012), and they mostly remain as a theoretical tool to achieve rate adaptivity.

## 4.6 Further mathematical analysis of modified density estimators

### *4.6.1 Asymptotic error for sample point variable bandwidth estimators*

To obtain the MSE of the sample point variable density estimator in Equation (4.1) with an unconstrained bandwidth matrix, we proceed with a multivariate generalisation of Jones et al. (1994).

**Theorem 4** *Suppose that (A1)–(A3) in Conditions A hold. Further suppose that the following conditions hold:*

**(A1')** *The density function $f$ is four times differentiable, with all of its fourth order partial derivatives bounded, continuous and square integrable.*

**(A2')** *The kernel $K$ has finite fourth order moment $\boldsymbol{m}_4(K) = \int_{\mathbb{R}^d} \boldsymbol{x}^{\otimes 4} K(\boldsymbol{x})\, d\boldsymbol{x}$.*

*For a non-random point $\boldsymbol{x}$ where $f(\boldsymbol{x}) \neq 0$, the mean squared error of the sample point estimator with the Abramson bandwidth function $\mathbf{H}_{\text{Ab}}(\boldsymbol{x}) = f(\boldsymbol{x})^{-1}\mathbf{H}$, is*

$$\begin{aligned}\text{MSE}\{\hat{f}_{\text{SP}}(\boldsymbol{x}; \mathbf{H}_{\text{Ab}})\} = \{&n^{-1}|\mathbf{H}|^{-1/2} R(K) f(\boldsymbol{x})^{1+d/2} \\ &+ \tfrac{1}{576}(\boldsymbol{m}_4(K)^\top)^{\otimes 2}(\mathbf{H}^{1/2})^{\otimes 8}(\boldsymbol{\lambda}(\boldsymbol{x}))^{\otimes 2}\}\{1+o(1)\}\end{aligned}$$

*where* $\boldsymbol{\lambda}(\boldsymbol{x}) = 24 f(\boldsymbol{x})^{-5}(\mathsf{D}f(\boldsymbol{x}))^{\otimes 4} - 36 f(\boldsymbol{x})^{-4}(\mathsf{D}^{\otimes 2}f(\boldsymbol{x}) \otimes (\mathsf{D}f(\boldsymbol{x}))^{\otimes 2} + f(\boldsymbol{x})^{-3}[8(\mathsf{D}^{\otimes 3}f(\boldsymbol{x}) \otimes \mathsf{D}f(\boldsymbol{x})) + 6(\mathsf{D}^{\otimes 2}f(\boldsymbol{x}))^{\otimes 2}] - f(\boldsymbol{x})^{-2}\mathsf{D}^{\otimes 4}f(\boldsymbol{x})$.

**Proof (Proof of Theorem 4)** We begin by establishing the $\text{MSE}\{\hat{f}_{\text{SP}}(\boldsymbol{x}; \boldsymbol{\Gamma}\}$ for a general $\boldsymbol{\Gamma}(\boldsymbol{x}) = \gamma(\boldsymbol{x})\mathbf{H}$. The pointwise expected value at a point $\boldsymbol{x}$ is

$$\mathbb{E}\{\hat{f}_{\text{SP}}(\boldsymbol{x}; \boldsymbol{\Gamma})\} = \int_{\mathbb{R}^d} |\mathbf{H}|^{-1/2}\gamma(\boldsymbol{y})^{-d/2} f(\boldsymbol{y}) K(\gamma(\boldsymbol{y})^{-1/2}\mathbf{H}^{-1/2}(\boldsymbol{x}-\boldsymbol{y}))\, d\boldsymbol{y}.$$

With the change of variables $\boldsymbol{w} = \gamma(\boldsymbol{y})^{-1/2}\mathbf{H}^{-1/2}(\boldsymbol{x}-\boldsymbol{y})$, then $\boldsymbol{y} = \boldsymbol{x} - \gamma(\boldsymbol{y})^{1/2}\mathbf{H}^{1/2}\boldsymbol{w} = \boldsymbol{x} - \gamma(\boldsymbol{x})^{1/2}\mathbf{H}^{1/2}\boldsymbol{w}\{1+o(1)\}$, and the expected value can be expanded as follows:

$$\begin{aligned}\mathbb{E}\{\hat{f}_{\text{SP}}(\boldsymbol{x}; \boldsymbol{\Gamma})\} &= \int_{\mathbb{R}^d} K(\boldsymbol{w}) f(\boldsymbol{x} - \gamma(\boldsymbol{x})^{1/2}\mathbf{H}^{1/2}\boldsymbol{w})\{1+o(1)\}\, d\boldsymbol{w} \\ &= f(\boldsymbol{x}) + \tfrac{1}{2}m_2(K)(\text{vec}^\top \mathbf{H})\mathsf{D}^{\otimes 2}[f(\boldsymbol{x})\gamma(\boldsymbol{x})] \\ &\quad + \tfrac{1}{24}\boldsymbol{m}_4(K)^\top(\mathbf{H}^{1/2})^{\otimes 4}\mathsf{D}^{\otimes 4}[f(\boldsymbol{x})\gamma(\boldsymbol{x})^2]\{1+o(1)\}d\boldsymbol{w}\end{aligned}$$

where the odd order terms in the Taylor expansion are omitted as they are identically zero due to the symmetry of the kernel $K$. The pointwise variance is $\text{Var}\{\hat{f}_{\text{SP}}(\boldsymbol{x}; \boldsymbol{\Gamma})\} = n^{-1}\,\text{Var}\{K_{\boldsymbol{\Gamma}(\boldsymbol{X})}(\boldsymbol{x}-\boldsymbol{X})\}$, so we require

$$\begin{aligned}&\mathbb{E}\{\{\gamma(\boldsymbol{X})^{d/2}|\mathbf{H}|^{-1/2}K(\gamma(\boldsymbol{X})^{1/2}\mathbf{H}^{-1/2}(\boldsymbol{x}-\boldsymbol{X}))\}^2\} \\ &= \int_{\mathbb{R}^d} |\mathbf{H}|^{-1}\gamma(\boldsymbol{y})^{-d} f(\boldsymbol{y}) K(\gamma(\boldsymbol{y})^{1/2}\mathbf{H}^{-1/2}(\boldsymbol{x}-\boldsymbol{y}))^2\, d\boldsymbol{y} \\ &= |\mathbf{H}|^{-1/2}\gamma(\boldsymbol{x})^{-d/2} f(\boldsymbol{x}) R(K)\{1+o(1)\}\, d\boldsymbol{w}\end{aligned}$$

which dominates the $O(1)$ term $\{\mathbb{E}\{\gamma(\boldsymbol{X})^{d/2}|\mathbf{H}|^{-1/2}K(\gamma(\boldsymbol{X})^{1/2}\mathbf{H}^{-1/2}(\boldsymbol{x} - \boldsymbol{X}))\}\}^2$. Hence $\text{Var}\{\hat{f}_{\text{SP}}(\boldsymbol{x}; \boldsymbol{\Gamma})\} = n^{-1}|\mathbf{H}|^{-1/2}R(K)\gamma(\boldsymbol{x})^{-d/2}f(\boldsymbol{x})\{1+o(1)\}$.

For the Abramsom bandwidth function $\mathbf{H}_{\text{Ab}}$, $\gamma(\boldsymbol{x}) \equiv f(\boldsymbol{x})^{-1}$ then the second order term in the Taylor expansion of $\mathbb{E}\{\hat{f}_{\text{SP}}(\boldsymbol{x};\mathbf{H}_{\text{Ab}})\}$ is identically zero, so the latter reduces to

$$\mathbb{E}\{\hat{f}_{\text{SP}}(\boldsymbol{x};\mathbf{H}_{\text{Ab}})\} = f(\boldsymbol{x}) + \tfrac{1}{24}\boldsymbol{m}_4(K)^\top(\mathbf{H}^{1/2})^{\otimes 4}\mathsf{D}^{\otimes 4}[f(\boldsymbol{x})^{-1}]\{1+o(1)\}.$$

To expand this fourth order derivative, we repeatedly apply the differential $d(f(\boldsymbol{x})^{-a}) = -af(\boldsymbol{x})^{-a-1}\mathsf{D}f(\boldsymbol{x})^\top d\boldsymbol{x}$. Starting with $a=1$, then $d(f(\boldsymbol{x})^{-1}) = -f(\boldsymbol{x})^{-2}\mathsf{D}f(\boldsymbol{x})^\top d\boldsymbol{x}$, i.e., the first derivative is $\mathsf{D}[f(\boldsymbol{x})^{-1}] = -f(\boldsymbol{x})^{-2}\mathsf{D}f(\boldsymbol{x})$. Continuing, the fourth derivative is

$$\begin{aligned}\mathsf{D}^{\otimes 4}[f(\boldsymbol{x})^{-1}] &= 24f(\boldsymbol{x})^{-5}(\mathsf{D}f(\boldsymbol{x}))^{\otimes 4} - 6f(\boldsymbol{x})^{-4}[\mathbf{I}_d\otimes\mathbf{K}_{d^2,d} + \mathbf{I}_{d^2}\otimes\mathbf{K}_{d,d}\\ &\quad + \mathbf{K}_{d^2,d^2} + \boldsymbol{\Lambda}\otimes\mathbf{I}_d](\mathsf{D}^{\otimes 2}f(\boldsymbol{x})\otimes(\mathsf{D}f(\boldsymbol{x}))^{\otimes 2})\\ &\quad + 2f(\boldsymbol{x})^{-3}[(\boldsymbol{\Lambda}\otimes\mathbf{I}_d)(\mathbf{I}_{d^2}\otimes\mathbf{K}_{d,d}) + \mathbf{I}_{d^4}](\mathsf{D}^{\otimes 3}f(\boldsymbol{x})\otimes\mathsf{D}f(\boldsymbol{x}))\\ &\quad + 2f(\boldsymbol{x})^{-3}(\boldsymbol{\Lambda}\otimes\mathbf{I}_d)(\mathbf{K}_{d,d^2}\otimes\mathbf{I}_d)(\mathsf{D}^{\otimes 2}f(\boldsymbol{x}))^{\otimes 2} - f(\boldsymbol{x})^{-2}\mathsf{D}^{\otimes 4}f(\boldsymbol{x})\end{aligned}$$

where $\boldsymbol{\Lambda} = \mathbf{I}_d\otimes\mathbf{K}_{d,d} + \mathbf{K}_{d,d^2} + \mathbf{I}_{d^3}$, with $\mathbf{K}_{m,n}$ denoting the $(mn)\times(mn)$-commutation matrix. See Appendix B or Magnus & Neudecker (1999, Section 3.7) for the definition and main properties of the commutation matrix. Since the action of these combinations of identity and commutation matrices does not affect the value of the inner product with $(\mathbf{H}^{1/2})^{\otimes 4}\boldsymbol{m}_4(K)$, the expected value simplifies to $\mathbb{E}\{\hat{f}_{\text{SP}}(\boldsymbol{x};\mathbf{H}_{\text{Ab}})\} = f(\boldsymbol{x}) + \frac{1}{24}\boldsymbol{m}_4(K)^\top(\mathbf{H}^{1/2})^{\otimes 4}\boldsymbol{\lambda}(\boldsymbol{x})\{1+o(1)\}$. The variance is $\text{Var}\{\hat{f}_{\text{SP}}(\boldsymbol{x};\mathbf{H}_{\text{Ab}})\} = n^{-1}|\mathbf{H}|^{-1/2}R(K)f(\boldsymbol{x})^{1+d/2}\{1+o(1)\}$. ■

### 4.6.2 *Asymptotic error for linear boundary density estimators*

To compute the coefficients of the linear boundary kernel $K_{\text{LB}}(\boldsymbol{x}) = (\boldsymbol{a}_0 + \boldsymbol{a}_1^\top)K(\boldsymbol{x})$ in Equation (4.4), we generalise the procedure of Hazelton & Marshall (2009). We set the zeroth and first moment at $\boldsymbol{x}$ of $K_{\text{LB}}$ to be

$$\begin{aligned}m_0(\boldsymbol{x};K_{\text{LB}}) &= \int_{\mathcal{B}(\boldsymbol{x})}(a_0 + \boldsymbol{a}_1^\top\boldsymbol{x})K(\boldsymbol{x})\,d\boldsymbol{x} = a_0 m_0(\boldsymbol{x}) + \boldsymbol{a}_1^\top\boldsymbol{m}_1(\boldsymbol{x}) = 1\\ \boldsymbol{m}_1(\boldsymbol{x};K_{\text{LB}}) &= \int_{\mathcal{B}(\boldsymbol{x})}(a_0 + \boldsymbol{a}_1^\top\boldsymbol{x})\boldsymbol{x}K(\boldsymbol{x})\,d\boldsymbol{x} = a_0\boldsymbol{m}_1(\boldsymbol{x}) + \mathbf{M}_2(\boldsymbol{x})\boldsymbol{a}_1 = \mathbf{0}.\end{aligned}$$

For brevity, we abbreviate the moments of $K$ as $\boldsymbol{m}_1(\boldsymbol{x}) \equiv \boldsymbol{m}_1(\mathcal{B}(\boldsymbol{x});K)$ etc. by omitting the explicit dependence on $\mathcal{B}$ and $K$. Assuming the invertibility of $\mathbf{M}_2(\boldsymbol{x})$, the second equation yields $\boldsymbol{a}_1 = -a_0\mathbf{M}_2(\boldsymbol{x})^{-1}\boldsymbol{m}_1(\boldsymbol{x})$. Substituting this value of $\boldsymbol{a}_1$ into the first equation yields $a_0 = [1 -$

$\boldsymbol{a}_1^\top \boldsymbol{m}_1(\boldsymbol{x})]/m_0(\boldsymbol{x}) = [1 + a_0\boldsymbol{m}_1(\boldsymbol{x})^\top \mathbf{M}_2(\boldsymbol{x})^{-1}\boldsymbol{m}_1(\boldsymbol{x})]/m_0(\boldsymbol{x})$, which has $a_0 = 1/[m_0(\boldsymbol{x}) - \boldsymbol{m}_1(\boldsymbol{x})^\top \mathbf{M}_2(\boldsymbol{x})^{-1}\boldsymbol{m}_1(\boldsymbol{x})]$ as its solution.

We derive the MSE of the linear boundary density estimator in Equation (4.5) with an unconstrained bandwidth matrix here. Suppose that (A1)–(A3) in Conditions A and (A2') in Theorem 4 hold. The pointwise expected value of $\hat{f}_{\mathrm{LB}}(\boldsymbol{x};\mathbf{H})$, for $\boldsymbol{x}$ such that $m_0(\boldsymbol{x}) \neq \boldsymbol{m}_1(\boldsymbol{x})^\top \mathbf{M}_2(\boldsymbol{x})^{-1}\boldsymbol{m}_1(\boldsymbol{x})$, is

$$
\begin{aligned}
&\mathbb{E}\{\hat{f}_{\mathrm{LB}}(\boldsymbol{x};\mathbf{H})\} \\
&\quad = \int_{\mathcal{B}(\boldsymbol{x})} K_{\mathrm{LB}}(\boldsymbol{w})[f(\boldsymbol{x}) - \boldsymbol{w}^\top \mathbf{H}^{1/2}\mathsf{D}f(\boldsymbol{x}) + \tfrac{1}{2}(\boldsymbol{w}^\top \mathbf{H}^{1/2})^{\otimes 2}\mathsf{D}^{\otimes 2}f(\boldsymbol{x})]\,d\boldsymbol{w} \\
&\qquad \times \{1 + o(1)\} \\
&\quad = m_0(\mathcal{B}(\boldsymbol{x});K_{\mathrm{LB}})f(\boldsymbol{x}) + \tfrac{1}{2}\boldsymbol{m}_2(\mathcal{B}(\boldsymbol{x});K_{\mathrm{LB}})^\top (\mathbf{H}^{1/2})^{\otimes 2}\mathsf{D}^{\otimes 2}f(\boldsymbol{x})\{1+o(1)\} \\
&\quad = f(\boldsymbol{x}) + \tfrac{1}{2}\{a_0\boldsymbol{m}_2(\boldsymbol{x})^\top + \boldsymbol{m}_3(\boldsymbol{x})^\top(\boldsymbol{a}_1 \otimes \mathbf{I}_{d^2})\}(\mathbf{H}^{1/2})^{\otimes 2}\mathsf{D}^{\otimes 2}f(\boldsymbol{x})\{1+o(1)\}
\end{aligned}
$$

since $m_0(\mathcal{B}(\boldsymbol{x});K_{\mathrm{LB}}) = 1$ and $\boldsymbol{m}_1(\mathcal{B}(\boldsymbol{x});K_{\mathrm{LB}}) = \mathbf{0}$, and $\boldsymbol{m}_2(\mathcal{B}(\boldsymbol{x});K_{\mathrm{LB}}) = \int_{\mathcal{B}(\boldsymbol{x})}(a_0 + \boldsymbol{a}_1^\top \boldsymbol{x})\boldsymbol{x}^{\otimes 2}K(\boldsymbol{x})\,d\boldsymbol{x} = a_0\boldsymbol{m}_2(\boldsymbol{x}) + (\boldsymbol{a}_1^\top \otimes \mathbf{I}_{d^2})\boldsymbol{m}_3(\boldsymbol{x}) \neq 0$.

We also have that

$$
\begin{aligned}
&\mathbb{E}\{K_{\mathrm{LB},\mathbf{H}}(\boldsymbol{x}-\boldsymbol{X})^2\} \\
&\quad = |\mathbf{H}|^{-1/2}\int_{\mathcal{B}(\boldsymbol{x})} K_{\mathrm{LB}}(\boldsymbol{w})^2 f(\boldsymbol{x} - \mathbf{H}^{1/2}\boldsymbol{w})\,d\boldsymbol{w} \\
&\quad = |\mathbf{H}|^{-1/2}[a_0^2 m_0(\mathcal{B}(\boldsymbol{x});K^2) + 2a_0\boldsymbol{a}_1^\top \boldsymbol{m}_1(\mathcal{B}(\boldsymbol{x});K^2) + (\boldsymbol{a}_1^\top)^{\otimes 2}\boldsymbol{m}_2(\mathcal{B}(\boldsymbol{x});K^2)] \\
&\qquad \times \{1+o(1)\}.
\end{aligned}
$$

As for the infinite support case, $\mathbb{E}\{K_{\mathrm{LB},\mathbf{H}}(\boldsymbol{x}-\boldsymbol{X})^2\}$ dominates $[\mathbb{E}\{K_{\mathrm{LB},\mathbf{H}}(\boldsymbol{x}-\boldsymbol{X})\}]^2$. The variance is therefore $\mathrm{Var}\{\hat{f}_{\mathrm{LB}}(\boldsymbol{x};\mathbf{H})\} = n^{-1}\,\mathrm{Var}\{K_{\mathrm{LB},\mathbf{H}}(\boldsymbol{x}-\boldsymbol{X})\} = n^{-1}\,\mathbb{E}\{K_{\mathrm{LB},\mathbf{H}}(\boldsymbol{x}-\boldsymbol{X})^2\}\{1+o(1)\}$.

Chapter 5

# Density derivative estimation

The density estimators examined in the previous chapters form a subset of a wider class of curve estimators. Crucial information about the structure of the underlying target density is not revealed by examining solely its values, and is only discerned via its derivatives. For example, the local minima/maxima are characterised as locations where the first derivative is identically zero and the Hessian matrix is positive/negative definite. So there is great interest in complementing the density estimators with the density derivative estimators.

The presentation of this chapter follows closely that of Chapters 2–3 on density estimation. For brevity, we have deliberately omitted certain details that are exact analogues of the zeroth derivative case (that is, the estimation of the density itself): the reader is urged to become familiar with these earlier chapters before perusing the current chapter. Sections 5.1–5.4 introduce estimators of the derivatives of the density function and their practical bandwidth selectors, focusing on the first and second derivatives (correspond to Sections 2.2–2.4). Section 5.5 sets up a mathematical framework for optimal bandwidth selection, akin to Sections 2.5–2.7. Sections 5.6–5.7 present automatic bandwidth selectors for density derivative estimation and summarise their convergence rates, as natural but non-trivial extensions of those in Chapter 3. As a case study, Section 5.8 focuses on obtaining explicit results for the case of a normal density, which are required to compute data-based selectors. Section 5.9 fills in the previously omitted mathematical details of the considered density derivative estimators.

## 5.1 Kernel density derivative estimators

The vector notation for the $r$-th derivative of a multivariate function introduced earlier in Section 2.6 is a crucial tool to manage higher order Taylor expansions. Recall that for a function $f\colon \mathbb{R}^d \to \mathbb{R}$, the expression $\mathsf{D}^{\otimes r} f(\boldsymbol{x})$ denotes a vector of length $d^r$ containing all the $r$-th order partial derivatives, arranged in the order determined by the formal $r$-fold Kronecker power of the gradient operator $\mathsf{D} = (\partial/\partial x_1, \ldots, \partial/\partial x_d)$.

The purpose of this chapter is to study the estimation of the $r$-th derivative $\mathsf{D}^{\otimes r} f$ of a density function. The kernel estimator of $\mathsf{D}^{\otimes r} f$ was introduced by Chacón et al. (2011) as the $r$-th derivative of the kernel density estimator,

$$\widehat{\mathsf{D}^{\otimes r} f}(\boldsymbol{x};\mathbf{H}) = \mathsf{D}^{\otimes r}\hat{f}(\boldsymbol{x};\mathbf{H}) = n^{-1}\sum_{i=1}^{n} \mathsf{D}^{\otimes r} K_{\mathbf{H}}(\boldsymbol{x}-\boldsymbol{X}_i). \tag{5.1}$$

It is a multivariate generalisation of the kernel estimator first considered in Bhattachatya (1967). Different variants in the univariate set up have been suggested by Jones (1994).

The derivative of the kernel function is taken after the scaling with the bandwidth matrix is applied, so a more explicit formula for each term in the summation is

$$\mathsf{D}^{\otimes r} K_{\mathbf{H}}(\boldsymbol{x}-\boldsymbol{X}_i) = |\mathbf{H}|^{-1/2}(\mathbf{H}^{-1/2})^{\otimes r}\mathsf{D}^{\otimes r} K(\mathbf{H}^{-1/2}(\boldsymbol{x}-\boldsymbol{X}_i)).$$

This expression greatly assists in implementing these estimators since it separate the roles of $K$ and $\mathbf{H}$. It is possible to propose such a direct estimator because the kernel density estimator inherits its differentiability from the underlying kernel function. This is a primary advantage of kernel estimators over histograms as the discontinuities of $\hat{f}_{\text{hist}}$ at the edges of the bins imply that $\mathsf{D}^{\otimes r}\hat{f}_{\text{hist}}$ is not well-defined.

The most important special cases of the general problem are those of the density gradient ($r = 1$) and the density Hessian ($r = 2$) estimation, so these are explored in more detail, before returning to the general case.

### *5.1.1 Density gradient estimators*

The density gradient is denoted as $\mathsf{D}f$ and its kernel estimator in Equation (5.1) simplifies to

$$\mathsf{D}\hat{f}(\boldsymbol{x};\mathbf{H}) = n^{-1}|\mathbf{H}|^{-1/2}(\mathbf{H}^{-1/2})\sum_{i=1}^{n} \mathsf{D}K(\mathbf{H}^{-1/2}(\boldsymbol{x}-\boldsymbol{X}_i)). \tag{5.2}$$

Equation (5.2) is the generalisation using an unconstrained bandwidth of the kernel estimator introduced in Fukunaga & Hostetler (1975).

To complement the compact notation for the vectorised total derivative, we denote a single partial derivative indexed by $\boldsymbol{r} = (r_1, \ldots, r_d)$ as

$$f^{(\boldsymbol{r})}(\boldsymbol{x}) = \frac{\partial^{|\boldsymbol{r}|} f(\boldsymbol{x})}{\partial x_1^{r_1}\cdots\partial x_d^{r_d}}$$

where $|\boldsymbol{r}| = r_1 + \cdots r_d$. For example, we utilise this notation for visualisation purposes since we can usually only display one partial derivative at a time.

Since the density gradient can be both negative and positive, we compute 'quasi-probability' contours by applying the definition of probability contours separately to the positive and negative parts of the density gradient i.e., $f_+^{(\boldsymbol{r})}(\boldsymbol{x}) = f^{(\boldsymbol{r})}(\boldsymbol{x})\mathbf{1}\{f^{(\boldsymbol{r})}(\boldsymbol{x}) \geq 0\}$ and $f_-^{(\boldsymbol{r})}(\boldsymbol{x}) = f^{(\boldsymbol{r})}(\boldsymbol{x})\mathbf{1}\{f^{(\boldsymbol{r})}(\boldsymbol{x}) < 0\}$. Since $f_+^{(\boldsymbol{r})}, f_-^{(\boldsymbol{r})}$ are not proper density functions, the probabilistic interpretation is no longer valid as for the contours of $f$, but as this procedure adapts to any range of density gradient values, it provides a useful choice for the contour levels for visualisation.

**Example 5.1** The partial density gradient estimates $\hat{f}^{(1,0)}, \hat{f}^{(0,1)}$ for the daily temperature data, with the plug-in bandwidth $[1.04, 0.98; 0.98, 1.69]$, are shown in Figure 5.1(a)–(b). The orange regions indicate the quasi-quartile contours of positive gradients $\hat{f}_+^{(1,0)}, \hat{f}_+^{(0,1)}$, the purple regions the negative gradients $\hat{f}_-^{(1,0)}, \hat{f}_-^{(0,1)}$, and the white regions represent the zones where the positive and negative gradients are below the first quasi-quartile (i.e., not far from zero). Darker orange colours indicate larger positive gradients and darker purple colours larger negative gradients. This is an example of a divergent colour scale (Zeileis et al., 2009).

As zero gradients coincide with local extrema, we focus on these regions. Around (20°C, 30°C), the gradient contours are composed of the two coloured regions with steep contours, separated only by a thin white region in which a local mode resides. Around (10°C, 20°C) there is a larger white region surrounded by gradual increments in the gradients in which a local anti-mode resides. This anti-mode separates the upper mode from the lower mode at around (5°C, 15°C).

As simultaneously interpreting the contour plots of the partial density gradients can be difficult, an alternative is the quiver or velocity plot in Figure 5.1(c), where the direction and length of the arrows is determined by the gradient vector. Longer arrows with larger heads indicate steeper gradients, shorter arrows with smaller heads, flatter gradients. These arrows follow the gradient ascent so they subtend a trajectory towards the local modes. The density estimate $\hat{f}$ in Figure 2.4 was computed with bandwidth $[0.67, 0.60; 0.60, 1.04]$ which is smaller than the bandwidth $[1.04, 0.98; 0.98, 1.69]$ for $\mathsf{D}\hat{f}$. We defer the reasoning on how to calculate the latter bandwidth matrix and the reasons why it is different from that for density estimation to Sections 5.5 and 5.6. □

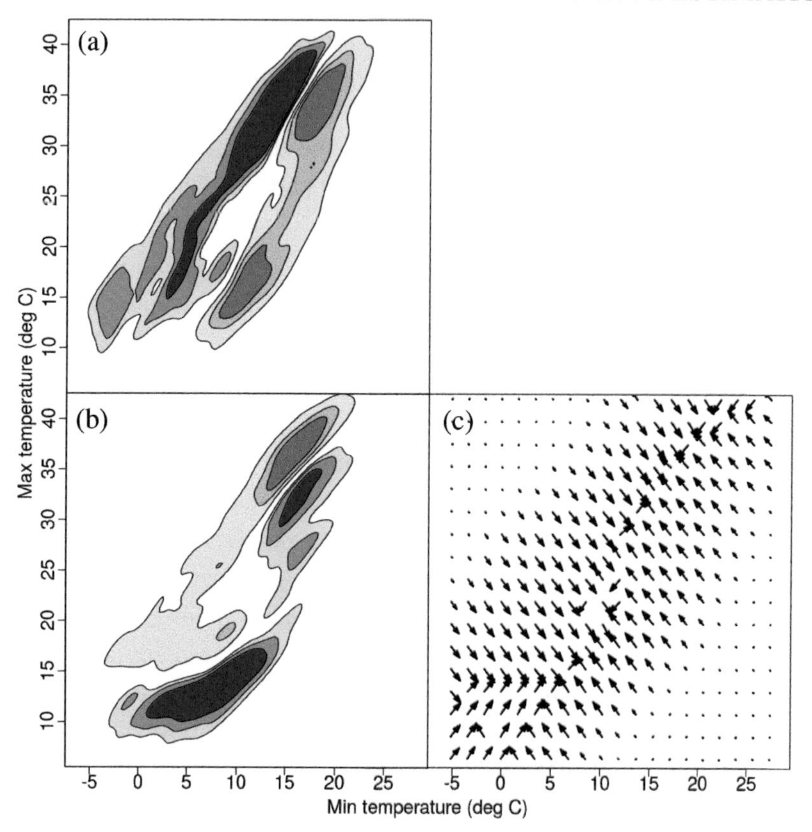


Figure 5.1 *Density gradient estimates for the daily temperature data. The horizontal axis is the daily minimum (°C) and the vertical axis is the daily maximum temperature (°C). (a) Partial density gradient estimate $\hat{f}^{(1,0)}$. (b) Partial density gradient estimate $\hat{f}^{(0,1)}$. The quasi-quartile contours are purple for positive gradients, orange for negative gradients. (c) Quiver plot of the density gradient estimate $\mathsf{D}\hat{f}$. The density gradient estimates are computed with the bandwidth* $[1.04, 0.98; 0.98, 1.69]$.

### *5.1.2 Density Hessian estimators*

As the density second derivative is denoted $\mathsf{D}^{\otimes 2} f$, its kernel estimator is

$$\mathsf{D}^{\otimes 2}\hat{f}(\boldsymbol{x};\mathbf{H}) = n^{-1}|\mathbf{H}|^{-1/2}(\mathbf{H}^{-1/2})^{\otimes 2}\sum_{i=1}^{n}\mathsf{D}^{\otimes 2}K(\mathbf{H}^{-1/2}(\boldsymbol{x}-\boldsymbol{X}_i)). \quad (5.3)$$

Visualising the individual mixed partial derivatives in the density second derivative concurrently is not trivial, and as the quiver plot representation is not easily adapted to the vectorised second derivative matrix, we search for a suitable alternative. As a local mode in $f$ requires that its Hessian matrix $\mathsf{H}f$

be negative definite, then let

$$s(\boldsymbol{x}) = -\mathbf{1}\{\mathsf{H}f(\boldsymbol{x}) < 0\} \operatorname{abs}(|\mathsf{H}f(\boldsymbol{x})|) \tag{5.4}$$

where $\operatorname{abs}(|\mathsf{H}f|)$ is the absolute value of the determinant of $\mathsf{H}f$. Since $\operatorname{vec}\mathsf{H}f = \mathsf{D}^{\otimes 2}f$, it is always possible to reconstruct the Hessian matrix exactly from its vectorised form.

An important goal in exploratory data analysis is the identification of data-rich regions, and these correspond to local modes in the density function. The quantity in Equation (5.4) thus focuses on local modes, rather than other local extrema like anti-modes or saddle points.

**Example 5.2** Figure 5.2(a)–(c) shows the partial density Hessian estimates of the daily temperature data corresponding to $\hat{f}^{(2,0)}, \hat{f}^{(0,2)}, \hat{f}^{(1,1)}$, with the plug-in bandwidth $[1.44, 1.42; 1.42, 2.46]$, with the analogous quasi-probability contours and colour scale from Figure 5.1. The dark orange regions $\hat{f}^{(2,0)}, \hat{f}^{(0,2)}$ in Figure 5.2(a)–(b) indicate large negative values of the density curvature surrounded by purple regions with positive curvature, which indicate that a local mode is located in the former. On the other hand, it is less easy to interpret the mixed partial derivative $\hat{f}^{(1,1)}$ in Figure 5.2(c). As it is also difficult to interpret all three partial density curvature plots simultaneously, the quasi-quartile contours of the summary curvature $\hat{s}$ are displayed in Figure 5.2(d). This allows for a clearer visualisation as the orange regions surround local modes in the data density and are hence the high data density regions. □

### *5.1.3 General density derivative estimators*

Whilst the density gradient and Hessian are the most commonly used derivatives of the density function, we have already encountered a number of situations where higher order derivatives of the density are required. For example, in the asymptotic MISE expression in Equation (3.10) involves the fourth order derivative, and the sixth and eighth order derivatives will appear in the 2-stage plug-in and smooth cross validation bandwidth selectors in Sections 5.6.4–5.6.5.

A more thorough theoretical analysis of the general density derivative estimators is deferred to Section 5.5, but one immediate difficulty in denoting these estimators is the computation of the kernel estimator in Equation (5.1).

The normal kernel $\phi(\boldsymbol{x}) = (2\pi)^{-d/2}\exp(-\frac{1}{2}\boldsymbol{x}^\top\boldsymbol{x})$ is the most common choice for density estimation in the multivariate context, and it enjoys further

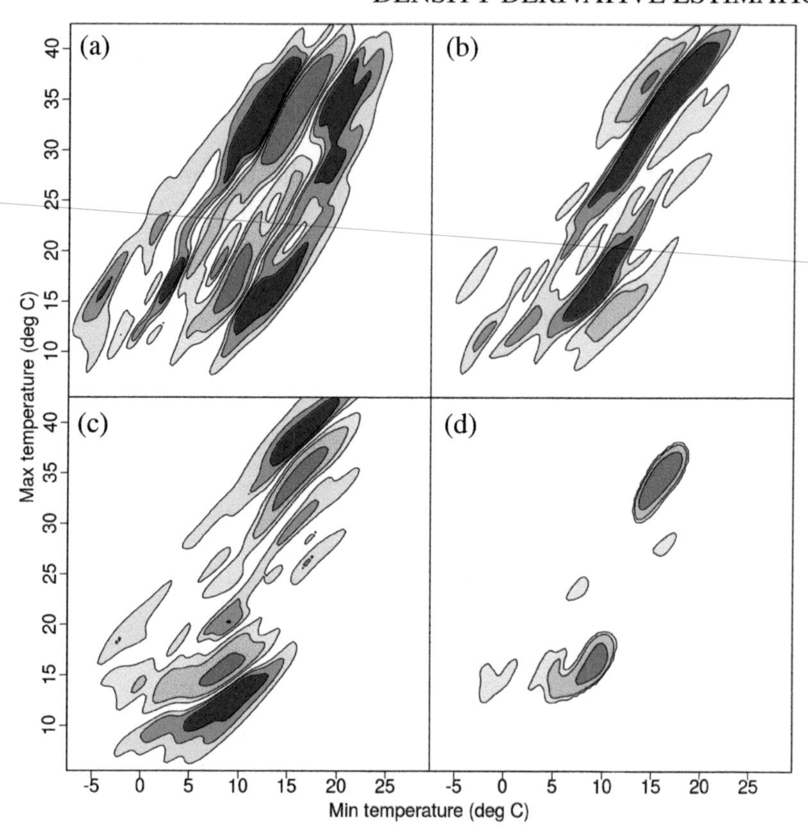


Figure 5.2 *Density curvature estimates for the daily temperature data. The horizontal axis is the daily minimum (°C) and the vertical axis is the daily maximum temperature (°C). (a) Partial density Hessian estimate* $\hat{f}^{(2,0)}$. *(b) Partial density Hessian estimate* $\hat{f}^{(0,2)}$. *(c) Partial density Hessian estimate* $\hat{f}^{(1,1)}$. *(d) Summary density curvature estimate* $\hat{s}$. *The quasi-quartile contours are purple for positive, orange for negative values. The curvature estimates are computed with the bandwidth* $[1.44, 1.42; 1.42, 2.46]$.

advantages for density derivative estimation. Its first two derivatives are

$$\mathsf{D}\phi(\boldsymbol{x}) = -\phi(\boldsymbol{x})\boldsymbol{x},$$
$$\mathsf{D}^{\otimes 2}\phi(\boldsymbol{x}) = \phi(\boldsymbol{x})(\boldsymbol{x}^{\otimes 2} - \operatorname{vec}\mathbf{I}_d).$$

More complicated expressions result for higher order derivatives, due to the fact that Kronecker products are not commutative. Fortunately Holmquist (1996*a*) showed that there exists a closed and compact formulation to express the derivatives of arbitrary order of the multivariate normal density function:

$$\mathsf{D}^{\otimes r}\phi_{\mathbf{\Sigma}}(\boldsymbol{x}) = (-1)^r(\mathbf{\Sigma}^{-1})^{\otimes r}\mathcal{H}_r(\boldsymbol{x};\mathbf{\Sigma})\phi_{\mathbf{\Sigma}}(\boldsymbol{x})$$

where $\mathcal{H}_r(\boldsymbol{x};\mathbf{\Sigma})$ is the $r$-th order vector Hermite polynomial in $\boldsymbol{x}$, given by

$$\mathcal{H}_r(\boldsymbol{x};\mathbf{\Sigma}) = r!\mathcal{S}_{d,r}\sum_{j=0}^{\lfloor r/2\rfloor}\frac{(-1)^j}{j!(r-2j)!2^j}\{\boldsymbol{x}^{\otimes(r-2j)}\otimes(\operatorname{vec}\mathbf{\Sigma})^{\otimes j}\} \tag{5.5}$$

with $\lfloor a\rfloor$ denoting the greatest integer that is less than or equal to $a$. Here, an important factor is the $d^r\times d^r$ symmetriser matrix $\mathcal{S}_{d,r}$, defined as

$$\mathcal{S}_{d,r} = \frac{1}{r!}\sum_{i_1,i_2,\ldots,i_r=1}^{d}\sum_{\sigma\in\mathcal{P}_r}\bigotimes_{\ell=1}^{r}\boldsymbol{e}_{i_\ell}\boldsymbol{e}_{i_{\sigma(\ell)}}^\top \tag{5.6}$$

with $\mathcal{P}_r$ denoting for the group of permutations of order $r$ and $\boldsymbol{e}_i$ for the $i$-th column of $\mathbf{I}_d$. The symmetriser matrix $\mathcal{S}_{d,r}$ maps the product $\bigotimes_{i=1}^r\boldsymbol{x}_i$ to an equally weighted linear combination of products of all possible permutations of $\boldsymbol{x}_1,\ldots,\boldsymbol{x}_r$. For instance, for $r=3$ the action of this symmetriser matrix on a 3-fold product results in

$$\begin{aligned}\mathcal{S}_{d,3}(\boldsymbol{x}_1\otimes\boldsymbol{x}_2\otimes\boldsymbol{x}_3) = \tfrac{1}{6}(&\boldsymbol{x}_1\otimes\boldsymbol{x}_2\otimes\boldsymbol{x}_3+\boldsymbol{x}_1\otimes\boldsymbol{x}_3\otimes\boldsymbol{x}_2+\boldsymbol{x}_2\otimes\boldsymbol{x}_1\otimes\boldsymbol{x}_3\\ &+\boldsymbol{x}_2\otimes\boldsymbol{x}_3\otimes\boldsymbol{x}_1+\boldsymbol{x}_3\otimes\boldsymbol{x}_1\otimes\boldsymbol{x}_2+\boldsymbol{x}_3\otimes\boldsymbol{x}_2\otimes\boldsymbol{x}_1).\end{aligned}$$

Simple explicit expressions for this matrix only exist for $r\le 2$, e.g., $\mathcal{S}_{d,0}=1$, $\mathcal{S}_{d,1}=\mathbf{I}_d$ and $\mathcal{S}_{d,2}=\frac{1}{2}(\mathbf{I}_{d^2}+\mathbf{K}_{d,d})$, where $\mathbf{K}_{d,d}$ is the $d^2\times d^2$ commutation matrix (see Appendix B). Chacón & Duong (2015) derived efficient algorithms to compute this symmetriser matrix and normal density derivatives (the latter is explored in more detail in Section 8.3). With these algorithms at hand, we are well-placed to compute $\mathsf{D}^{\otimes r}\hat{f}$.

The symmetriser matrix is also involved in defining other quantities related to the normal distribution, which play a significant role in the analysis of kernel smoothers for derivative estimation. For instance, for the $2r$-th moment of a standard normal random variable $\boldsymbol{Z}$, denoted as $\boldsymbol{\mu}_{2r}=\mathbb{E}(\boldsymbol{Z}^{\otimes 2r})$, Holmquist (1988) showed that an explicit formula in terms of the symmetriser matrix is given by

$$\boldsymbol{\mu}_{2r} = \mathrm{OF}(2r)\mathcal{S}_{d,2r}(\operatorname{vec}\mathbf{I}_d)^{\otimes r} \tag{5.7}$$

where $\mathrm{OF}(2r)=(2r-1)(2r-3)\cdots 5\cdot 3\cdot 1=(2r)!/(r!2^r)$ denotes the odd factorial of an even number $2r$.

Moreover, we repeatedly make use of the expectation of products of quadratic forms in normal random variables

$$\begin{aligned} \nu_r(\mathbf{A};\boldsymbol{\mu},\boldsymbol{\Sigma}) &= \mathbb{E}\{(\boldsymbol{Y}^\top\mathbf{A}\boldsymbol{Y})^r\} \\ \nu_{r,s}(\mathbf{A},\mathbf{B};\boldsymbol{\mu},\boldsymbol{\Sigma}) &= \mathbb{E}\{(\boldsymbol{Y}^\top\mathbf{A}\boldsymbol{Y})^r(\boldsymbol{Y}^\top\mathbf{B}\boldsymbol{Y})^s\} \end{aligned} \tag{5.8}$$

where $\boldsymbol{Y} \sim N(\boldsymbol{\mu},\boldsymbol{\Sigma})$ and $\mathbf{A}$, $\mathbf{B}$ are symmetric matrices. For a standard normal random variable we tend to omit the mean and variance and write $\nu_r(\mathbf{A})$, $\nu_{r,s}(\mathbf{A},\mathbf{B})$ for brevity. Holmquist (1996*b*) showed that closed, explicit formulas for these expectations can be expressed in terms of the symmetriser matrix as

$$\begin{aligned} \nu_r(\mathbf{A}) &= \mathrm{OF}(2r)(\mathrm{vec}^\top\mathbf{A})^{\otimes r}\mathcal{S}_{d,2r}(\mathrm{vec}\,\mathbf{I}_d)^{\otimes r} \\ \nu_{r,s}(\mathbf{A},\mathbf{B}) &= \mathrm{OF}(2r+2s)\big\{(\mathrm{vec}^\top\mathbf{A})^{\otimes r}\otimes(\mathrm{vec}^\top\mathbf{B})^{\otimes s}\big\}\mathcal{S}_{d,2r+2s}(\mathrm{vec}\,\mathbf{I}_d)^{\otimes(r+s)}. \end{aligned} \tag{5.9}$$

In addition to providing a compact notation for such moments, these $\nu$ functionals play an important role in the efficient computation of bandwidth selectors as outlined in Section 8.4.

## 5.2 Gains from unconstrained bandwidth matrices

To examine the effect of unconstrained bandwidth matrices on the estimators of density derivatives, we return to the *Grevillea* data and the dumbbell density.

**Example 5.3** The quiver plots for the kernel gradient estimates of the *Grevillea* data with an unconstrained bandwidth $[0.043,-0.028;-0.028,0.061]$ is in Figure 5.3(a), and with a diagonal bandwidth $\mathrm{diag}(0.025,0.031)$ in Figure 5.3(b). The summary curvature plots for the kernel curvature estimates with an unconstrained bandwidth $[0.053,-0.036;-0.036,0.076]$ are shown in Figure 5.3(c), and with a diagonal bandwidth $\mathrm{diag}(0.030,0.038)$ in Figure 5.3(d). With the unconstrained matrices, the oblique contours suggest a directionality in the geographical distribution. □

**Example 5.4** Regarding the dumbbell density, Figures 5.4(a)–(c) show the quiver plots for the target density gradient, the kernel estimate with unconstrained bandwidth and the kernel estimate with diagonal bandwidth. The unimodality of the target density is confirmed with the quiver plots for the unconstrained bandwidth but not for the diagonal one. The gradient arrows for the target density in Figure 5.4(a) fall away from the 45° line in the central region, which is also the case in Figure 5.4(b), whereas in Figure 5.4(c), these

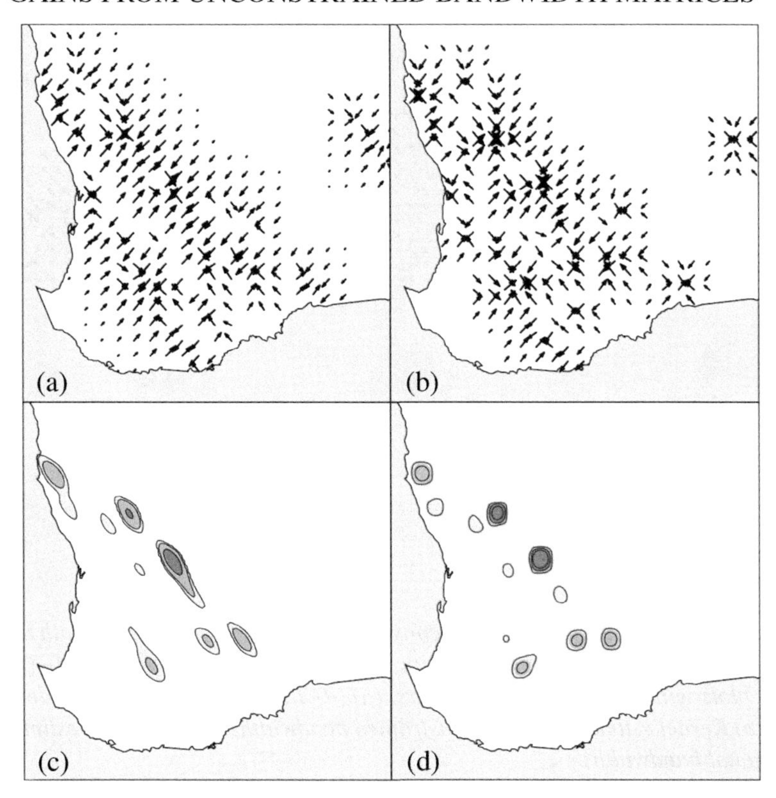


Figure 5.3 *Potential gains of an unconstrained over a diagonal bandwidth matrix for the density derivatives of the* Grevillea *data. (a)–(b) Gradient quiver plots. (c)–(d) Summary curvature plots with quasi-quartile contours. Kernel estimates with unconstrained bandwidths (a)* $[0.043, -0.028; -0.028, 0.061]$, *(c)* $[0.053, -0.036; -0.036, 0.076]$. *Kernel estimates with diagonal bandwidths (b)* $\text{diag}(0.025, 0.031)$, *(d)* $\text{diag}(0.030, 0.038)$.

arrows converge to two or three separate modes. The corresponding summary curvature plots are displayed in Figure 5.4(d)–(f). The unconstrained bandwidth estimate reproduces fairly accurately the target contours, whilst the contours for the diagonal bandwidth estimate are too circular in shape. □

To supplement these heuristic observations, we quantify the performance gain in using an unconstrained matrix from class $\mathcal{F}$ with respect to a diagonal matrix from class $\mathcal{D}$, via the asymptotic relative efficiency (ARE)

$$\text{ARE}(\mathcal{F}, \mathcal{D}) = \left[\frac{\min_{\mathbf{H}\in\mathcal{F}} \text{AMISE}\{\mathsf{D}^{\otimes r}\hat{f}(\cdot;\mathbf{H})\}}{\min_{\mathbf{H}\in\mathcal{D}} \text{AMISE}\{\mathsf{D}^{\otimes r}\hat{f}(\cdot;\mathbf{H})\}}\right]^{(d+2r+4)/4}$$

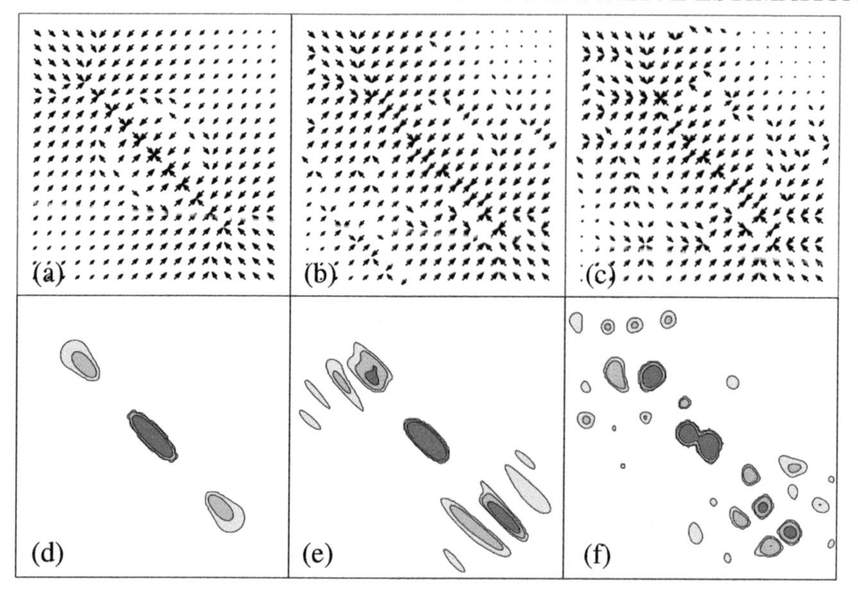


Figure 5.4 *Potential gains of an unconstrained over a diagonal bandwidth matrix for the derivatives of the dumbbell density. (a)–(c) Gradient quiver plots. (d)–(f) Curvature plots with quasi-quartile contours. (a),(d) Target dumbbell density derivatives. (b),(e) Kernel estimates with unconstrained bandwidths. (c),(f) Kernel estimates with diagonal bandwidths.*

as introduced by Chacón et al. (2011). The ARE for comparing the $\mathcal{F}$ class to the scalar $\mathcal{A}$ class follows analogously by suitably replacing the denominator.

Here, $\min_{\mathbf{H}\in\mathcal{F}} \text{AMISE}\{\mathsf{D}^{\otimes r}\hat{f}(\cdot;\mathbf{H})\}$ denotes the minimal asymptotic MISE achievable with a bandwidth matrix in class $\mathcal{F}$ (and analogously for classes $\mathcal{D}$ and $\mathcal{A}$). The MISE for density derivative estimators will be formally defined in Section 5.5 as the equivalent of the MISE for the density case. As $0 \leq \text{ARE}(\mathcal{F},\mathcal{D}) \leq 1$, low values (close to zero) of the ARE indicate that the unconstrained matrix produces a much lower minimal AMISE than a constrained matrix, so the former is strongly preferred; and high values (close to one) of the ARE indicate that the unconstrained matrix produces a similar minimal AMISE as the constrained matrix, so there is a only modest gain in using the former. An alternative interpretation of the $\text{ARE}(\mathcal{F},\mathcal{D})$ is that, for large $n$, the minimal achievable error using diagonal matrices with $n$ observations can be replicated using only $\text{ARE}(\mathcal{F},\mathcal{D})n$ observations with unconstrained matrices.

**Example 5.5** As the ARE does not have a closed form for a general density $f$, we numerically evaluate it for the special case of a bivariate normal density with variance matrix $\mathbf{\Sigma} = [\sigma_1^2, \rho\sigma_1\sigma_2; \rho\sigma_1\sigma_2, \sigma_2^2]$. In Figure 5.5(a), we have an isotropic variance $\sigma_1 = \sigma_2 = 1$. In this case, the ARE$(\mathcal{F}, \mathcal{D})$ and ARE$(\mathcal{F}, \mathcal{A})$ coincide exactly as $\mathbf{H}_{\text{AMISE},\mathcal{D}} = \mathbf{H}_{\text{AMISE},\mathcal{A}}$ whenever $\sigma_1 = \sigma_2$. In Figure 5.5(b), we have anisotropic variance $\sigma_1 = 1, \sigma_2 = 5$. The set of solid curves are the ARE$(\mathcal{F}, \mathcal{D})$ and of dashed curves are the ARE$(\mathcal{F}, \mathcal{A})$. With these ARE curves, we can observe the evolution from when the coordinate variables are perfectly linearly correlated ($\rho = \pm 1$) to when they are perfectly uncorrelated ($\rho = 0$). All the ARE curves tend to 0 as $|\rho|$ tends to 1, indicating that for highly correlated data, the unconstrained matrix is preferred. For a fixed $r$, the ARE$(\mathcal{F}, \mathcal{D}) \geq$ ARE$(\mathcal{F}, \mathcal{A})$ uniformly, i.e., the diagonal matrix class $\mathcal{D}$ performs uniformly better than the scalar class $\mathcal{A}$. As $r$ increases, the decrease in the ARE is steeper as $|\rho|$ tends to 1, demonstrating that the gains in AMISE performance for unconstrained matrices can be greater for $r > 0$ than for $r = 0$. The ARE$(\mathcal{F}, \mathcal{D})$ remain unchanged from Figure 5.5(a) to (b), as the diagonal matrix class handles anisotropy correctly, whereas the ARE$(\mathcal{F}, \mathcal{A})$ in Figure 5.5(b) are rather flat curves with values close to 0, implying that the scalar matrix class is inadequate for smoothing data with highly different dispersions. □

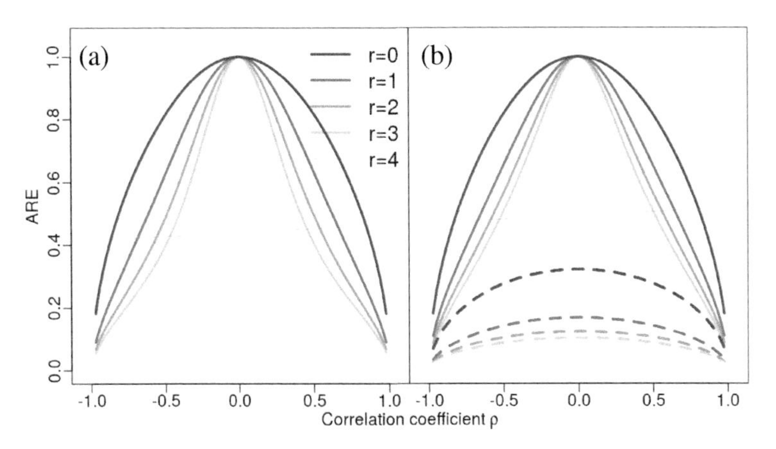


Figure 5.5 *Asymptotic relative errors (ARE) for the derivatives of the density of* $N((0,0), [\sigma_1^2, \rho\sigma_1\sigma_2; \rho\sigma_1\sigma_2, \sigma_2^2])$, *as a function of the correlation coefficient* $\rho$. *The horizontal axis is* $\rho$, *and the vertical axis is the ARE. (a)* $\sigma_1 = 1, \sigma_2 = 1$. *(b)* $\sigma_1 = 1, \sigma_2 = 5$. *The solid curves are the* ARE$(\mathcal{F}, \mathcal{D})$ *and the dashed curves the* ARE$(\mathcal{F}, \mathcal{A})$. *The colour changes from purple to grey as r increases from 0 to 4.*

## 5.3 Advice for practical bandwidth selection

In Section 2.4, we illustrated the density estimates resulting from different bandwidth selectors. Given the earlier advice, the possible selectors are reduced to the most 'promising' ones: (a) normal scale, (b) unbiased cross validation, (c) plug-in and (d) smoothed cross validation. These will be exposited in detail in Section 5.6, so we only mention them briefly here.

**Example 5.6** For the *Grevillea* data, Figures 5.6–5.7 for the density gradient and curvature estimates are the equivalents of Figure 2.11 for the density estimates.

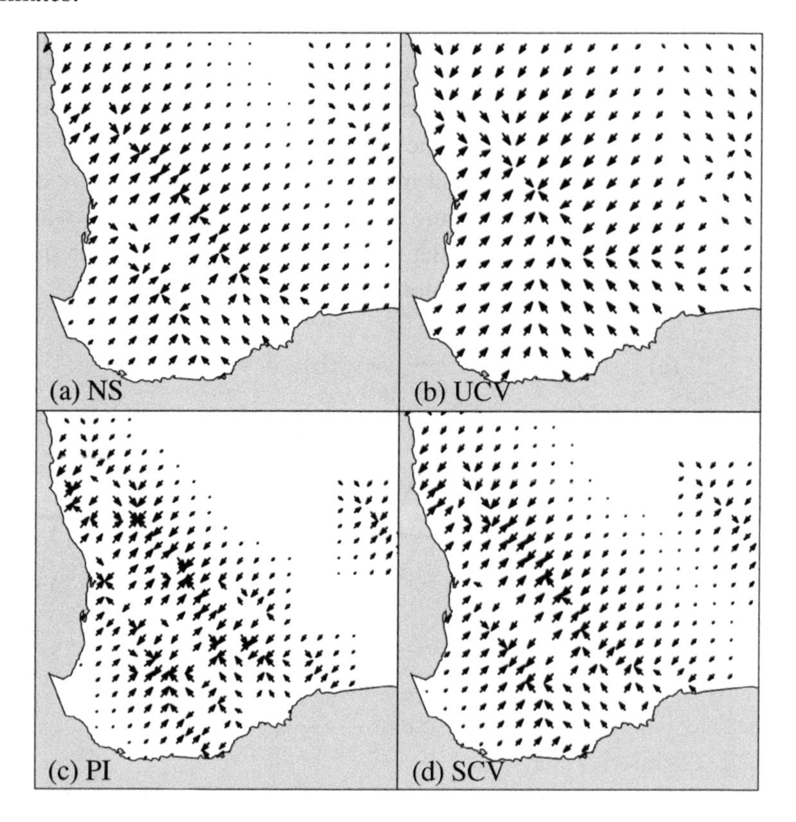


Figure 5.6 *Different bandwidth selectors for the density gradient estimates of the* Grevillea *data. Quiver plots. (a) Normal scale* $\hat{\mathbf{H}}_{\text{NS},1}$. *(b) Unbiased cross validation* $\hat{\mathbf{H}}_{\text{UCV},1}$. *(c) Plug-in* $\hat{\mathbf{H}}_{\text{PI},1}$. *(d) Smoothed cross validation* $\hat{\mathbf{H}}_{\text{SCV},1}$.

The normal scale selector is efficient to compute, and tends to produce oversmoothed density estimates, so it is useful for a quick visualisation of the overall trends in the data, but generally is not as accurate as the cross validation and the plug-in selectors. Unbiased cross validation, as it does not rely

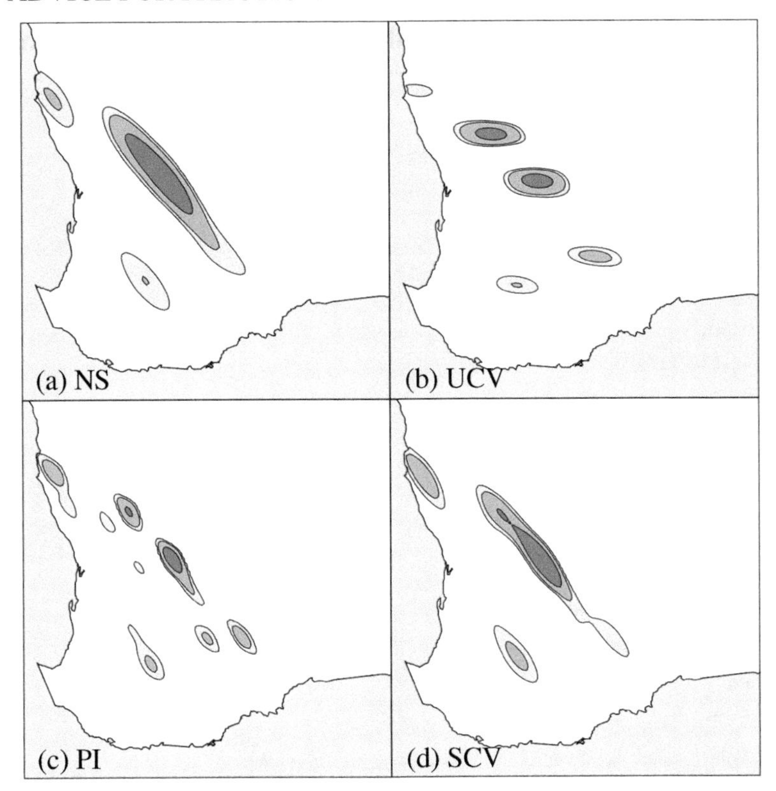


Figure 5.7 *Different bandwidth selectors for the density curvature estimates of the* Grevillea *data. Summary curvature plots. (a) Normal scale* $\hat{\mathbf{H}}_{\mathrm{NS},2}$. *(b) Unbiased cross validation* $\hat{\mathbf{H}}_{\mathrm{UCV},2}$. *(c) Plug-in* $\hat{\mathbf{H}}_{\mathrm{PI},2}$. *(d) Smoothed cross validation* $\hat{\mathbf{H}}_{\mathrm{SCV},2}$.

on asymptotic expansions, can be less biased than other selectors, though this smaller bias tends to give a wider variability (see Section 3.8). This leads to oversmoothed density derivative estimates for the *Grevillea* data, in contrast to the undersmoothed density estimates in Figure 2.11. The wide variability in the UCV selector is also well-documented in Chacón & Duong (2013). □

**Example 5.7** We repeat this comparison of these different bandwidth selectors for the stem cell data in Figure 5.8 for the summary curvature plots since the quiver arrow plots are not currently available for 3-dimensional visualisations. The normal scale $\hat{\mathbf{H}}_{\mathrm{NS},2}$ and unbiased cross validation $\hat{\mathbf{H}}_{\mathrm{UCV},2}$ yield similar density estimates. The quartile contours for these three selectors are ellipsoidal and evenly spaced, and tend to indicate oversmoothed estimates. For the $\hat{\mathbf{H}}_{\mathrm{PI},2}$ and smoothed cross validation $\hat{\mathbf{H}}_{\mathrm{SCV},2}$ selectors, more details of the data structure are visible due to the irregularly shaped and spaced contours shells, and that the upper left mode is more peaked. □

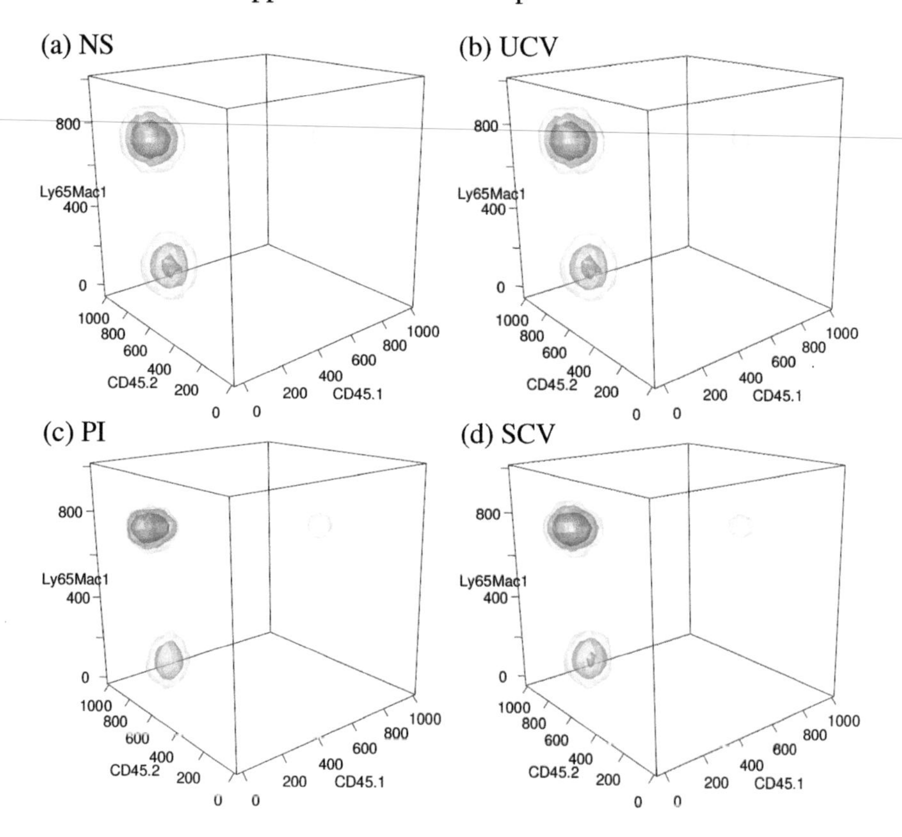


Figure 5.8 *Different bandwidth selectors for the density curvature estimates of the stem cell data. Summary curvature plots. (a) Normal scale* $\hat{\mathbf{H}}_{\text{NS},2}$*. (b) Unbiased cross validation* $\hat{\mathbf{H}}_{\text{UCV},2}$*. (c) Plug-in* $\hat{\mathbf{H}}_{\text{PI},2}$*. (d) Smoothed cross validation* $\hat{\mathbf{H}}_{\text{SCV},2}$*.*

The smoothed cross validation $\hat{\mathbf{H}}_{\text{SCV}}$ and plug-in $\hat{\mathbf{H}}_{\text{PI}}$ (and their slight variants) are the most widely recommended bandwidth selectors, with a small advantage to the latter as it is computationally more efficient. Whilst it is inadvisable to base these general recommendations on the sole analysis of the *Grevillea* and the stem cell data as presented here, they have been indeed verified for a wider range of target density shapes (Chacón & Duong, 2013).

## 5.4 Empirical comparison of bandwidths of different derivative orders

We asserted earlier in Example 5.1 that the optimal bandwidths for density derivative estimation are larger than those for density estimation. Even though we defer the definition of the latter until Section 5.6, we present here an empirical comparison of the advantages of smoothing density derivatives with these bandwidths rather than those developed for the density function.

Density #10 from Chacón (2009) is defined as the density of the normal mixture $\frac{1}{2}N((0,0),\mathbf{I}_2) + \frac{1}{10}N((0,0),\frac{1}{16}\mathbf{I}_2) + \frac{1}{10}N((-1,-1),\frac{1}{16}\mathbf{I}_2) + \frac{1}{10}N((-1,1),\frac{1}{16}\mathbf{I}_2) + \frac{1}{10}N((1,-1),\frac{1}{16}\mathbf{I}_2) + \frac{1}{10}N((1,1),\frac{1}{16}\mathbf{I}_2)$. It shows an intricate multimodal structure that can be challenging to recover, and this is perhaps even more so for its derivatives. The quartile contours of $f$ are displayed in Figure 5.9(a) and the quasi-quartile contours of the second order partial derivative $f^{(0,2)}$ in Figure 5.9(b). In Figure 5.9(c) is the estimate $\hat{f}^{(0,2)}(\cdot;\hat{\mathbf{H}}_{\text{PI}})$ with a representative data sample and with the plug-in selector $\hat{\mathbf{H}}_{\text{PI}} = [0.037, 0.001; 0.001, 0.039]$ which is optimal for the density. In Figure 5.9(d) is the estimate $\hat{f}^{(0,2)}(\cdot;\hat{\mathbf{H}}_{\text{PI},2})$ with the plug-in selector which is optimal for the density Hessian $\hat{\mathbf{H}}_{\text{PI},2} = [0.054, 0.001; 0.001, 0.057]$. An inspection of the $\hat{\mathbf{H}}_{\text{PI}}$ and $\hat{\mathbf{H}}_{\text{PI},2}$ reveals that the latter contains larger terms on the main diagonal, which lead to the estimate with $\hat{\mathbf{H}}_{\text{PI},2}$ having smoother contours which more closely resemble the target contours than that with $\hat{\mathbf{H}}_{\text{PI}}$.

## 5.5 Squared error analysis

The mean squared error and its related error measures can be developed analogously to those for the density estimator $\hat{f}$. The MSE of the density derivative estimator admits the variance plus squared bias decomposition

$$\begin{aligned}\text{MSE}\{\mathsf{D}^{\otimes r}\hat{f}(\boldsymbol{x};\mathbf{H})\} &= \mathbb{E}\left\{\|\mathsf{D}^{\otimes r}\hat{f}(\boldsymbol{x};\mathbf{H}) - \mathsf{D}^{\otimes r}f(\boldsymbol{x})\|^2\right\}\\ &= \operatorname{tr}\text{Var}\{\mathsf{D}^{\otimes r}\hat{f}(\boldsymbol{x};\mathbf{H})\} + \|\text{Bias}\{\mathsf{D}^{\otimes r}\hat{f}(\boldsymbol{x};\mathbf{H})\}\|^2,\end{aligned}$$

where $\|\boldsymbol{v}\|^2$ denotes the squared norm of a vector $\boldsymbol{v}$ and $\operatorname{tr}\mathbf{A}$ denotes the trace of a square matrix $\mathbf{A}$. We employ the usual convention that the MSE is always a scalar whereas the expected value of a random $d$-vector is a $d$-vector and its variance is a $d \times d$ matrix. This implies that $\text{Bias}\{\mathsf{D}^{\otimes r}\hat{f}(\boldsymbol{x};\mathbf{H})\} = \mathbb{E}\{\mathsf{D}^{\otimes r}\hat{f}(\boldsymbol{x};\mathbf{H})\} - \mathsf{D}^{\otimes r}f(\boldsymbol{x}) \in \mathbb{R}^{d^r}$ and that $\text{Var}\{\mathsf{D}^{\otimes r}\hat{f}(\boldsymbol{x};\mathbf{H})\} = \mathbb{E}\left\{[\mathsf{D}^{\otimes r}\hat{f}(\boldsymbol{x};\mathbf{H}) - \mathbb{E}\{\mathsf{D}^{\otimes r}\hat{f}(\boldsymbol{x};\mathbf{H})\}][\mathsf{D}^{\otimes r}\hat{f}(\boldsymbol{x};\mathbf{H}) - \mathbb{E}\{\mathsf{D}^{\otimes r}\hat{f}(\boldsymbol{x};\mathbf{H})\}]^\top\right\} \in \mathcal{M}_{d^r \times d^r}$.

Given sufficient regularity, we obtain the MISE as

$$\text{MISE}\{\mathsf{D}^{\otimes r}\hat{f}(\cdot;\mathbf{H})\} = \mathbb{E}\int_{\mathbb{R}^d}\|\mathsf{D}^{\otimes r}\hat{f}(\boldsymbol{x};\mathbf{H}) - \mathsf{D}^{\otimes r}f(\boldsymbol{x})\|^2\,d\boldsymbol{x}.$$

The variance-squared bias decomposition of the MSE now leads to a decomposition of the MISE into the IV and ISB, namely

$$\text{MISE}\{\mathsf{D}^{\otimes r}\hat{f}(\cdot;\mathbf{H})\} = \text{IV}\{\mathsf{D}^{\otimes r}\hat{f}(\cdot;\mathbf{H})\} + \text{ISB}\{\mathsf{D}^{\otimes r}\hat{f}(\cdot;\mathbf{H})\}$$

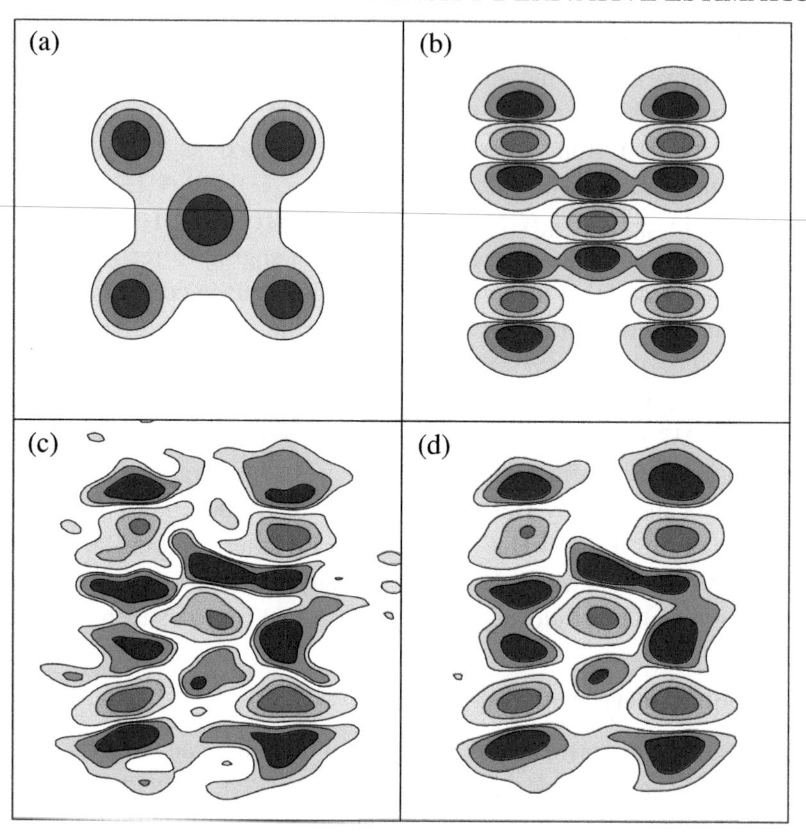


Figure 5.9 *Plug-in selectors of different derivative orders for the fountain density #10 data. (a) Target density $f$. (a) Target second order partial density derivative $f^{(0,2)}$. (c) Partial derivative estimate $\hat{f}^{(0,2)}(\cdot;\hat{\mathbf{H}}_{\mathrm{PI}})$ with plug-in bandwidth $\hat{\mathbf{H}}_{\mathrm{PI}} =$* $[0.037, 0.001; 0.001, 0.039]$. *(d) Partial derivative estimate $\hat{f}^{(0,2)}(\cdot;\hat{\mathbf{H}}_{\mathrm{PI},2})$ with plug-in bandwidth $\hat{\mathbf{H}}_{\mathrm{PI},2} =$* $[0.054, 0.001; 0.001, 0.057]$.

where

$$\mathrm{IV}\{\mathsf{D}^{\otimes r}\hat{f}(\cdot;\mathbf{H})\} = \int_{\mathbb{R}^d} \operatorname{tr}\operatorname{Var}\{\mathsf{D}^{\otimes r}\hat{f}(\boldsymbol{x};\mathbf{H})\}\,d\boldsymbol{x},$$

$$\mathrm{ISB}\{\mathsf{D}^{\otimes r}\hat{f}(\cdot;\mathbf{H})\} = \int_{\mathbb{R}^d} \|\mathrm{Bias}\{\mathsf{D}^{\otimes r}\hat{f}(\boldsymbol{x};\mathbf{H})\}\|^2\,d\boldsymbol{x}.$$

The MISE is a non-stochastic quantity that describes the performance of the kernel estimator with respect to a typical sample drawn from $f$. In contrast, the integrated squared error $\mathrm{ISE}\{\mathsf{D}^{\otimes r}\hat{f}(\cdot;\mathbf{H})\} = \int_{\mathbb{R}^d}\|\mathsf{D}^{\otimes r}\hat{f}(\boldsymbol{x};\mathbf{H}) - \mathsf{D}^{\otimes r}f(\boldsymbol{x})\|^2\,d\boldsymbol{x}$ is a stochastic discrepancy measure depending on the data at hand.

The expected value and the variance of the kernel density derivative estimator can be written in convolution form as

$$\begin{aligned}\mathbb{E}\{\mathsf{D}^{\otimes r}\hat{f}(\boldsymbol{x};\mathbf{H})\} &= (K_{\mathbf{H}} * \mathsf{D}^{\otimes r}f)(\boldsymbol{x})\\ \mathrm{Var}\{\mathsf{D}^{\otimes r}\hat{f}(\boldsymbol{x};\mathbf{H})\} &= n^{-1}\{[\{(\mathsf{D}^{\otimes r}K_{\mathbf{H}})(\mathsf{D}^{\otimes r}K_{\mathbf{H}})^\top\} * f](\boldsymbol{x})\\ &\quad - (K_{\mathbf{H}} * \mathsf{D}^{\otimes r}f)(\boldsymbol{x})(K_{\mathbf{H}} * \mathsf{D}^{\otimes r}f)(\boldsymbol{x})^\top\} \qquad (5.10)\end{aligned}$$

(Chacón et al., 2011, Theorem 4), where convolutions with vector-valued functions are applied in a component-wise manner. Combining the two formulas in Equation (5.10) we obtain a more explicit formula for the MSE

$$\begin{aligned}\mathrm{MSE}\{\mathsf{D}^{\otimes r}\hat{f}(\boldsymbol{x};\mathbf{H})\} &= n^{-1}\{(\|\mathsf{D}^{\otimes r}K_{\mathbf{H}}\|^2 * f)(\boldsymbol{x}) - \|(K_{\mathbf{H}} * \mathsf{D}^{\otimes r}f)(\boldsymbol{x})\|^2\}\\ &\quad + \|(K_{\mathbf{H}} * \mathsf{D}^{\otimes r}f)(\boldsymbol{x}) - \mathsf{D}^{\otimes r}f(\boldsymbol{x})\|^2.\end{aligned}$$

Integrating over $\boldsymbol{x}$, it follows that

$$\begin{aligned}&\mathrm{MISE}\{\mathsf{D}^{\otimes r}\hat{f}(\cdot;\mathbf{H})\}\\ &\quad = \{n^{-1}|\mathbf{H}|^{-1/2}\operatorname{tr}((\mathbf{H}^{-1})^{\otimes r}\mathbf{R}(\mathsf{D}^{\otimes r}K)) - n^{-1}\operatorname{tr}\mathbf{R}^*(K_{\mathbf{H}} * K_{\mathbf{H}}, \mathsf{D}^{\otimes r}f)\}\\ &\qquad + \{\operatorname{tr}\mathbf{R}^*(K_{\mathbf{H}} * K_{\mathbf{H}}, \mathsf{D}^{\otimes r}f) - 2\operatorname{tr}\mathbf{R}^*(K_{\mathbf{H}}, \mathsf{D}^{\otimes r}f) + \operatorname{tr}\mathbf{R}(\mathsf{D}^{\otimes r}f)\}\\ &\hspace{25em}(5.11)\end{aligned}$$

where $\mathbf{R}(\boldsymbol{a}) = \int_{\mathbb{R}^d} \boldsymbol{a}(\boldsymbol{x})\boldsymbol{a}(\boldsymbol{x})^\top d\boldsymbol{x}$ and $\mathbf{R}^*(b,\boldsymbol{a}) = \int_{\mathbb{R}^d} b * \boldsymbol{a}(\boldsymbol{x})\boldsymbol{a}(\boldsymbol{x})^\top d\boldsymbol{x}$ for suitable functions $\boldsymbol{a}\colon \mathbb{R}^d \to \mathbb{R}^p$ and $b\colon \mathbb{R}^d \to \mathbb{R}$. The first set of braces in Equation (5.11) contains the expression for the IV, the second the ISB.

To demonstrate the effect of the bandwidth more apparently, it is useful to derive an asymptotic MISE formula, which satisfies $\mathrm{MISE}\{\mathsf{D}^{\otimes r}\hat{f}(\cdot;\mathbf{H})\} = \mathrm{AMISE}\{\mathsf{D}^{\otimes r}\hat{f}(\cdot;\mathbf{H})\}\{1 + o(1)\}$. Since the density derivatives are vector-valued functions, the main mathematical tool required to obtain the desired asymptotic approximation in this case is Taylor's theorem for vector-valued functions. First it is necessary to clarify the notion of a derivative that we employ for a vector-valued function. If $\boldsymbol{f}\colon \mathbb{R}^d \to \mathbb{R}^p$ is a vector-valued function of a vector variable, with components $\boldsymbol{f} = (f_1, \ldots, f_p)$, then we define $\mathsf{D}^{\otimes r}\boldsymbol{f}(\boldsymbol{x}) \in \mathbb{R}^{pd^r}$ to be

$$\mathsf{D}^{\otimes r}\boldsymbol{f}(\boldsymbol{x}) = \begin{bmatrix} \mathsf{D}^{\otimes r}f_1(\boldsymbol{x}) \\ \hline \vdots \\ \hline \mathsf{D}^{\otimes r}f_p(\boldsymbol{x}) \end{bmatrix}.$$

Observe that, using this notation, we have $\mathsf{D}(\mathsf{D}^{\otimes r}f) = \mathsf{D}^{\otimes (r+1)}f$ due to the formal algebraic properties of the Kronecker product. Then, our version of Taylor's theorem reads as follows.

**Theorem 5** *Let $\boldsymbol{f}: \mathbb{R}^d \to \mathbb{R}^p$ be an r-times continuously differentiable vector-valued function. For $\boldsymbol{x}, \boldsymbol{a} \in \mathbb{R}^d$, the Taylor expansion of $\boldsymbol{f}$ at a point $\boldsymbol{x}+\boldsymbol{a}$ for a small perturbation $\boldsymbol{a}$ is*

$$\boldsymbol{f}(\boldsymbol{x}+\boldsymbol{a}) = \sum_{j=0}^{r} \frac{1}{j!}\{\mathbf{I}_p \otimes (\boldsymbol{a}^\top)^{\otimes j}\}\mathsf{D}^{\otimes j}\boldsymbol{f}(\boldsymbol{x}) + Re(\boldsymbol{a}).$$

*The remainder $Re(\boldsymbol{a})$ is a vector whose norm is $o(\|\boldsymbol{a}\|^r)$.*

Since this is a nonstandard formulation of Taylor's theorem, its proof is given in Section 5.9. In order to use Taylor expansions to find an asymptotic expression for the MISE we are required to make the following assumptions:

**Conditions B**

**(B1)** The density derivative function $\mathsf{D}^{\otimes r} f$ is square integrable and $\mathsf{D}^{\otimes (r+2)} f$ exists, with all of its $(r+2)$-th order partial derivatives bounded, continuous and square integrable.

**(B2)** The kernel $K$ is spherically symmetric and with finite second order moment i.e., $\int_{\mathbb{R}^d} \boldsymbol{z}K(\boldsymbol{z})d\boldsymbol{z} = \mathbf{0}$ and that $\int_{\mathbb{R}^d} \boldsymbol{z}^{\otimes 2}K(\boldsymbol{z})d\boldsymbol{z} = m_2(K)\operatorname{vec}\mathbf{I}_d$ with $m_2(K) = \int_{\mathbb{R}^d} z_i^2 K(\boldsymbol{z})d\boldsymbol{z}$ for all $i = 1, \ldots, d$; and all of its partial derivatives up to order $r$ are square integrable.

**(B3)** The bandwidth matrices $\mathbf{H} = \mathbf{H}_n$ form a sequence of positive definite, symmetric matrices such that $\operatorname{vec}\mathbf{H} \to 0$ and $n^{-1}|\mathbf{H}|^{-1/2}\operatorname{vec}((\mathbf{H}^{-1})^{\otimes r}) \to 0$ as $n \to \infty$.

We begin by developing an asymptotic form for the bias. After a change of variables, the expected value of $\mathsf{D}^{\otimes r}\hat{f}(\boldsymbol{x};\mathbf{H})$ can be written as

$$\mathbb{E}\{\mathsf{D}^{\otimes r}\hat{f}(\boldsymbol{x};\mathbf{H})\} = \int_{\mathbb{R}^d} K_{\mathbf{H}}(\boldsymbol{x}-\boldsymbol{y})\mathsf{D}^{\otimes r}f(\boldsymbol{y})\,d\boldsymbol{y} = \int_{\mathbb{R}^d} K(\boldsymbol{z})\mathsf{D}^{\otimes r}f(\boldsymbol{x}-\mathbf{H}^{1/2}\boldsymbol{z})\,d\boldsymbol{z}. \tag{5.12}$$

By applying Theorem 5 we obtain

$$\begin{aligned}\mathsf{D}^{\otimes r}f(\boldsymbol{x}-\mathbf{H}^{1/2}\boldsymbol{z}) &= \mathsf{D}^{\otimes r}f(\boldsymbol{x}) - \{\mathbf{I}_{d^r} \otimes (\boldsymbol{z}^\top\mathbf{H}^{1/2})\}\mathsf{D}^{\otimes(r+1)}f(\boldsymbol{x}) \\ &\quad + \tfrac{1}{2}\{\mathbf{I}_{d^r} \otimes (\boldsymbol{z}^\top\mathbf{H}^{1/2})^{\otimes 2}\}\mathsf{D}^{\otimes(r+2)}f(\boldsymbol{x}) + o(\|\operatorname{vec}\mathbf{H}\|)\mathbf{1}_{d^r}\end{aligned}$$

where $\mathbf{1}_{d^r}$ denotes the vector in $\mathbb{R}^{d^r}$ with one as all of its elements. Taking into account condition (B2) on $K$, substituting the expansion of $\mathsf{D}^{\otimes r}f(\boldsymbol{x}-\mathbf{H}^{1/2}\boldsymbol{z})$

into Equation (5.12) yields

$$\begin{aligned}
&\mathbb{E}\{\mathsf{D}^{\otimes r}\hat{f}(\boldsymbol{x};\mathbf{H})\} \\
&\quad = \mathsf{D}^{\otimes r}f(\boldsymbol{x}) + \tfrac{1}{2}[\mathbf{I}_{d^r}\otimes\{m_2(K)(\operatorname{vec}^\top\mathbf{I}_d)(\mathbf{H}^{1/2})^{\otimes 2}\}]\mathsf{D}^{\otimes(r+2)}f(\boldsymbol{x}) \\
&\qquad + o(\|\operatorname{vec}\mathbf{H}\|)\mathbf{1}_{d^r} \\
&\quad = \mathsf{D}^{\otimes r}f(\boldsymbol{x}) + \tfrac{1}{2}m_2(K)(\mathbf{I}_{d^r}\otimes\operatorname{vec}^\top\mathbf{H})\mathsf{D}^{\otimes(r+2)}f(\boldsymbol{x}) + o(\|\operatorname{vec}\mathbf{H}\|)\mathbf{1}_{d^r}.
\end{aligned}$$

From this, squaring and integrating with respect to $\boldsymbol{x}$ we obtain

$$\begin{aligned}
\operatorname{ISB}\{\mathsf{D}^{\otimes r}\hat{f}(\cdot;\mathbf{H})\} = {}& \tfrac{1}{4}m_2(K)^2\operatorname{tr}\big\{(\mathbf{I}_{d^r}\otimes\operatorname{vec}^\top\mathbf{H})\mathbf{R}(\mathsf{D}^{\otimes(r+2)}f)(\mathbf{I}_{d^r}\otimes\operatorname{vec}\mathbf{H})\big\} \\
& + o(\|\operatorname{vec}\mathbf{H}\|^2).
\end{aligned} \tag{5.13}$$

Alternative forms for the trace in the dominant term are

$$\begin{aligned}
&\operatorname{tr}\big\{(\mathbf{I}_{d^r}\otimes\operatorname{vec}^\top\mathbf{H})\mathbf{R}(\mathsf{D}^{\otimes(r+2)}f)(\mathbf{I}_{d^r}\otimes\operatorname{vec}\mathbf{H})\big\} \\
&\qquad\qquad = \operatorname{tr}\big[\{\mathbf{I}_{d^r}\otimes(\operatorname{vec}\mathbf{H}\operatorname{vec}^\top\mathbf{H})\}\mathbf{R}(\mathsf{D}^{\otimes(r+2)}f)\big] \\
&\qquad\qquad = \big\{\operatorname{vec}^\top\mathbf{R}(\mathsf{D}^{\otimes(r+2)}f)\big\}\big\{\operatorname{vec}\mathbf{I}_{d^r}\otimes(\operatorname{vec}\mathbf{H})^{\otimes 2}\big\}.
\end{aligned}$$

Regarding the IV, starting from Equation (5.10) and integrating with respect to $\boldsymbol{x}$ we have, for the first term,

$$\begin{aligned}
&n^{-1}\operatorname{tr}\int_{\mathbb{R}^d}[\{(\mathsf{D}^{\otimes r}K_{\mathbf{H}})(\mathsf{D}^{\otimes r}K_{\mathbf{H}})^\top\}*f](\boldsymbol{x})\,d\boldsymbol{x} \\
&\quad = n^{-1}\operatorname{tr}\int_{\mathbb{R}^d}\int_{\mathbb{R}^d}\mathsf{D}^{\otimes r}K_{\mathbf{H}}(\boldsymbol{x}-\boldsymbol{y})\mathsf{D}^{\otimes r}K_{\mathbf{H}}(\boldsymbol{x}-\boldsymbol{y})^\top f(\boldsymbol{y})\,d\boldsymbol{y}d\boldsymbol{x} \\
&\quad = n^{-1}|\mathbf{H}|^{-1/2}\operatorname{tr}\int_{\mathbb{R}^d}\int_{\mathbb{R}^d}(\mathbf{H}^{-1/2})^{\otimes r}\mathsf{D}^{\otimes r}K(\boldsymbol{z})\mathsf{D}^{\otimes r}K(\boldsymbol{z})^\top(\mathbf{H}^{-1/2})^{\otimes r} \\
&\qquad \times f(\boldsymbol{x}-\mathbf{H}^{1/2}\boldsymbol{z})\,d\boldsymbol{z}d\boldsymbol{x} \\
&\quad = n^{-1}|\mathbf{H}|^{-1/2}\operatorname{tr}\big\{(\mathbf{H}^{-1})^{\otimes r}\mathbf{R}(\mathsf{D}^{\otimes r}K)\big\}.
\end{aligned} \tag{5.14}$$

For the second term in Equation (5.10) we can reuse the calculations for $\mathbb{E}\{\mathsf{D}^{\otimes r}\hat{f}(\boldsymbol{x};\mathbf{H})\}$ to obtain $n^{-1}\operatorname{tr}\int_{\mathbb{R}^d}(K_{\mathbf{H}}*\mathsf{D}^{\otimes r}f)(\boldsymbol{x})(K_{\mathbf{H}}*(\mathsf{D}^{\otimes r}f)^\top)(\boldsymbol{x})\,d\boldsymbol{x} = n^{-1}\operatorname{tr}\mathbf{R}(\mathsf{D}^{\otimes r}f)+o(n^{-1})$, so that this second term in the IV is of a smaller order than the first one. Therefore,

$$\begin{aligned}
\operatorname{IV}\{\hat{f}(\cdot;\mathbf{H})\} = {}& n^{-1}|\mathbf{H}|^{-1/2}\operatorname{tr}\big\{(\mathbf{H}^{-1})^{\otimes r}\mathbf{R}(\mathsf{D}^{\otimes r}K)\big\} \\
& + o(n^{-1}|\mathbf{H}|^{-1/2}\|\operatorname{vec}\mathbf{H}^{-1}\|^r).
\end{aligned} \tag{5.15}$$

Combining Equations (5.13)–(5.15), it follows that an asymptotic approximation to the MISE can be written as

$$\begin{aligned}
\operatorname{AMISE}\{\mathsf{D}^{\otimes r}\hat{f}(\cdot;\mathbf{H})\} = {}& n^{-1}|\mathbf{H}|^{-1/2}\operatorname{tr}\big\{(\mathbf{H}^{-1})^{\otimes r}\mathbf{R}(\mathsf{D}^{\otimes r}K)\big\} \\
& + \tfrac{1}{4}m_2(K)^2\operatorname{tr}\big\{(\mathbf{I}_{d^r}\otimes\operatorname{vec}^\top\mathbf{H})\mathbf{R}(\mathsf{D}^{\otimes(r+2)}f)(\mathbf{I}_{d^r}\otimes\operatorname{vec}\mathbf{H})\big\}.
\end{aligned} \tag{5.16}$$

Examining Equation (5.16), if the bandwidth has small entries, the bias tends to be small as well whilst the variance is inflated. If the bandwidth is large, in the sense of having a large determinant, the bias tends also to be large with a correspondingly diminishing variance.

We note that the leading term $n^{-1}|\mathbf{H}|^{-1/2}\operatorname{tr}\{(\mathbf{H}^{-1})^{\otimes r}\mathbf{R}(\mathsf{D}^{\otimes r}K)\}$ of the IV is virtually indistinguishable asymptotically from its exact form for any density $f$. Thus there is no effect on bandwidth selection, at least in an asymptotic sense, in changing the exact variance for its leading term. Whereas the accuracy of the asymptotic ISB depends highly on the structure of $f$. Finally, observe that if we impose the condition (B3) on the bandwidth sequence we obtain that $\text{AMISE}\{\mathsf{D}^{\otimes r}\hat{f}(\cdot;\mathbf{H})\} \to 0$. That is, the kernel density derivative estimator is consistent for the target density derivative as $n \to \infty$. Thus $\mathsf{D}^{\otimes r}\hat{f}$ is asymptotically unbiased and consistent, which assures that, at least in the limiting case, kernel estimators perform well.

From an asymptotic point of view, the optimal bandwidth is $\mathbf{H}_{\text{AMISE},r} = \operatorname{argmin}_{\mathbf{H}\in\mathcal{F}}\{\text{AMISE}\{\mathsf{D}^{\otimes r}\hat{f}(\cdot;\mathbf{H})\}$. Under the assumptions (B1)–(B3) in Conditions B, by balancing the terms of order $n^{-1}|\mathbf{H}|^{-1/2}\operatorname{tr}((\mathbf{H}^{-1})^{\otimes r})$ and $\|\operatorname{vec}\mathbf{H}\|^2$ which comprise the AMISE, this leads to that $\mathbf{H}_{\text{AMISE},r}$ is of order $n^{-2/(d+2r+4)}$. This implies that the minimal MISE rate is $\inf_{\mathbf{H}\in\mathcal{F}}\text{MISE}\{\mathsf{D}^{\otimes r}\hat{f}(\cdot;\mathbf{H})\} = O(n^{-4/(d+2r+4)})$. This rate is uniformly slower than the parametric rate $O(n^{-1})$ for all $d$, implying that non-parametric kernel estimation is a more difficult problem than its parametric counterpart. This $O(n^{-4/(d+2r+4)})$ rate also implies that as the dimension $d$ and the derivative order $r$ increase, kernel estimation becomes increasingly difficult: a unit increase in the derivative order leads to a difficulty that is asymptotically equivalent to a two-fold unit increase in the dimension $d$.

## 5.6 Bandwidth selection for density derivative estimators

The oracle optimal bandwidth is the minimiser of the MISE, $\mathbf{H}_{\text{MISE},r} = \operatorname{argmin}_{\mathbf{H}\in\mathcal{F}}\text{MISE}\{\mathsf{D}^{\otimes r}\hat{f}(\cdot;\mathbf{H})\}$. As noted above, the simpler proxy $\mathbf{H}_{\text{AMISE},r}$ can be equally considered an asymptotic oracle bandwidth. As they require knowledge of the target density derivative $\mathsf{D}^{\otimes r}f$ to be computed, the usual approach is to estimate the (A)MISE from the data to get a data-based criterion $\widehat{\text{(A)MISE}}$ whose minimiser yields a data-based bandwidth selector, denoted as $\hat{\mathbf{H}}_r = \operatorname{argmin}_{\mathbf{H}\in\mathcal{F}}\widehat{\text{(A)MISE}}\{\mathsf{D}^{\otimes r}\hat{f}(\cdot;\mathbf{H})\}$.

### 5.6.1 *Normal scale bandwidths*

To obtain a normal scale selector, the unknown target density $f$ is replaced by a normal density with mean $\boldsymbol{\mu}$ and variance $\boldsymbol{\Sigma}$ in the error formulas, and the kernel $K$ is taken to be the normal kernel. Due to the neat mathematical properties of normal densities, the MISE and AMISE of $\mathsf{D}^{\otimes r}\hat{f}$ become

$$\begin{aligned}\mathrm{MISE}_{\mathrm{NS}}\{\mathsf{D}^{\otimes r}\hat{f}(\cdot;\mathbf{H})\} &= 2^{-r}(4\pi)^{-d/2}\{n^{-1}|\mathbf{H}|^{-1/2}\nu_r(\mathbf{H}^{-1})\\ &\quad + (1-n^{-1})|\mathbf{H}+\boldsymbol{\Sigma}|^{-1/2}\nu_r((\mathbf{H}+\boldsymbol{\Sigma})^{-1})\\ &\quad - 2^{(d+2r+2)/2}|\mathbf{H}+2\boldsymbol{\Sigma}|^{-1/2}\nu_r((\mathbf{H}+2\boldsymbol{\Sigma})^{-1}) + |\boldsymbol{\Sigma}|^{-1/2}\nu_r(\boldsymbol{\Sigma}^{-1})\}\\ \mathrm{AMISE}_{\mathrm{NS}}\{\mathsf{D}^{\otimes r}\hat{f}(\cdot;\mathbf{H})\} &= 2^{-r}(4\pi)^{-d/2}\{n^{-1}|\mathbf{H}|^{-1/2}\nu_r(\mathbf{H}^{-1})\\ &\quad + \tfrac{1}{16}|\boldsymbol{\Sigma}|^{-1/2}\nu_{r,2}(\boldsymbol{\Sigma}^{-1},\boldsymbol{\Sigma}^{-1/2}\mathbf{H}\boldsymbol{\Sigma}^{-1/2})\},\end{aligned}$$

as shown in Section 5.8, and where $\nu_r$ and $\nu_{r,2}$ refer to the functionals related to quadratic forms in normal variables introduced in Equations (5.8)–(5.9).

There is no explicit formula for the minimiser of $\mathrm{MISE}_{\mathrm{NS}}$, whereas the minimiser of the $\mathrm{AMISE}_{\mathrm{NS}}$ is given by

$$\mathbf{H}_{\mathrm{NS},r} = \{4/(d+2r+2)\}^{2/(d+2r+4)}n^{-2/(d+2r+4)}\boldsymbol{\Sigma} \tag{5.17}$$

from Chacón et al. (2011). If we replace the population variance $\boldsymbol{\Sigma}$ with an estimator, usually the sample variance $\mathbf{S}$, then we obtain a data-based, normal scale bandwidth

$$\hat{\mathbf{H}}_{\mathrm{NS},r} = \{4/(d+2r+2)\}^{2/(d+2r+4)}n^{-2/(d+2r+4)}\mathbf{S}. \tag{5.18}$$

As the normal density is amongst the smoothest densities available, the normal scale selector tends to yield bandwidths which lead to oversmoothing for non-normal data.

As for the case of the density, it can be useful to have explicit forms for the constrained minimizers of $\mathrm{AMISE}_{\mathrm{NS}}\{\mathsf{D}^{\otimes r}\hat{f}(\cdot;\mathbf{H})\}$. Within class $\mathcal{A}$, it is shown in Section 5.8 that the bandwidth that minimizes the AMISE can be written as $h^2_{\mathrm{NS},r}\mathbf{I}_d$ with

$$h_{\mathrm{NS},r} = \big\{4\nu_{r+1}(\mathbf{I}_d)|\boldsymbol{\Sigma}|^{1/2}/\nu_{r+2}(\boldsymbol{\Sigma}^{-1})\big\}^{1/(d+2r+4)}n^{-1/(d+2r+4)}. \tag{5.19}$$

From the definition of the $\nu_r$ functionals it follows that $\nu_{r+1}(\mathbf{I}_d)$ is the $(r+1)$-th raw moment of the chi-squared distribution, and its explicit form is $\nu_{r+1}(\mathbf{I}_d) = 2^{r+1}\Gamma(r+1+d/2)/\Gamma(d/2) = \prod_{j=0}^{r}(d+2j)$, see Johnson et al. (1994, Chapter 18). On the other hand, an explicit formula for $\nu_{r+2}(\boldsymbol{\Sigma}^{-1})$ is given in Equation (5.9), and computationally efficient ways to obtain it will be provided in Section 8.4. For the most common cases $r = 0,1,2$ the required quantities are $\nu_2(\mathbf{\Sigma}^{-1}) = 2\,\mathrm{tr}(\mathbf{\Sigma}^{-2}) + \mathrm{tr}^2(\mathbf{\Sigma}^{-1})$, $\nu_3(\mathbf{\Sigma}^{-1}) = 8\,\mathrm{tr}(\mathbf{\Sigma}^{-3}) + 6\,\mathrm{tr}(\mathbf{\Sigma}^{-2})\,\mathrm{tr}(\mathbf{\Sigma}) + \mathrm{tr}^3(\mathbf{\Sigma}^{-1})$, and $\nu_4(\mathbf{\Sigma}^{-1}) = 48\,\mathrm{tr}(\mathbf{\Sigma}^{-4}) + 32\,\mathrm{tr}(\mathbf{\Sigma}^{-3})\,\mathrm{tr}(\mathbf{\Sigma}^{-1}) + 12\,\mathrm{tr}(\mathbf{\Sigma}^{-2})\,\mathrm{tr}^2(\mathbf{\Sigma}^{-1}) + 12\,\mathrm{tr}^2(\mathbf{\Sigma}^{-2}) + \mathrm{tr}^4(\mathbf{\Sigma}^{-1})$.

The problem of finding an explicit formula for the optimal diagonal bandwidth in the normal case remains open, although by comparison with the bandwidth Equation (3.4) for $r = 0$, there is numerical evidence that the diagonal bandwidth whose $i$-th diagonal entry is the square of

$$h_{\mathrm{NS},r,i} = \{4\nu_{r+1}(\mathbf{I}_d)|\mathbf{\Delta}|^{1/2}/\nu_{r+2}(\mathbf{\Delta}^{-1})\}^{1/(d+2r+4)}\sigma_i n^{-1/(d+2r+4)} \tag{5.20}$$

provides a reasonable approximation, where $\sigma_i^2$ denotes the variance of the $i$-th coordinate, and $\mathbf{\Delta} = (\mathrm{diag}\,\mathbf{\Sigma})^{-1}\mathbf{\Sigma}$ with $\mathrm{diag}\,\mathbf{\Sigma} = \mathrm{diag}(\sigma_1^2,\ldots,\sigma_d^2)$.

### 5.6.2 *Normal mixture bandwidths*

The normal scale selector can be extended to a more general case of a normal mixture $\sum_{\ell=1}^q w_\ell \phi_{\mathbf{\Sigma}_\ell}(\cdot - \boldsymbol{\mu}_\ell)$. The corresponding MISE and AMISE formulas, from Theorems 7–8 in Chacón et al. (2011), are

$$\begin{aligned}\mathrm{MISE}_{\mathrm{NM}}\{\mathsf{D}^{\otimes r}\hat f(\cdot;\mathbf{H})\} &= 2^{-r}(4\pi)^{-d/2}n^{-1}|\mathbf{H}|^{-1/2}\nu_r(\mathbf{H}^{-1})\\ &\quad + \boldsymbol{w}^\top[(1-n^{-1})\mathbf{\Omega}_{r,2} - 2\mathbf{\Omega}_{r,1} + \mathbf{\Omega}_{r,0}]\boldsymbol{w}\\ \mathrm{AMISE}_{\mathrm{NM}}\{\mathsf{D}^{\otimes r}\hat f(\cdot;\mathbf{H})\} &= 2^{-r}(4\pi)^{-d/2}n^{-1}|\mathbf{H}|^{-1/2}\nu_r(\mathbf{H}^{-1}) + \tfrac{1}{4}\boldsymbol{w}^\top\mathbf{\Xi}_r\boldsymbol{w}\end{aligned}$$

where $\mathbf{\Omega}_{r,a}$ and $\mathbf{\Xi}_r$ are $q\times q$ matrices whose $(\ell,\ell')$-th entries are given, respectively, by

$$\begin{aligned}[\mathbf{\Omega}_{r,a}]_{\ell,\ell'} &= (-1)^r(\mathrm{vec}^\top\mathbf{I}_{d^r})\mathsf{D}^{\otimes 2r}\phi_{a\mathbf{H}+\mathbf{\Sigma}_\ell+\mathbf{\Sigma}_{\ell'}}(\boldsymbol{\mu}_\ell - \boldsymbol{\mu}_{\ell'}),\\ [\mathbf{\Xi}_r]_{\ell,\ell'} &= (-1)^r\,\mathrm{vec}^\top[\mathbf{I}_{d^r}\otimes(\mathrm{vec}\,\mathbf{H}\,\mathrm{vec}^\top\mathbf{H})]\mathsf{D}^{\otimes 2r+4}\phi_{\mathbf{\Sigma}_\ell+\mathbf{\Sigma}_{\ell'}}(\boldsymbol{\mu}_\ell - \boldsymbol{\mu}_{\ell'}).\end{aligned}$$

These authors also showed that these expressions for $r = 0$ reduce to those introduced in Section 3.3. With this closed form AMISE expression, it is possible to develop a normal mixture selector as:

$$\mathrm{NM}_r(\mathbf{H}) = 2^{-r}n^{-1}(4\pi)^{-d/2}|\mathbf{H}|^{-1/2}\nu_r(\mathbf{H}^{-1}) + \tfrac{1}{4}\hat{\boldsymbol{w}}^\top\hat{\mathbf{\Xi}}_r\hat{\boldsymbol{w}}$$

and thus $\hat{\mathbf{H}}_{\mathrm{NM},r} = \mathrm{argmin}_{\mathbf{H}\in\mathcal{F}}\mathrm{NM}_r(\mathbf{H})$. The NM selector has a straightforward implementation in Algorithm 6.

**Algorithm 6** Normal mixture bandwidth selector for density derivative estimation

**Input:** $\{\boldsymbol{X}_1, \dots, \boldsymbol{X}_n\}$
**Output:** $\hat{\mathbf{H}}_{\text{NM}}$
1: Fit a normal mixture model with $\hat{q}$ components to the data
2: $\hat{\mathbf{H}}_{\text{NM},r} :=$ minimiser of $\text{NM}_r(\mathbf{H})$

### 5.6.3 *Unbiased cross validation bandwidths*

The basis of unbiased cross validation is the formula for the integrated squared error. For $r = 0$, the ISE can be expanded as $\text{ISE}\{\hat{f}(\cdot;\mathbf{H})\} = \int_{\mathbb{R}^d} \hat{f}(\boldsymbol{x};\mathbf{H})^2 d\boldsymbol{x} - 2\mathbb{E}\{\hat{f}(\boldsymbol{X};\mathbf{H})|\boldsymbol{X}_1, \dots, \boldsymbol{X}_n\} + R(f)$ where $\boldsymbol{X} \sim f$. For $r \geq 0$, the equivalent expansion is

$$\begin{aligned}\text{ISE}\{\mathsf{D}^{\otimes r}\hat{f}(\cdot;\mathbf{H})\} = (-1)^r(\text{vec}^\top \mathbf{I}_{d^r})\Big[n^{-2}\sum_{i,j=1}^{n} \mathsf{D}^{\otimes 2r} K_{\mathbf{H}} * K_{\mathbf{H}}(\boldsymbol{X}_i - \boldsymbol{X}_j) \\ - 2\mathbb{E}\{\mathsf{D}^{\otimes 2r}\hat{f}(\boldsymbol{X};\mathbf{H})|\boldsymbol{X}_1, \dots, \boldsymbol{X}_n\}\Big] + \text{tr}\,\mathbf{R}(\mathsf{D}^{\otimes r} f).\end{aligned}$$

Leaving out the third term as it does not involve the bandwidth, and noting that an estimator of the conditional expectation is $n^{-1}\sum_{i=1}^{n} \mathsf{D}^{\otimes 2r}\hat{f}_{\mathbf{H},-i}(\boldsymbol{X}_i)$ where $\mathsf{D}^{\otimes 2r}\hat{f}_{\mathbf{H},-i}$ denotes the kernel estimator based on the sample with the $i$-th observation deleted, this leads to the UCV criterion (Chacón & Duong, 2013)

$$\begin{aligned}\text{UCV}_r(\mathbf{H}) = (-1)^r(\text{vec}^\top \mathbf{I}_{d^r})\Big[n^{-2}\sum_{i,j=1}^{n} \mathsf{D}^{\otimes 2r} K_{\mathbf{H}} * K_{\mathbf{H}}(\boldsymbol{X}_i - \boldsymbol{X}_j) \\ - 2\{n(n-1)\}^{-1}\sum_{\substack{i,j=1 \\ j \neq i}}^{n} \mathsf{D}^{\otimes 2r} K_{\mathbf{H}}(\boldsymbol{X}_i - \boldsymbol{X}_j)\Big]\end{aligned}$$

as the second summation is equal to $n^{-1}\sum_{i=1}^{n} \mathsf{D}^{\otimes 2r}\hat{f}_{\mathbf{H},-i}(\boldsymbol{X}_i)$. Extracting the terms in the first double sum with an identically zero argument, we have

$$\begin{aligned}\text{UCV}_r(\mathbf{H}) &= n^{-1}|\mathbf{H}|^{-1/2}\text{tr}\{(\mathbf{H}^{-1})^{\otimes r}\mathbf{R}(\mathsf{D}^{\otimes r} K)\} \\ &\quad + (-1)^r\{n(n-1)\}^{-1}(\text{vec}^\top \mathbf{I}_{d^r}) \\ &\quad \times \sum_{\substack{i,j=1 \\ j \neq i}}^{n} \{(1-n^{-1})\mathsf{D}^{\otimes 2r} K_{\mathbf{H}} * K_{\mathbf{H}} - 2\mathsf{D}^{\otimes 2r} K_{\mathbf{H}}\}(\boldsymbol{X}_i - \boldsymbol{X}_j). \quad (5.21)\end{aligned}$$

From this, $\hat{\mathbf{H}}_{\mathrm{UCV},r} = \mathrm{argmin}_{\mathbf{H}\in\mathcal{F}} \mathrm{UCV}_r(\mathbf{H})$. Examining Equation (5.21), the expected value of the UCV criterion is

$$\begin{aligned}\mathbb{E}\{\mathrm{UCV}_r(\mathbf{H})\} &= n^{-1}|\mathbf{H}|^{-1/2}\,\mathrm{tr}\left\{(\mathbf{H}^{-1})^{\otimes r}\mathbf{R}(\mathsf{D}^{\otimes r}K)\right\} + \mathrm{tr}\,\mathbf{R}^*(K_{\mathbf{H}} * K_{\mathbf{H}}, \mathsf{D}^{\otimes r}f)\\ &\quad - 2\,\mathrm{tr}\,\mathbf{R}^*(K_{\mathbf{H}}, \mathsf{D}^{\otimes r}f)\\ &= \mathrm{MISE}\{\hat{f}(\cdot;\mathbf{H})\} - \mathrm{tr}\,\mathbf{R}(\mathsf{D}^{\otimes r}f)\end{aligned}$$

from Equation (5.11). Hence the UCV is an unbiased estimator of the MISE, ignoring the $\mathrm{tr}\,\mathbf{R}(\mathsf{D}^{\otimes r}f)$ constant which does not involve the bandwidth.

If the normal kernel $K = \phi$ is used, then the UCV can be rewritten in a more computationally efficient form:

$$\begin{aligned}\mathrm{UCV}_r(\mathbf{H}) = (-1)^r\Big\{ & n^{-2}\sum_{i,j=1}^{n} \eta_{2r}(\boldsymbol{X}_i - \boldsymbol{X}_j; 2\mathbf{H})\\ & - 2[n(n-1)]^{-1}\sum_{\substack{i,j=1\\ j\neq i}}^{n} \eta_{2r}(\boldsymbol{X}_i - \boldsymbol{X}_j; \mathbf{H})\Big\}\end{aligned} \qquad (5.22)$$

where $\eta_{2r}(\boldsymbol{x};\boldsymbol{\Sigma}) = (\mathrm{vec}^\top \mathbf{I}_d)^{\otimes r}\mathsf{D}^{\otimes 2r}\phi_{\boldsymbol{\Sigma}}(\boldsymbol{x})$ as introduced by Chacón & Duong (2015). The reasons why these $\eta$ functionals allow for more efficient computation are detailed in Section 8.4.

The UCV selector has a straightforward implementation in Algorithm 7, which contains only one step for the numerical minimisation of the UCV criterion.

**Algorithm 7** UCV bandwidth selector for density derivative estimation

**Input:** $\{\boldsymbol{X}_1,\ldots,\boldsymbol{X}_n\}, r$
**Output:** $\hat{\mathbf{H}}_{\mathrm{UCV},r}$
1: $\hat{\mathbf{H}}_{\mathrm{UCV},r} :=$ minimiser of $\mathrm{UCV}_r(\mathbf{H})$

### 5.6.4 *Plug-in bandwidths*

We can rewrite the AMISE formula in Equation (5.16) as

$$\begin{aligned}\mathrm{AMISE}_r(\mathbf{H}) &= n^{-1}|\mathbf{H}|^{-1/2}\,\mathrm{tr}\left\{(\mathbf{H}^{-1})^{\otimes r}\mathbf{R}(\mathsf{D}^{\otimes r}K)\right\}\\ &\quad + (-1)^r\tfrac{1}{4}m_2(K)^2\boldsymbol{\psi}_{2r+4}^\top\{\mathrm{vec}\,\mathbf{I}_{d^r}\otimes(\mathrm{vec}\,\mathbf{H})^{\otimes 2}\}\end{aligned} \qquad (5.23)$$

where we recall from Section 3.6 that $\boldsymbol{\psi}_{2r+4}$ corresponds to the case $s = r+2$ of the functional

$$\boldsymbol{\psi}_{2s} = (-1)^s\,\mathrm{vec}\,\mathbf{R}(\mathsf{D}^{\otimes s}f) = \int_{\mathbb{R}^d}\mathsf{D}^{\otimes 2s}f(\boldsymbol{x})f(\boldsymbol{x})\,d\boldsymbol{x},$$

whose estimator had been denoted in Equation (3.11) as $\hat{\boldsymbol{\psi}}_{2s} \equiv \hat{\boldsymbol{\psi}}_{2s}(\mathbf{G})$ for a pilot bandwidth $\mathbf{G}$. This leads immediately to the plug-in criterion

$$\begin{aligned}\mathrm{PI}_r(\mathbf{H};\mathbf{G}) &= n^{-1}|\mathbf{H}|^{-1/2}\operatorname{tr}\{(\mathbf{H}^{-1})^{\otimes r}\mathbf{R}(\mathsf{D}^{\otimes r}K)\}\\ &\quad + (-1)^r\tfrac{1}{4}m_2(K)^2\hat{\boldsymbol{\psi}}_{2r+4}(\mathbf{G})^\top\{\operatorname{vec}\mathbf{I}_{d^r}\otimes(\operatorname{vec}\mathbf{H})^{\otimes 2}\} \qquad (5.24)\end{aligned}$$

as introduced by Chacón & Duong (2013). The resulting plug-in selector is $\hat{\mathbf{H}}_{\mathrm{PI},r} = \operatorname{argmin}_{\mathbf{H}\in\mathcal{F}}\mathrm{PI}_r(\mathbf{H};\mathbf{G})$.

If the normal kernels $K = L = \phi$ are used, then the plug-in criterion can be rewritten in a more computationally efficient form:

$$\begin{aligned}\mathrm{PI}_r(\mathbf{H};\mathbf{G}) &= 2^{-r}(4\pi)^{-d/2}n^{-1}|\mathbf{H}|^{-1/2}\nu_r(\mathbf{H}^{-1})\\ &\quad + (-1)^r(2n)^{-2}\sum_{i,j=1}^{n}\eta_{r,2}(\boldsymbol{X}_i - \boldsymbol{X}_j;\mathbf{H},\mathbf{G})\end{aligned}$$

where $\eta_{r,s}(\boldsymbol{x};\mathbf{A},\boldsymbol{\Sigma}) = [(\operatorname{vec}^\top\mathbf{I}_d)^{\otimes r}\otimes(\operatorname{vec}^\top\mathbf{A})^{\otimes s}]\mathsf{D}^{\otimes 2r+2s}\phi_{\boldsymbol{\Sigma}}(\boldsymbol{x})$.

As the use of a pilot bandwidth $\mathbf{G}$ is the key for good performance for the class of plug-in selectors, we set up an analogous optimality criterion for selecting $\mathbf{G}$ as

$$\begin{aligned}\mathrm{MSE}(\hat{\mathbf{H}}_{\mathrm{PI},r}) &= \mathbb{E}\{\|\operatorname{vec}(\hat{\mathbf{H}}_{\mathrm{PI},r} - \mathbf{H}_{\mathrm{AMISE},r})\|^2\}\\ &= \mathrm{const}\cdot\mathrm{MSE}\{\hat{\boldsymbol{\psi}}_{2r+4}(\mathbf{G})\}\{1+o(1)\} \qquad (5.25)\end{aligned}$$

where the constant does not involve $\mathbf{G}$ or the data, and so can be ignored when optimising the $\mathrm{MSE}(\hat{\mathbf{H}}_{\mathrm{PI},r})$ with respect to $\mathbf{G}$. The leading asymptotic term of the $\mathrm{MSE}\{\hat{\boldsymbol{\psi}}_{2r+4}(\mathbf{G})\}$ is

$$\begin{aligned}\mathrm{AMSE}\{\hat{\boldsymbol{\psi}}_{2r+4}(\mathbf{G})\} &= \|n^{-1}|\mathbf{G}|^{-1/2}(\mathbf{G}^{-1/2})^{\otimes 2r+4}\mathsf{D}^{\otimes 2r+4}L(\mathbf{0})\\ &\quad + \tfrac{1}{2}m_2(L)(\operatorname{vec}^\top\mathbf{G}\otimes\mathbf{I}_{d^{2r+4}})\boldsymbol{\psi}_{2r+6}\|^2 \qquad (5.26)\end{aligned}$$

as developed by Chacón & Duong (2010, Theorem 1).

The asymptotically optimal $\mathbf{G}$ is the minimiser of this AMSE in Equation (5.26). As it contains the unknown quantity $\boldsymbol{\psi}_{2r+6}$, we construct an estimate $\widehat{\mathrm{AMSE}}\{\hat{\boldsymbol{\psi}}_{2r+4}(\mathbf{G})\}$ of the AMSE by replacing $\boldsymbol{\psi}_{2r+6}$ with $\hat{\boldsymbol{\psi}}_{2r+6}(\mathbf{G}_{\mathrm{NS},2r+6})$, where this time we use a normal scale pilot bandwidth $\hat{\mathbf{G}}_{\mathrm{NS},2r+6} = \{2/(d+2r+6)\}^{2/(d+2r+8)}2\mathbf{S}n^{-2/(d+2r+8)}$ rather than another stage of numerical optimisation, see Chacón & Duong (2010, Equation (8)) and also Section 5.8.5 below.

This insensitivity of the pilot bandwidth to an imprecise estimation can be extended to afford a further reduction in complexity without overly compromising the accuracy of $\hat{f}$ by using the scalar bandwidth class $\mathbf{G} \in \mathcal{A}$, as there is an analytic expression for the minimiser of $\widehat{\text{AMSE}}\{\hat{\boldsymbol{\psi}}_{2r+4}(g^2\mathbf{I}_d)\}$,

$$\hat{g}_{2r+4} = [2A_1/\{-A_2 + (A_2^2 + 4A_1A_3)^{1/2}\}]^{1/(d+2r+6)} n^{-1/(d+2r+6)} \quad (5.27)$$

where the constants are $A_1 = (2d+4r+8)\|\mathsf{D}^{\otimes 2r+4}L(\mathbf{0})\|^2, A_2 = (d+2r+2)m_2(L)[\text{vec}\,\mathbf{I}_d \otimes \mathsf{D}^{\otimes 2r+4}L(\mathbf{0})]^\top \hat{\boldsymbol{\psi}}_{2r+6}(\hat{g}^2_{\text{NS},2r+6}\mathbf{I}_d), A_3 = m_2(L)^2\|(\text{vec}^\top \mathbf{I}_d \otimes \mathbf{I}_{d^{2r+4}})\hat{\boldsymbol{\psi}}_{2r+6}(\hat{g}^2_{\text{NS},2r+6}\mathbf{I}_d)\|^2$. The normal scale scalar pilot is

$$\hat{g}_{\text{NS},2r+6} = [2B_1/\{-B_2 + (B_2^2 + 4B_1B_3)^{1/2}\}]^{1/(d+2r+8)} n^{-1/(d+2r+8)}$$

where $B_1 = 2(2\pi)^{-d}\text{OF}(2r+6)\prod_{j=0}^{r+3}(d+2j), B_2 = -(d+2r+4)2^{-d/2-r+2} \times (2\pi)^{-d}\text{OF}(2r+6)|\mathbf{S}|^{-1/2}\nu_{r+4}(\mathbf{S}^{-1})$, and $B_3 = \hat{\boldsymbol{\psi}}^\top_{\text{NS},2r+8}(\text{vec}\,\mathbf{I}_d\,\text{vec}^\top\,\mathbf{I}_d \otimes \mathbf{I}_{d^{2r+6}})\hat{\boldsymbol{\psi}}_{\text{NS},2r+8}$. A pilot bandwidth of class $\mathcal{A}$ is sufficient for most cases if the data are pre-scaled to have the same marginal variance, though for cases where unconstrained pilots of class $\mathcal{F}$ are more appropriate, see Chacón & Duong (2010).

This 2-stage plug-in selector is given in Algorithm 8, with the option in steps 1 and 3 of an unconstrained or a scalar pilot. The result is $\hat{\mathbf{H}}_{\text{PI},r}$ which is a generalisation of $\hat{\mathbf{H}}_{\text{PI}}$.

---

**Algorithm 8** Two-stage plug-in bandwidth selector for density derivative estimation

---

**Input:** $\{\boldsymbol{X}_1,\ldots,\boldsymbol{X}_n\}, r$
**Output:** $\hat{\mathbf{H}}_{\text{PI},r}$

1: Compute the $(2r+6)$-th order normal scale pilot bandwidth
 (a) Unconstrained pilot $\hat{\mathbf{G}}_{2r+6} := \hat{\mathbf{G}}_{\text{NS},2r+6}$
 (b) Scalar pilot $\hat{\mathbf{G}}_{2r+6} := \hat{g}^2_{\text{NS},2r+6}\mathbf{I}_d$
2: Compute the $(2r+6)$-th order kernel functional estimate $\hat{\boldsymbol{\psi}}_{2r+6}(\hat{\mathbf{G}}_{2r+6})$
 /* Stage 1 */
3: Plug $\hat{\boldsymbol{\psi}}_{2r+6}(\hat{\mathbf{G}}_{2r+6})$ into the formula for the pilot bandwidth $\hat{\mathbf{G}}_{2r+4}$
 (a) Unconstrained pilot $\hat{\mathbf{G}}_{2r+4} :=$ minimiser of $\widehat{\text{AMSE}}\{\hat{\boldsymbol{\psi}}_{2r+4}(\mathbf{G})\}$
 (b) Scalar pilot $\hat{\mathbf{G}}_{2r+4} := \hat{g}^2_{2r+4}\mathbf{I}_d$
4: Compute the $(2r+4)$-th order kernel functional estimate $\hat{\boldsymbol{\psi}}_{2r+4}(\hat{\mathbf{G}}_{2r+4})$
 /* Stage 2 */
5: $\hat{\mathbf{H}}_{\text{PI},r} :=$ minimiser of $\text{PI}_r(\mathbf{H};\hat{\mathbf{G}}_{2r+4})$

---

### 5.6.5 Smoothed cross validation bandwidths

Instead of estimating the asymptotic ISB, the smoothed cross validation criterion aims to estimate the exact ISB by replacing the true density $f$ by a pilot kernel density estimator $\tilde{f}(\boldsymbol{x};\mathbf{G})$ to obtain

$$\begin{aligned}\widehat{\mathrm{ISB}}_r(\mathbf{H};\mathbf{G}) &= \int_{\mathbb{R}^d} \|\{K_{\mathbf{H}} * \mathsf{D}^{\otimes r}\tilde{f}(\cdot;\mathbf{G})\}(\boldsymbol{x}) - \mathsf{D}^{\otimes r}\tilde{f}(\boldsymbol{x};\mathbf{G})\|^2\, d\boldsymbol{x} \\ &= (-1)^r n^{-2}(\mathrm{vec}^\top \mathbf{I}_{d^r}) \sum_{i,j=1}^{n} (\bar{K}_{\mathbf{H}} - 2K_{\mathbf{H}} + K_{\mathbf{0}}) * \mathsf{D}^{\otimes 2r}\bar{L}_{\mathbf{G}}(\boldsymbol{X}_i - \boldsymbol{X}_j),\end{aligned}$$

where $\bar{K} = K * K$, $\bar{L} = L * L$ and $K_{\mathbf{0}}$ denotes the Dirac delta function. Adding the dominant term of the IV to this yields

$$\begin{aligned}\mathrm{SCV}_r(\mathbf{H};\mathbf{G}) &= n^{-1}|\mathbf{H}|^{-1/2}\,\mathrm{tr}\{(\mathbf{H}^{-1})^{\otimes r}\mathbf{R}(\mathsf{D}^{\otimes r}K)\} \\ &\quad + (-1)^r n^{-2}(\mathrm{vec}^\top \mathbf{I}_{d^r}) \sum_{i,j=1}^{n} (\bar{K}_{\mathbf{H}} - 2K_{\mathbf{H}} + K_{\mathbf{0}}) * \mathsf{D}^{\otimes 2r}\bar{L}_{\mathbf{G}}(\boldsymbol{X}_i - \boldsymbol{X}_j)\end{aligned} \tag{5.28}$$

as the SCV estimator of the (A)MISE. Then, the SCV selector is $\hat{\mathbf{H}}_{\mathrm{SCV},r} = \mathrm{argmin}_{\mathbf{H}\in\mathcal{F}}\,\mathrm{SCV}_r(\mathbf{H};\mathbf{G})$.

In contrast to the plug-in estimator of the AMISE in Equation (5.24), the integrated density functional $\boldsymbol{\psi}_{2r+4}$ is not required be estimated in Equation (5.28), with the trade-off of the more computationally intensive double sums. If there are no duplicates in the data $\boldsymbol{X}_1,\ldots,\boldsymbol{X}_n$, then the unbiased cross validation is a special case of the smoothed cross validation as $\mathrm{SCV}_r(\mathbf{H};\mathbf{0}) \equiv \mathrm{UCV}_r(\mathbf{H})$, i.e., the SCV pre-smooths the data differences $\boldsymbol{X}_i - \boldsymbol{X}_j$ by $\bar{L}_{\mathbf{G}}$.

If, instead of using the dominant term of the IV, the exact IV is employed, the resulting estimate of the MISE is

$$\begin{aligned}\mathrm{BMISE}_r(\mathbf{H};\mathbf{G}) &= n^{-1}|\mathbf{H}|^{-1/2}\,\mathrm{tr}\{(\mathbf{H}^{-1})^{\otimes r}\mathbf{R}(\mathsf{D}^{\otimes r}K)\} \\ &\quad + (-1)^r n^{-2}(\mathrm{vec}^\top \mathbf{I}_{d^r}) \\ &\quad \times \sum_{i,j=1}^{n} \{(1-n^{-1})\bar{K}_{\mathbf{H}} - 2K_{\mathbf{H}} + K_{\mathbf{0}})\} * \mathsf{D}^{\otimes 2r}\bar{L}_{\mathbf{G}}(\boldsymbol{X}_i - \boldsymbol{X}_j).\end{aligned}$$

The only difference with the SCV in Equation (5.28) is the $1 - n^{-1}$ factor in the first term of the double sum, hence they are asymptotically equivalent.

If normal kernels $K = L = \phi$ are used, then the SCV has a computationally efficient form in Equation (5.29):

$$\begin{aligned}\mathrm{SCV}_r(\mathbf{H}) &= 2^{-r}(4\pi)^{-d/2}n^{-1}|\mathbf{H}|^{-1/2}\nu_r(\mathbf{H}^{-1}) \\ &\quad + (-1)^r n^{-2}\sum_{i,j=1}^{n}\{\eta_r(\boldsymbol{X}_i-\boldsymbol{X}_j;2\mathbf{H}+2\mathbf{G}) \\ &\quad - 2\eta_r(\boldsymbol{X}_i-\boldsymbol{X}_j;\mathbf{H}+2\mathbf{G})+\eta_r(\boldsymbol{X}_i-\boldsymbol{X}_j;2\mathbf{G})\}. \qquad (5.29)\end{aligned}$$

The analogous pilot bandwidth selection problem for the SCV is slightly different from the PI selector. We begin with the $\mathrm{MSE}(\hat{\mathbf{H}}_{\mathrm{SCV},r}) = \mathbb{E}\{\|\mathrm{vec}(\hat{\mathbf{H}}_{\mathrm{SCV},r}-\mathbf{H}_{\mathrm{MISE},r})\|^2\}$ which has the leading term $\mathrm{AMSE}(\hat{\mathbf{H}}_{\mathrm{SCV},r}) = \mathrm{const}\cdot\mathrm{AMSE}^*\{\hat{\boldsymbol{\psi}}_{2r+4}(\mathbf{G})\}$ where the constant does not involve $\mathbf{G}$, and

$$\begin{aligned}\mathrm{AMSE}^*\{\hat{\boldsymbol{\psi}}_{2r+4}(\mathbf{G})\} &= \|n^{-1}|\mathbf{G}|^{-1/2}(\mathbf{G}^{-1/2})^{\otimes 2r+4}\mathsf{D}^{\otimes 2r+4}\bar{L}(\mathbf{0}) \\ &\quad + \tfrac{1}{2}m_2(\bar{L})(\mathrm{vec}^\top\mathbf{G}\otimes\mathbf{I}_{d^{2r+4}})\boldsymbol{\psi}_{2r+6}\|^2 \qquad (5.30)\end{aligned}$$

which is the $\mathrm{AMSE}\{\hat{\boldsymbol{\psi}}_{2r+4}(\mathbf{G})\}$ in Equation (5.26) except that the convolved kernel $\bar{L}$ rather than $L$ is used. Whilst no explicit computation of $\hat{\boldsymbol{\psi}}_{2r+4}(\mathbf{G})$ is required for the SCV, we are still required to compute the corresponding matrix $\hat{\mathbf{G}}_{2r+4}$. As for the PI bandwidth, for a 2-stage SCV selector this $\hat{\mathbf{G}}_{2r+4}$ is obtained by minimising $\widehat{\mathrm{AMSE}}^*\{\hat{\boldsymbol{\psi}}_{2r+4}(\mathbf{G})\}$, which is an estimate of the AMSE$^*$ constructed by replacing $\boldsymbol{\psi}_{2r+6}$ in the AMSE$^*$ formula with a kernel estimate $\hat{\boldsymbol{\psi}}_{2r+6}(\hat{\mathbf{G}}^*_{\mathrm{NS},2r+6})$ using a normal scale bandwidth $\hat{\mathbf{G}}^*_{\mathrm{NS},2r+6} = \{2/(d+2r+6)\}^{2/(d+2r+8)}\mathbf{S}n^{-2/(d+2r+8)}$ rather than another stage of numerical optimisation, see Section 5.8.5.

Proceeding as for the PI selector, a similar scalar pilot bandwidth may be derived. The scalar minimiser of $\widehat{\mathrm{AMSE}}^*\{\hat{\boldsymbol{\psi}}_{2r+2s}(g^2\mathbf{I}_d)\}$ is

$$\hat{g}^*_{2r+4} = [2A_1/\{-A_2+(A_2^2+4A_1A_3)^{1/2}\}^{1/(d+2r+8)}n^{-1/(d+2r+8)} \qquad (5.31)$$

where the constants are $A_1 = (2d+4r+8)\|\mathsf{D}^{\otimes 2r+4}\bar{L}(\mathbf{0})\|^2, A_2 = (d+2r+2)m_2(\bar{L})[\mathrm{vec}\,\mathbf{I}_d\otimes\mathsf{D}^{\otimes 2r+4}\bar{L}(\mathbf{0})]^\top\hat{\boldsymbol{\psi}}_{2r+6}(\hat{g}^{*2}_{\mathrm{NS},2r+6}\mathbf{I}_d), A_3 = m_2(\bar{L})^2\|(\mathrm{vec}^\top\mathbf{I}_d\otimes\mathbf{I}_{d^{2r+4}})\hat{\boldsymbol{\psi}}_{2r+6}(\hat{g}^{*2}_{\mathrm{NS},2r+6}\mathbf{I}_d)\|^2$. The normal scale scalar pilot is

$$\hat{g}^*_{\mathrm{NS},2r+6} = \{2B_1/[-B_2+(B_2^2+4B_1B_3)^{1/2}]\}^{1/(d+2r+8)}n^{-1/(d+2r+8)}$$

where $B_1 = 2^{-2r-5}(4\pi)^{-d}\mathrm{OF}(2r+6)\prod_{j=0}^{r+3}(d+2j), B_2 = -(d+2r+4)\times 2^{-2r-6}(4\pi)^{-d}\mathrm{OF}(2r+6)|\mathbf{S}|^{-1/2}\nu_{r+4}(\mathbf{S}^{-1}), B_3 = 4\hat{\boldsymbol{\psi}}^\top_{\mathrm{NS},2r+8}(\mathrm{vec}\,\mathbf{I}_d\,\mathrm{vec}^\top\mathbf{I}_d\otimes\mathbf{I}_{d^{2r+6}})\hat{\boldsymbol{\psi}}_{\mathrm{NS},2r+8}$.

The 2-stage SCV selector is given in Algorithm 9, with the option in steps 1 and 3 to use unconstrained or scalar pilots. The result is $\hat{\mathbf{H}}_{\mathrm{SCV},r}$ which is a generalisation of $\hat{\mathbf{H}}_{\mathrm{SCV}}$.

**Algorithm 9** Two-stage SCV bandwidth selector for density derivative estimation

**Input:** $\{\boldsymbol{X}_1,\ldots,\boldsymbol{X}_n\}, r$
**Output:** $\hat{\mathbf{H}}_{\mathrm{SCV},r}$
1: Compute $(2r+6)$-th order normal scale pilot bandwidth
(a) Unconstrained pilot $\hat{\mathbf{G}}_{2r+6} := \hat{\mathbf{G}}^*_{\mathrm{NS},2r+6}$
(b) Scalar pilot $\hat{\mathbf{G}}_{2r+6} := \hat{g}^{*2}_{\mathrm{NS},2r+6}\mathbf{I}_d$
2: Compute $(2r+6)$-th order kernel functional estimate $\hat{\boldsymbol{\psi}}_{2r+6}(\hat{\mathbf{G}}_{2r+6})$
/* Stage 1 */
3: Plug $\hat{\boldsymbol{\psi}}_{2r+6}(\hat{\mathbf{G}}_{2r+6})$ into formula for pilot bandwidth $\hat{\mathbf{G}}_{2r+4}$
(a) Unconstrained pilot $\hat{\mathbf{G}}_{2r+4} :=$ minimiser of $\widehat{\mathrm{AMSE}}^*\{\hat{\boldsymbol{\psi}}_{2r+4}(\mathbf{G})\}$
(b) Scalar pilot $\hat{\mathbf{G}}_{2r+4} := \hat{g}^{*2}_{2r+4}\mathbf{I}_d$
4: $\mathbf{H}_{\mathrm{SCV},r} :=$ minimiser of $\mathrm{SCV}_r(\mathbf{H};\hat{\mathbf{G}}_{2r+4})$ /* Stage 2 */

## 5.7 Relative convergence rates of bandwidth selectors

Whilst we do not have in general explicit expressions for the unconstrained data-based selectors, or even the oracle (A)MISE selectors, we are able to establish that all these selectors are of order $n^{-2/(d+2r+4)}$. Analogously we can show that optimal pilot selector $\hat{\mathbf{G}}_{2r+6}$ for both the plug-in and SCV is of order $n^{-2/(d+2r+6)}$. Since the pilot and final selectors do not share a common asymptotic order, this indicates that the functionally independent pilot selector is crucial to the solid performance of the plug-in and SCV selectors, and the reduced performance for the UCV where there is no pilot bandwidth.

The analogous generalisation of Equation (3.21) is that a bandwidth selector $\hat{\mathbf{H}}_r$ converges to the oracle bandwidth $\mathbf{H}_{\mathrm{AMISE}_r}$ at relative rate $n^{-\alpha}$, $\alpha > 0$, when $\operatorname{vec}(\hat{\mathbf{H}}_r - \mathbf{H}_{\mathrm{AMISE},r}) = O_P(\mathbf{J}_{d^2}n^{-\alpha}) \operatorname{vec}\mathbf{H}_{\mathrm{AMISE},r}$, where $\mathbf{J}_{d^2}$ is the $d^2 \times d^2$ matrix of all ones. The rates for the unconstrained selectors are summarised in Table 5.1, which complement those in Table 3.1. The rates in the former table are taken from Chacón & Duong (2013, Theorem 1), except that these authors established the convergence rate to $\mathbf{H}_{\mathrm{MISE},r}$ rather than $\mathbf{H}_{\mathrm{AMISE},r}$ . This does not change the stated rates, apart from for $\hat{\mathbf{H}}_{\mathrm{UCV},r}$, since the convergence rate of $\mathbf{H}_{\mathrm{AMISE},r}$ to $\mathbf{H}_{\mathrm{MISE},r}$ is $O(n^{-2/(d+2r+4)})$.

The performance decreases with increasing dimension and with increasing derivative order. For $d \le 3$, the UCV selector rate of $n^{-\min\{d,4\}/(2d+4r+8)}$ is slower than the SCV and PI selectors rate, which is $n^{-2/(d+2r+6)}$ for $d \ge 2$ and $n^{-5/(4r+14)}$ for $d = 1$, due to bias annihilation, as shown in Jones (1992). This swaps over for $d \ge 4$, as the choice of an independent pilot bandwidth decreases the convergence rates of the latter. The convergence rates remain the same for constrained matrices for the cross validation selectors, though the constrained plug-in selectors have a faster rate convergence $n^{-\min\{d+4,8\}/(2d+4r+12)}$ (again from Jones, 1992) indicating that selecting the off-diagonal elements of the bandwidth matrix, which determine the orientation of the kernel, is the most difficult aspect of unconstrained plug-in selection.

A unit increase in the derivative order leads to an equivalent decrease in the convergence as twice the increase in the dimension, indicating that density derivative estimation raises more difficulties than increasing the data dimension, at least asymptotically. This difficulty partially explains the scarcity of theoretical and practical results for the former in comparison to density estimators.

| **Selector** | **Class** | **Convergence rate to $\mathbf{H}_{\text{AMISE}}$** |
|---|---|---|
| $\hat{\mathbf{H}}_{\text{UCV},r}$ | $\mathcal{A},\mathcal{D},\mathcal{F}$ | $n^{-\min\{d,4\}/(2d+4r+8)}$ |
| $\hat{\mathbf{H}}_{\text{SCV},r}$ | $\mathcal{A},\mathcal{D},\mathcal{F}$ | $n^{-2/(d+2r+6)}$ |
| $\hat{\mathbf{H}}_{\text{PI},r}$ | $\mathcal{A}$ | $n^{-\min\{d+4,8\}/(2d+4r+12)}$ |
| $\hat{\mathbf{H}}_{\text{PI},r}$ | $\mathcal{D},\mathcal{F}$ | $n^{-2/(d+2r+6)}$ |

Table 5.1 *Convergence rates to $\mathbf{H}_{\text{AMISE},r}$ for the unbiased cross validation $\hat{\mathbf{H}}_{\text{UCV},r}$, smoothed cross validation $\hat{\mathbf{H}}_{\text{SCV},r}$ and plug-in $\hat{\mathbf{H}}_{\text{PI},r}$ selectors, for $d > 1$.*

An equivalent treatment of the relative convergence rate of a data-based pilot bandwidth $\hat{\mathbf{G}}_{2r+4}$ to $\mathbf{G}_{\text{AMSE},2r+4}$ follows directly, and which we conjecture would be $O_P(n^{-2/(d+2r+8)})$.

## 5.8 Case study: The normal density

In this section we provide a detailed collation of numerous results for the normal case, i.e., where $f$ is the density of the $N(\boldsymbol{\mu},\boldsymbol{\Sigma})$ distribution and $K$ is taken to be the standard normal kernel. This is an important case, since it is useful both for better understanding of the AMISE approximation and for providing normal scale rules as a starting point for developing more advanced bandwidth selectors.

### 5.8.1 *Exact MISE*

As noted in Equation (2.4) for the density case or in Equation (5.11) for the density derivative case, the MISE can be written as the sum of four terms. In the normal case, the computations are greatly simplified because all these four terms are closely related to each other. To determine them, we first note that an element-wise application of formula (A.3) in Wand & Jones (1993) leads to

$$\int_{\mathbb{R}^d} \mathsf{D}^{\otimes r}\phi_{\mathbf{\Sigma}}(\boldsymbol{x}-\boldsymbol{\mu}) \otimes \mathsf{D}^{\otimes r'}\phi_{\mathbf{\Sigma}'}(\boldsymbol{x}-\boldsymbol{\mu}')d\boldsymbol{x} = (-1)^r \mathsf{D}^{\otimes(r+r')}\phi_{\mathbf{\Sigma}+\mathbf{\Sigma}'}(\boldsymbol{\mu}-\boldsymbol{\mu}') \tag{5.32}$$

which implies that we have $\operatorname{tr}\int_{\mathbb{R}^d} \mathsf{D}^{\otimes r}\phi_{\mathbf{\Sigma}}(\boldsymbol{x}-\boldsymbol{\mu})\mathsf{D}^{\otimes r'}\phi_{\mathbf{\Sigma}'}(\boldsymbol{x}-\boldsymbol{\mu}')^\top d\boldsymbol{x} = (-1)^r(\operatorname{vec}^\top \mathbf{I}_{d^{r+r'}})\mathsf{D}^{\otimes(r+r')}\phi_{\mathbf{\Sigma}+\mathbf{\Sigma}'}(\boldsymbol{\mu}-\boldsymbol{\mu}')$. This can be further simplified for $\boldsymbol{\mu} = \boldsymbol{\mu}' = \mathbf{0}$ and $r = r'$, since using the formulas for the derivatives of the normal density given in Holmquist (1996*a*) yields

$$\operatorname{tr}\int_{\mathbb{R}^d} \mathsf{D}^{\otimes r}\phi_{\mathbf{\Sigma}}(\boldsymbol{x})\mathsf{D}^{\otimes r}\phi_{\mathbf{\Sigma}'}(\boldsymbol{x})^\top d\boldsymbol{x} = (2\pi)^{-d/2}|\mathbf{\Sigma}+\mathbf{\Sigma}'|^{-1/2}\nu_r\big((\mathbf{\Sigma}+\mathbf{\Sigma}')^{-1}\big)$$

where, as introduced in Section 5.1.3, for a standard normal variable $\boldsymbol{Z}$, $\nu_r(\mathbf{A}) = \mathrm{OF}(2r)(\operatorname{vec}^\top \mathbf{A})^{\otimes r}\mathcal{S}_{d,2r}(\operatorname{vec}\mathbf{I}_d)^{\otimes r} = \mathbb{E}\{(\boldsymbol{Z}^\top\mathbf{A}\boldsymbol{Z})^r\}$.

Observing also that $(\mathsf{D}^{\otimes r}\phi_{\mathbf{\Sigma}}) * \phi_{\mathbf{\Sigma}'} = \mathsf{D}^{\otimes r}\phi_{\mathbf{\Sigma}+\mathbf{\Sigma}'}$, the four terms involved in the exact $\mathrm{MISE}_{\mathrm{NS}}$ formula in Section 5.6.1 are

$$\begin{aligned}
|\mathbf{H}|^{-1/2}\operatorname{tr}\big\{(\mathbf{H}^{-1})^{\otimes r}\mathbf{R}(\mathsf{D}^{\otimes r}K)\big\} &= \operatorname{tr}\int_{\mathbb{R}^d}\mathsf{D}^{\otimes r}\phi_{\mathbf{H}}(\boldsymbol{x})\mathsf{D}^{\otimes r}\phi_{\mathbf{H}}(\boldsymbol{x})^\top d\boldsymbol{x}\\
&= (2\pi)^{-d/2}|2\mathbf{H}|^{-1/2}\nu_r\big((2\mathbf{H})^{-1}\big)\\
\operatorname{tr}\mathbf{R}^*(K_{\mathbf{H}} * K_{\mathbf{H}}, \mathsf{D}^{\otimes r}f) &= \operatorname{tr}\int_{\mathbb{R}^d}\mathsf{D}^{\otimes r}\phi_{2\mathbf{H}+\mathbf{\Sigma}}(\boldsymbol{x})\mathsf{D}^{\otimes r}\phi_{\mathbf{\Sigma}}(\boldsymbol{x})^\top d\boldsymbol{x}\\
&= (2\pi)^{-d/2}|2\mathbf{H}+2\mathbf{\Sigma}|^{-1/2}\nu_r\big((2\mathbf{H}+2\mathbf{\Sigma})^{-1}\big)\\
\operatorname{tr}\mathbf{R}^*(K_{\mathbf{H}}, \mathsf{D}^{\otimes r}f) &= \operatorname{tr}\int_{\mathbb{R}^d}\mathsf{D}^{\otimes r}\phi_{\mathbf{H}+\mathbf{\Sigma}}(\boldsymbol{x})\mathsf{D}^{\otimes r}\phi_{\mathbf{\Sigma}}(\boldsymbol{x})^\top d\boldsymbol{x}\\
&= (2\pi)^{-d/2}|\mathbf{H}+2\mathbf{\Sigma}|^{-1/2}\nu_r\big((\mathbf{H}+2\mathbf{\Sigma})^{-1}\big)\\
\operatorname{tr}\mathbf{R}(\mathsf{D}^{\otimes r}f) &= \operatorname{tr}\int_{\mathbb{R}^d}\mathsf{D}^{\otimes r}\phi_{\mathbf{\Sigma}}(\boldsymbol{x})\mathsf{D}^{\otimes r}\phi_{\mathbf{\Sigma}}(\boldsymbol{x})^\top d\boldsymbol{x}\\
&= (2\pi)^{-d/2}|2\mathbf{\Sigma}|^{-1/2}\nu_r\big((2\mathbf{\Sigma})^{-1}\big).
\end{aligned}$$

### 5.8.2 *Curvature matrix*

To find an explicit form for the AMISE requires a formula for the matrix $\mathbf{R}(\mathsf{D}^{\otimes s}\phi_{\mathbf{\Sigma}}) = \int_{\mathbb{R}^d}\mathsf{D}^{\otimes s}\phi_{\mathbf{\Sigma}}(\boldsymbol{x})\mathsf{D}^{\otimes s}\phi_{\mathbf{\Sigma}}(\boldsymbol{x})^\top d\boldsymbol{x} \in \mathcal{M}_{d^s\times d^s}$ or, equivalently, for $\boldsymbol{\psi}_{\mathrm{NS},2s} = (-1)^s\operatorname{vec}\mathbf{R}(\mathsf{D}^{\otimes s}\phi_{\mathbf{\Sigma}})$, for different values of $s$. To begin, observe that from $\mathsf{D}^{\otimes s}\phi_{\mathbf{\Sigma}}(\boldsymbol{x}) = |\mathbf{\Sigma}|^{-1/2}(\mathbf{\Sigma}^{-1/2})^{\otimes s}\mathsf{D}^{\otimes s}\phi(\mathbf{\Sigma}^{-1/2}\boldsymbol{x})$, after a change of variables, it follows that

$$\mathbf{R}(\mathsf{D}^{\otimes s}\phi_{\mathbf{\Sigma}}) = |\mathbf{\Sigma}|^{-1/2}(\mathbf{\Sigma}^{-1/2})^{\otimes s}\mathbf{R}(\mathsf{D}^{\otimes s}\phi)(\mathbf{\Sigma}^{-1/2})^{\otimes s} \tag{5.33}$$

or $\boldsymbol{\psi}_{\mathrm{NS},2s} = (-1)^s|\boldsymbol{\Sigma}|^{-1/2}(\boldsymbol{\Sigma}^{-1/2})^{\otimes 2s}\operatorname{vec}\mathbf{R}(\mathsf{D}^{\otimes s}\phi)$, so it suffices to find a formula for the standard normal distribution.

For the base case of density estimation, we require a formula for $\mathbf{R}(\mathsf{D}^{\otimes 2}\phi)$. Since $\mathsf{D}^{\otimes 2}\phi(\mathbf{z}) = \phi(\mathbf{z})(\mathbf{z}^{\otimes 2} - \operatorname{vec}\mathbf{I}_d)$, then

$$\mathsf{D}^{\otimes 2}\phi(\mathbf{z})\mathsf{D}^{\otimes 2}\phi(\mathbf{z})^\top = (4\pi)^{-d/2}\phi_{2^{-1}\mathbf{I}_d}(\mathbf{z})\{(\mathbf{z}\mathbf{z}^\top)\otimes(\mathbf{z}\mathbf{z}^\top) - \mathbf{z}^{\otimes 2}\operatorname{vec}^\top\mathbf{I}_d \\ - (\operatorname{vec}\mathbf{I}_d)\mathbf{z}^{\top\otimes 2} + \operatorname{vec}\mathbf{I}_d\operatorname{vec}^\top\mathbf{I}_d\}.$$

Therefore $\mathbf{R}(\mathsf{D}^{\otimes 2}\phi)$ can be computed as a combination of the moments of order 0, 2 and 4 of the $N(\mathbf{0}, 2^{-1}\mathbf{I}_d)$ distribution. From Theorem 9.20 in Schott (1996) it follows that $\int_{\mathbb{R}^d}\phi_{2^{-1}\mathbf{I}_d}(\mathbf{z})(\mathbf{z}\mathbf{z}^\top)\otimes(\mathbf{z}\mathbf{z}^\top)d\mathbf{z} = \frac{1}{4}(\mathbf{I}_{d^2} + \mathbf{K}_{d,d} + \operatorname{vec}\mathbf{I}_d\operatorname{vec}^\top\mathbf{I}_d)$, with $\mathbf{K}_{d,d}$ the $d^2\times d^2$ commutation matrix, and $\int_{\mathbb{R}^d}\phi_{2^{-1}\mathbf{I}_d}(\mathbf{z})\mathbf{z}^{\otimes 2}d\mathbf{z}\operatorname{vec}^\top\mathbf{I}_d = \frac{1}{2}\operatorname{vec}\mathbf{I}_d\operatorname{vec}^\top\mathbf{I}_d = \operatorname{vec}\mathbf{I}_d\int_{\mathbb{R}^d}\phi_{2^{-1}\mathbf{I}_d}(\mathbf{z})\mathbf{z}^{\top\otimes 2}d\mathbf{z}$. Thus we have

$$\mathbf{R}(\mathsf{D}^{\otimes 2}\phi) = \tfrac{1}{4}(4\pi)^{-d/2}(\mathbf{I}_{d^2} + \mathbf{K}_{d,d} + \operatorname{vec}\mathbf{I}_d\operatorname{vec}^\top\mathbf{I}_d). \tag{5.34}$$

More generally, Equation (5.32) immediately yields $\boldsymbol{\psi}_{\mathrm{NS},2s} = \mathsf{D}^{\otimes 2s}\phi_{2\boldsymbol{\Sigma}}(\mathbf{0})$, and the expression of the normal density derivative using the Hermite polynomial in Equation (5.5) leads to

$$\boldsymbol{\psi}_{\mathrm{NS},2s} = (-2)^{-s}(4\pi)^{-d/2}\mathrm{OF}(2s)|\boldsymbol{\Sigma}|^{-1/2}\mathcal{S}_{d,2s}(\operatorname{vec}\boldsymbol{\Sigma}^{-1})^{\otimes s}.$$

Analogous to Equation (5.7) in Section 5.1.3, the relationship with the moments of the multivariate normal distribution $\boldsymbol{\mu}_{2s} = \mathbb{E}(\mathbf{Z}^{\otimes 2s})$ is $\boldsymbol{\psi}_{\mathrm{NS},2s} = (-2)^{-s}(4\pi)^{-d/2}|\boldsymbol{\Sigma}|^{-1/2}(\boldsymbol{\Sigma}^{-1/2})^{\otimes 2s}\boldsymbol{\mu}_{2s}$. It then follows that, for every $s \geq 0$,

$$\mathbf{R}(\mathsf{D}^{\otimes s}\phi_{\boldsymbol{\Sigma}}) = 2^{-s}(4\pi)^{-d/2}|\boldsymbol{\Sigma}|^{-1/2}(\boldsymbol{\Sigma}^{-1/2})^{\otimes s}\mathbb{E}\{(\mathbf{Z}\mathbf{Z}^\top)^{\otimes s}\}(\boldsymbol{\Sigma}^{-1/2})^{\otimes s}. \tag{5.35}$$

### 5.8.3 *Asymptotic MISE*

We asserted that an expression for the asymptotic IV in Equation (5.15) is $n^{-1}|\mathbf{H}|^{-1/2}\operatorname{tr}\{(\mathbf{H}^{-1})^{\otimes r}\mathbf{R}(\mathsf{D}^{\otimes r}K)\} = 2^{-r}(4\pi)^{-d/2}n^{-1}|\mathbf{H}|^{-1/2}\nu_r(\mathbf{H}^{-1})$, so to find the AMISE in the normal case, it only remains to compute the ISB.

For the density base case, the IV reduces to $(4\pi)^{-d/2}n^{-1}|\mathbf{H}|^{-1/2}$ and, from Equations (5.33)–(5.34), the quadratic form that appears in the asymptotic ISB can be written as

$$\begin{aligned}
Q(\mathbf{H}) &= (\operatorname{vec}^\top\mathbf{H})\mathbf{R}(\mathsf{D}^{\otimes 2}\phi_{\boldsymbol{\Sigma}})\operatorname{vec}\mathbf{H} \\
&= \tfrac{1}{4}(4\pi)^{-d/2}|\boldsymbol{\Sigma}|^{-1/2}\{\operatorname{vec}^\top(\boldsymbol{\Sigma}^{-1/2}\mathbf{H}\boldsymbol{\Sigma}^{-1/2})\} \\
&\quad\times(\mathbf{I}_{d^2} + \mathbf{K}_{d,d} + \operatorname{vec}\mathbf{I}_d\operatorname{vec}^\top\mathbf{I}_d)\operatorname{vec}(\boldsymbol{\Sigma}^{-1/2}\mathbf{H}\boldsymbol{\Sigma}^{-1/2}) \\
&= \tfrac{1}{4}(4\pi)^{-d/2}|\boldsymbol{\Sigma}|^{-1/2}\{2\operatorname{tr}(\boldsymbol{\Sigma}^{-1/2}\mathbf{H}\boldsymbol{\Sigma}^{-1}\mathbf{H}\boldsymbol{\Sigma}^{-1/2}) + \operatorname{tr}^2(\boldsymbol{\Sigma}^{-1/2}\mathbf{H}\boldsymbol{\Sigma}^{-1/2})\} \\
&= \tfrac{1}{4}(4\pi)^{-d/2}|\boldsymbol{\Sigma}|^{-1/2}\{2\operatorname{tr}(\mathbf{H}\boldsymbol{\Sigma}^{-1}\mathbf{H}\boldsymbol{\Sigma}^{-1}) + \operatorname{tr}^2(\mathbf{H}\boldsymbol{\Sigma}^{-1})\}.
\end{aligned} \tag{5.36}$$

For the first equality, and in several occasions hereafter, we use the formula $(\mathbf{C}^\top \otimes \mathbf{A})\operatorname{vec}\mathbf{B} = \operatorname{vec}(\mathbf{ABC})$ for conformable matrices $\mathbf{A}, \mathbf{B}, \mathbf{C}$ (see Appendix B). The previous display, combined with Equation (2.10) and that $m_2(\phi) = 1$, results in the normal scale AMISE included in Section 3.1; that is, $\text{AMISE}_{\text{NS}}\{\hat{f}(\cdot;\mathbf{H})\} = n^{-1}|\mathbf{H}|^{-1/2}(4\pi)^{-d/2} + \frac{1}{16}(4\pi)^{-d/2}|\boldsymbol{\Sigma}|^{-1/2}\{2\operatorname{tr}(\mathbf{H}\boldsymbol{\Sigma}^{-1}\mathbf{H}\boldsymbol{\Sigma}^{-1}) + \operatorname{tr}^2(\mathbf{H}\boldsymbol{\Sigma}^{-1})\}$. This result was first asserted in Wand (1992), but the derivation here provided is different.

For density derivative estimation, noting that $(\boldsymbol{\Sigma}^{-1/2})^{\otimes 2}\operatorname{vec}\mathbf{H} = \operatorname{vec}\mathbf{A}$ with $\mathbf{A} = \boldsymbol{\Sigma}^{-1/2}\mathbf{H}\boldsymbol{\Sigma}^{-1/2}$, and making use of Equation (5.35) with $s = r+2$, then for the asymptotic ISB we have

$$\begin{aligned}
&\operatorname{tr}\left[(\mathbf{I}_{d^r} \otimes \operatorname{vec}^\top \mathbf{H})\mathbf{R}(\mathsf{D}^{\otimes(r+2)}\phi_{\boldsymbol{\Sigma}})(\mathbf{I}_{d^r} \otimes \operatorname{vec}\mathbf{H})\right] \\
&\quad = 2^{-(r+2)}(4\pi)^{-d/2}|\boldsymbol{\Sigma}|^{-1/2}\operatorname{tr}\left[\{(\boldsymbol{\Sigma}^{-1})^{\otimes r} \otimes (\operatorname{vec}\mathbf{A}\operatorname{vec}^\top \mathbf{A})\}\right. \\
&\quad\quad \left.\times\, \mathbb{E}\{(\mathbf{Z}\mathbf{Z}^\top)^{\otimes(r+2)}\}\right].
\end{aligned}$$

The trace on the right-hand side can be written as

$$\begin{aligned}
&\mathbb{E}\operatorname{tr}\{(\boldsymbol{\Sigma}^{-1}\mathbf{Z}\mathbf{Z}^\top)^{\otimes r} \otimes [(\operatorname{vec}\mathbf{A}\operatorname{vec}^\top \mathbf{A})(\mathbf{Z}\mathbf{Z}^\top)^{\otimes 2}]\} \\
&\qquad\qquad = \mathbb{E}\{\operatorname{tr}^r(\boldsymbol{\Sigma}^{-1}\mathbf{Z}\mathbf{Z}^T)\operatorname{tr}[\operatorname{vec}\mathbf{A}\operatorname{vec}^\top(\mathbf{Z}\mathbf{Z}^\top\mathbf{A}\mathbf{Z}\mathbf{Z}^\top)]\} \\
&\qquad\qquad = \mathbb{E}\{(\mathbf{Z}^\top\boldsymbol{\Sigma}^{-1}\mathbf{Z})^r(\mathbf{Z}^\top\mathbf{A}\mathbf{Z})^2\} = \nu_{r,2}(\boldsymbol{\Sigma}^{-1}, \mathbf{A})
\end{aligned}$$

thus yielding the normal scale AMISE as given in Section 5.6.1.

### 5.8.4 *Normal scale bandwidth*

The normal scale bandwidth is the minimiser of the AMISE in the normal case. For density estimation, an explicit formula for such a bandwidth is obtained by minimising $\text{AMISE}_{\text{NS}}\{\hat{f}(\cdot;\mathbf{H})\}$. A possible way to derive it is by computing the gradient of the AMISE with respect to $\operatorname{vec}\mathbf{H}$, as it is done in Wand (1992).

Here, an alternative path is followed. From Section 2.9.2, taking into account that $R(K) = (4\pi)^{-d/2}$ and $m_2(K) = 1$ for the normal kernel, it follows that

$$\mathbf{H}_{\text{NS}} = \{d(4\pi)^{-d/2}/Q(\mathbf{A}_0)\}^{2/(d+4)} n^{-2/(d+4)}\mathbf{A}_0, \tag{5.37}$$

where $\mathbf{A}_0 = \operatorname{argmin}_{\mathbf{A}\in\mathcal{F}, |\mathbf{A}|=1} Q(\mathbf{A})$ and $Q(\mathbf{A})$ is defined as in (5.36). Reasoning as in Theorem A.1 in Celeux & Govaert (1995) it is possible to show that, in this case, $\mathbf{A}_0 = |\boldsymbol{\Sigma}|^{-1/d}\boldsymbol{\Sigma}$ so that $Q(\mathbf{A}_0) = \frac{1}{4}(4\pi)^{-d/2}|\boldsymbol{\Sigma}|^{-(d+4)/(2d)}d(d+2)$ and, therefore, $\mathbf{H}_{\text{NS}} = \{4/(d+2)\}^{2/(d+4)}n^{-2/(d+4)}\boldsymbol{\Sigma}$, as in Equation (3.2).

The constrained normal scale bandwidths in Equations (3.3)–(3.4) can be obtained similarly, by finding the minimiser of $Q(\mathbf{A})$ within the appropriate class of matrices. The only matrix in $\mathcal{A}$ with unit determinant is $\mathbf{I}_d$, with $Q(\mathbf{I}_d) = \frac{1}{4}(4\pi)^{-d/2}|\mathbf{\Sigma}|^{-1/2}\{2\,\mathrm{tr}(\mathbf{\Sigma}^{-2}) + \mathrm{tr}^2(\mathbf{\Sigma}^{-1})\}$, which gives Equation (3.3). For class $\mathcal{D}$, now reasoning as in Corollary A.1 in Celeux & Govaert (1995), it follows that the minimiser of $Q(\mathbf{A})$ for $\mathbf{A} \in \mathcal{D}$ with $|\mathbf{A}| = 1$ is $\tilde{\mathbf{A}}_0 = |\mathrm{diag}\,\mathbf{\Sigma}|^{-1/d}\,\mathrm{diag}\,\mathbf{\Sigma}$, i.e., the (normalised) diagonal matrix having the same diagonal as $\mathbf{\Sigma}$. So writing $\mathbf{\Delta} = (\mathrm{diag}\,\mathbf{\Sigma})^{-1}\mathbf{\Sigma}$ as in Section 3.1, the attained minimum is $Q(\tilde{\mathbf{A}}_0) = \frac{1}{4}(4\pi)^{-d/2}|\mathbf{\Sigma}|^{-1/2}|\mathrm{diag}\,\mathbf{\Sigma}|^{-2/d}\{2\,\mathrm{tr}(\mathbf{\Delta}^{-2}) + \mathrm{tr}^2(\mathbf{\Delta}^{-1})\}$ which, once substituted into Equation (5.37) gives the optimal diagonal bandwidth for the normal case in Equation (3.4).

For density derivative estimation, the calculations are slightly more complicated. It is simplest to begin with the assumption that the normal scale bandwidth takes the form $\mathbf{H} = c^2\mathbf{\Sigma}$ for some $c > 0$ and then to minimise $\mathrm{AMISE}_{\mathrm{NS}}\{\mathsf{D}^{\otimes r}\hat{f}(\cdot;\mathbf{H})\}$ with respect to $c$. It is then necessary to prove that $\nu_{r,2}(\mathbf{\Sigma}^{-1}, \mathbf{I}_d) = (d+2r+2)(d+2r)\nu_r(\mathbf{\Sigma}^{-1})$ in order to obtain the normal scale bandwidth in Equation (5.18). The details of this non-trivial proof can be found in Section A.2.2 of Chacón et al. (2011). The scalar bandwidth in Equation (5.19) is then obtained by minimising $\mathrm{AMISE}_{\mathrm{NS}}\{\mathsf{D}^{\otimes r}\hat{f}(\cdot;\mathbf{H})\}$ for $\mathbf{H} = h^2\mathbf{I}_d$.

### 5.8.5 *Asymptotic MSE for curvature estimation*

Plug-in bandwidth selectors rely on the choice of a pilot bandwidth matrix $\mathbf{G}$ to estimate the curvature matrix, or more generally, the functional $\boldsymbol{\psi}_{2s} = (-1)^s\,\mathrm{vec}\,\mathbf{R}(\mathsf{D}^{\otimes s}f)$. Theorem 1 in Chacón & Duong (2010) shows that the dominant term of the MSE of the kernel estimator $\hat{\boldsymbol{\psi}}_{2s}(\mathbf{G}) = n^{-2}\sum_{i,j=1}^n \mathsf{D}^{\otimes 2s}L_{\mathbf{G}}(\boldsymbol{X}_i - \boldsymbol{X}_j)$ is given by

$$\begin{aligned}\mathrm{AMSE}\{\hat{\boldsymbol{\psi}}_{2s}(\mathbf{G})\} = \|&n^{-1}|\mathbf{G}|^{-1/2}(\mathbf{G}^{-1/2})^{\otimes 2s}\mathsf{D}^{\otimes 2s}L(\mathbf{0}) \\ &+ \tfrac{1}{2}m_2(L)(\mathrm{vec}^\top\mathbf{G} \otimes \mathbf{I}_{d^{2s}})\boldsymbol{\psi}_{2s+2}\|^2. \qquad (5.38)\end{aligned}$$

In order to obtain a normal scale rule for the choice of $\mathbf{G}$ it is necessary to find a formula for such AMSE and its minimiser in the normal case.

To begin with, using a normal kernel $L = \phi$ it follows from Section 5.1.3 that $\mathsf{D}^{\otimes 2s}\phi(\mathbf{0}) = (-1)^s(2\pi)^{-d/2}\boldsymbol{\mu}_{2s}$. From Section 5.8.2 we already have $\boldsymbol{\psi}_{\mathrm{NS},2s+2} = (-2)^{-(s+1)}(4\pi)^{-d/2}|\mathbf{\Sigma}|^{-1/2}(\mathbf{\Sigma}^{-1/2})^{\otimes 2s+2}\boldsymbol{\mu}_{2s+2}$, so that

$$\begin{aligned}\mathrm{AMSE}_{\mathrm{NS}}\{\hat{\boldsymbol{\psi}}_{2s}(\mathbf{G})\} = (2\pi)^{-d}\|&n^{-1}|\mathbf{G}|^{-1/2}(\mathbf{G}^{-1/2})^{\otimes 2s}\boldsymbol{\mu}_{2s} \\ &- 2^{-(d+2s+4)/2}|\mathbf{\Sigma}|^{-1/2}\{\mathrm{vec}^\top(\mathbf{\Sigma}^{-1/2}\mathbf{G}\mathbf{\Sigma}^{-1/2}) \otimes (\mathbf{\Sigma}^{-1/2})^{\otimes 2s}\}\boldsymbol{\mu}_{2s+2}\|^2.\end{aligned}$$

As before, we seek to minimise $\text{AMSE}_{\text{NS}}\{\hat{\boldsymbol{\psi}}_{2s}(\mathbf{G})\}$ within the class of matrices of the form $\mathbf{G} = c^2\boldsymbol{\Sigma}$ with $c > 0$, for which

$$\begin{aligned}\text{AMSE}_{\text{NS}}\{\hat{\boldsymbol{\psi}}_{2s}(c^2\boldsymbol{\Sigma})\} = (2\pi)^{-d}|\boldsymbol{\Sigma}|^{-1}\|n^{-1}c^{-d-2s}(\boldsymbol{\Sigma}^{-1/2})^{\otimes 2s}\boldsymbol{\mu}_{2s} \\ - 2^{-(d+2s+4)/2}c^2\{\text{vec}^\top \mathbf{I}_d \otimes (\boldsymbol{\Sigma}^{-1/2})^{\otimes 2s}\}\boldsymbol{\mu}_{2s+2}\|^2.\end{aligned}$$

This can be further simplified by taking into account the result in Section 5.9.2, which shows that $\mathbb{E}\{(\mathbf{Z}^\top\mathbf{Z})\mathbf{Z}^{\otimes 2s}\} = (d+2s)\boldsymbol{\mu}_{2s}$. Thus $\{\text{vec}^\top \mathbf{I}_d \otimes (\boldsymbol{\Sigma}^{-1/2})^{\otimes 2s}\}\boldsymbol{\mu}_{2s+2} = (\boldsymbol{\Sigma}^{1/2})^{\otimes 2s}\mathbb{E}\{(\mathbf{Z}^\top\mathbf{Z})\mathbf{Z}^{\otimes 2s}\} = (d+2s)(\boldsymbol{\Sigma}^{1/2})^{\otimes 2s}\boldsymbol{\mu}_{2s}$, and this implies that

$$\begin{aligned}\text{AMSE}_{\text{NS}}\{\hat{\boldsymbol{\psi}}_{2s}(c^2\boldsymbol{\Sigma})\} = (2\pi)^{-d}|\boldsymbol{\Sigma}|^{-1}\|(\boldsymbol{\Sigma}^{-1/2})^{\otimes 2s}\boldsymbol{\mu}_{2s}\|^2 \\ \times \left\{n^{-1}c^{-d-2s} - 2^{-(d+2s+4)/2}c^2(d+2s)\right\}^2.\end{aligned}$$

This is annihilated for $c = \{2/(d+2s)\}^{1/(d+2s+2)}2^{1/2}n^{-1/(d+2s+2)}$, leading to $\mathbf{G}_{\text{NS},2s} = \{2/(d+2s)\}^{2/(d+2s+2)}2\boldsymbol{\Sigma}n^{-2/(d+2s+2)}$, as asserted in Section 5.6.4.

This situation in which the dominant term of the MSE can be completely annihilated is a special property of the normal case, due to the fact that the two vectors that make up the bias are proportional for this particular choice of $L$ and $f$. On the other hand, for the general case, the AMSE in Equation (5.38) can only be minimised. In the univariate case, in contrast, Jones & Sheather (1991) noted that the leading term of the MSE can always be annihilated for this problem (subject to some sign restrictions which are easily fulfilled), thus leading to faster convergence rates than would be expected from the general multivariate analysis.

For the pilot bandwidth for SCV, the $\text{AMSE}^*_{\text{NS}}\{\hat{\boldsymbol{\psi}}_{2s}(\mathbf{G})\}$ is similar to $\text{AMSE}_{\text{NS}}\{\hat{\boldsymbol{\psi}}_{2s}(\mathbf{G})\}$ but with different coefficients of the same terms. Supposing that $\mathbf{G} = c^2\boldsymbol{\Sigma}$, we have

$$\begin{aligned}&\text{AMSE}^*_{\text{NS}}\{\hat{\boldsymbol{\psi}}_{2s}(c^2\boldsymbol{\Sigma})\} \\ &\quad = (2\pi)^{-d}|\boldsymbol{\Sigma}|^{-1}\|n^{-1}c^{-d-2s}(\boldsymbol{\Sigma}^{-1/2})^{\otimes 2s}\boldsymbol{\mu}_{2s} \\ &\qquad - \tfrac{1}{2}c^2\{\text{vec}^\top \mathbf{I}_d \otimes (\boldsymbol{\Sigma}^{-1/2})^{\otimes 2s}\}\boldsymbol{\mu}_{2s+2}\|^2 \\ &\quad = (2\pi)^{-d}|\boldsymbol{\Sigma}|^{-1}\|(\boldsymbol{\Sigma}^{-1/2})^{\otimes 2s}\boldsymbol{\mu}_{2s}\|^2\left\{n^{-1}c^{-d-2s} - \tfrac{1}{2}c^2(d+2s)\right\}^2.\end{aligned}$$

This is annihilated at $c = \{2/(d+2s)\}^{1/(d+2s+2)}n^{-1/(d+2s+2)}$, leading to $\mathbf{G}^*_{\text{NS},2s} = \{2/(d+2s)\}^{2/(d+2s+2)}\boldsymbol{\Sigma}n^{-2/(d+2s+2)}$, as asserted in Section 5.6.5.

## 5.9 Further mathematical analysis of density derivative estimators

### 5.9.1 Taylor expansions for vector-valued functions

It is not common to find Taylor's theorem for vector-valued functions in the literature, and if it is indeed provided, it is unlikely to be in the form of Theorem 5 that we require. The closest previously published form is in Theorem 1.4.8 of Kollo & von Rosen (2005). Given its importance in this chapter, we outline a brief proof.

**Proof (Proof of Theorem 5)** Writing a vector-valued function $\boldsymbol{f}\colon \mathbb{R}^d \to \mathbb{R}^p$ in terms of its components $\boldsymbol{f} = (f_1, \ldots, f_p)$, it is possible to apply the usual form of Taylor's theorem to each of the real-valued components, to obtain

$$f_i(\boldsymbol{x}+\boldsymbol{a}) = \sum_{j=0}^{r} \frac{1}{j!}(\boldsymbol{a}^\top)^{\otimes j}\mathsf{D}^{\otimes j} f_i(\boldsymbol{x}) + Re_i(\boldsymbol{a})$$

where $Re_i(\boldsymbol{a}) = o(\|\boldsymbol{a}\|^r)$ for every $i = 1, \ldots, p$, as stated in Equation (2.5). Then, all the component-wise expansions can be gathered together by noting that

$$\begin{aligned}\left[\begin{array}{c}(\boldsymbol{a}^\top)^{\otimes j}\mathsf{D}^{\otimes j} f_1(\boldsymbol{x}) \\ \hline \vdots \\ \hline (\boldsymbol{a}^\top)^{\otimes j}\mathsf{D}^{\otimes j} f_p(\boldsymbol{x})\end{array}\right] &= \left[\begin{array}{c|c|c}(\boldsymbol{a}^\top)^{\otimes j} & \cdots & \boldsymbol{0}_d^\top \\ \hline \vdots & \ddots & \vdots \\ \hline \boldsymbol{0}_d^\top & \cdots & (\boldsymbol{a}^\top)^{\otimes j}\end{array}\right] \cdot \left[\begin{array}{c}\mathsf{D}^{\otimes j} f_1(\boldsymbol{x}) \\ \hline \vdots \\ \hline \mathsf{D}^{\otimes j} f_p(\boldsymbol{x})\end{array}\right] \\ &= \{\mathbf{I}_p \otimes (\boldsymbol{a}^\top)^{\otimes j}\}\mathsf{D}^{\otimes j}\boldsymbol{f}(\boldsymbol{x}).\end{aligned}$$

■

### 5.9.2 Relationship between multivariate normal moments

A key fact to obtain the normal scale pilot bandwidth for curvature estimation in Section 5.8.5 was to notice that, for a $d$-variate standard normal random vector $\boldsymbol{Z} = (Z_1, \ldots, Z_d)$ and an even number $p = 2q$, the vectors $\boldsymbol{v}_1 = \mathbb{E}\{\boldsymbol{Z}^{\otimes p}\} \in \mathbb{R}^{d^p}$ and $\boldsymbol{v}_2 = \mathbb{E}\{(\boldsymbol{Z}^\top\boldsymbol{Z})\boldsymbol{Z}^{\otimes p}\} \in \mathbb{R}^{d^p}$ are proportional, i.e.,

$$\boldsymbol{v}_2 = (d+p)\boldsymbol{v}_1. \tag{5.39}$$

In the original work of Chacón & Duong (2010) a weaker version of Equation (5.39) was shown to hold: $(\boldsymbol{v}_1^\top\boldsymbol{v}_2)/(\boldsymbol{v}_1^\top\boldsymbol{v}_1) = d+p$. For completeness, we provide the proof for the stronger result.

For $d = 1$, taking into account the symmetry of the standard normal density $\phi(z) = (2\pi)^{-1/2}\exp\{-\frac{1}{2}z^2\}$ and making the change of variables $y = z^2$, we can write

$$\begin{aligned} v_1 &= \mathbb{E}(Z^p) = (2\pi)^{-1/2} \int_{-\infty}^{\infty} (z^2)^q \exp(-\tfrac{1}{2}z^2)\, dz \\ &= (2\pi)^{-1/2} 2 \int_0^{\infty} (z^2)^q \exp(-\tfrac{1}{2}z^2)\, dz = (2\pi)^{-1/2} \int_0^{\infty} y^{q-1/2} \exp(-\tfrac{1}{2}y)\, dy. \end{aligned}$$

Therefore, by changing $q$ for $q+1$ in the last formula, we obtain $v_2 = (2\pi)^{-1/2} \int_0^{\infty} y^{q+1/2} \exp(-\frac{1}{2}y)\, dy$. Integration by parts with $u = y^{q+1/2}$ and $dv = \exp(-\frac{1}{2}y)\, dy$, so that $du = (q+1/2)y^{q-1/2}dy$ and $v = -2\exp(-\frac{1}{2}y)$, yields

$$\begin{aligned} v_2 &= (2\pi)^{-1/2} \int_0^{\infty} y^{q+1/2} \exp(-\tfrac{1}{2}y)\, dy \\ &= 2(q+\tfrac{1}{2})(2\pi)^{-1/2} \int_0^{\infty} y^{q-1/2} \exp(-\tfrac{1}{2}y)\, dy = (p+1)v_1. \end{aligned}$$

A simpler proof in the univariate case can be obtained from the fact that $\mathbb{E}(Z^p) = \mathrm{OF}(p)$, as shown in Johnson et al. (1994, p. 89), but the above constructive reasoning is useful for the multivariate case.

For $d > 1$, the goal is to show that all the elements of the vectors $\boldsymbol{v}_1$ and $\boldsymbol{v}_2$ are proportional, with the same proportionality constant. All the elements of $\boldsymbol{v}_1$ have the form $\mathbb{E}(Z_{j_1} \cdots Z_{j_p})$ with $j_\ell \in \{1, \ldots, d\}$ for all $\ell = 1, \ldots, p$ and, since $\mathbf{Z}^\top \mathbf{Z} = \sum_{i=1}^d Z_i^2$, the corresponding coordinate in $\boldsymbol{v}_2$ is $\sum_{i=1}^d \mathbb{E}(Z_i^2 Z_{j_1} \cdots Z_{j_p})$. Equivalently, it is useful to express the coordinates of $\boldsymbol{v}_1$ as $v_{1;p_1,\ldots,p_d} = \mathbb{E}(Z_1^{p_1} \cdots Z_d^{p_d})$ with $0 \le p_i \le p$ for all $i = 1, \ldots, d$ and $\sum_{i=1}^d p_i = p$. The corresponding coordinates in $\boldsymbol{v}_2$ are given as $v_{2;p_1,\ldots,p_d} = \sum_{i=1}^d \mathbb{E}(Z_i^2 Z_1^{p_1} \cdots Z_d^{p_d})$. If any of the $p_i$ are odd, then both corresponding coordinates are zero, due to the symmetry of the normal distribution, so any proportionality constant is valid. Hence, for the remaining case, we assume that $p_i = 2q_i$, again with $\sum_{i=1}^d q_i = q$. For a fixed $i \in \{1, \ldots, d\}$, integrating first with respect to the $i$-th coordinate and applying the same reasoning as for the univariate case, we have $\mathbb{E}(Z_i^2 Z_1^{p_1} \cdots Z_d^{p_d}) = (p_i + 1)\, \mathbb{E}(Z_1^{p_1} \cdots Z_d^{p_d})$. This implies, as $\sum_{i=1}^d p_i = p$, that

$$v_{2;p_1,\ldots,p_d} = \sum_{i=1}^d (p_1+1) v_{1;p_1,\ldots,p_d} = (p+d) v_{1;p_1,\ldots,p_d},$$

which concludes the proof of Equation (5.39).

Chapter 6

# Applications related to density and density derivative estimation

In the previous chapters we focused on density and density derivative estimators in their own right as they play a directly important role for exploratory data analysis. In this chapter, we focus on more complex data analysis applications in which density and density derivative estimators are crucial components. Section 6.1 covers the estimation of the level sets of a density function: high threshold level sets are used in modal region estimation and bump-hunting, and low threshold sets for density support estimation. Section 6.2 examines cluster analysis where the clusters are identified as the basins of attractions of the density gradient to local data density modes. Section 6.3 introduces the ridges of the density function, since they are based on the eigenvector decomposition of the density Hessian, as a tool for analysing filamentary data. Section 6.4 places density curvature estimators into a formal inference framework to delimit significantly data dense regions, which offers an alternative to the level sets of the data density in Section 6.1 for modal region estimation.

## 6.1 Level set estimation

Since the densities and density derivatives are functions, then important related objects are level sets of these functions and their associated graphs. The level set of a function $f$ at level $c$ is defined as the set of points at which the values of $f$ are above $c$; that is, $\mathcal{L}(c) = \{\boldsymbol{x} : f(\boldsymbol{x}) \geq c\}$. We focus on the level sets of a density function, but our exposition applies equally to other functions; for example, in the previous chapter we explored the level sets of the summary curvature function $s$, and the level sets of regression functions are also of interest (see Polonik & Wang, 2005, and references therein).

Regarding density level sets, they possess a large potential for applications in diverse statistical problems. Hartigan (1975) considered the connected components of density level sets as high-density clusters of a distribution (see also Section 6.2), Hyndman (1996) explored the use of these highest density regions as a graphical representation of a distribution, and Klemelä (2004) proposed level set trees as a tool for visualisation of multivariate density estimates. Level sets at low level values can also be useful for outlier detection (Baíllo et al., 2001). More generally, by exploring all the level sets of a density function as the level varies, it is possible to recover more detailed topological features of the distribution (Bobrowski et al., 2017), as covered in a recently emerging field known as topological data analysis (Wasserman, 2018).

Given a fixed threshold $c$ and any estimator $\hat{f}$ of the density function $f$, the plug-in estimator of $\mathcal{L}(c) \equiv \mathcal{L}(f;c)$ is $\hat{\mathcal{L}}(c) \equiv \mathcal{L}(\hat{f};c) = \{\boldsymbol{x} : \hat{f}(\boldsymbol{x}) \geq c\}$. A natural choice is to take $\hat{f}$ as a kernel density estimator, as introduced in Dobrow (1992), but it is worth pointing out that there exist alternative estimators which target the level sets directly without the intermediate density estimation, or estimators based on the geometric properties of the level sets (see the recent exhaustive review of Saavedra-Nieves et al., 2014).

As noted earlier in Section 2.2.1, an alternative parameterisation may be considered so that, for a given $\alpha \in (0,1)$, the level $c = c_\alpha$ is chosen as the largest threshold such that $\mathcal{L}(c_\alpha)$ has probability content greater than or equal to $1-\alpha$. This threshold $c_\alpha$ also depends on the unknown density, so for practical data analysis it needs to be estimated from the data as well. Cadre (2006) explored the possibility of estimating $c_\alpha$ by the corresponding quantity defined in terms of the kernel level set estimator, but Cadre et al. (2013) noted that this is asymptotically equivalent to the (computationally more efficient) proposal of Hyndman (1996), which defines $\hat{c}_\alpha$ as the $\alpha$-quantile of $\hat{f}(\boldsymbol{X}_1;\mathbf{H}),\dots,\hat{f}(\boldsymbol{X}_n;\mathbf{H})$, where $\hat{f}(\cdot;\mathbf{H})$ is the kernel density estimator based on $\boldsymbol{X}_1,\dots,\boldsymbol{X}_n$. The resulting 'nested' plug-in estimator of the level set is $\hat{\mathcal{L}}(\hat{c}_\alpha) \equiv \mathcal{L}(\hat{f};\hat{c}_\alpha) = \{\boldsymbol{x} : \hat{f}(\boldsymbol{x};\mathbf{H}) \geq \hat{c}_\alpha\}$.

To evaluate the accuracy of level set estimators, as the involved entities are sets, it is necessary to use a distance between sets. This is usually based on a measure of the symmetric difference, defined for sets $A$ and $B$ as $A\Delta B = (A\cup B)\backslash(A\cap B) = (A\cap B^c)\cup(A^c\cap B)$. Some common choices for the measure $\mu$ are the Lebesgue measure $\mu = \lambda$ (so that $\lambda(A\Delta B)$ is the hyper-volume of $A\Delta B$), or the underlying probability measure $\mu = \mathbb{P}$. These are known as distance-in-measure. A different analysis stems from the use of the Hausdorff distance (Cuevas et al., 2006; Singh et al., 2009).

The asymptotic behaviour of the plug-in level set estimators has been the subject of intensive investigation, although these studies focus on the use of a constrained bandwidth of the form $\mathbf{H} = h^2\mathbf{I}_d$ for some $h > 0$. A general consistency result for the distance-in-measure was shown in Cuevas et al. (2006) and consistency for the Hausdorff measure can be deduced from the results in Molchanov (1991). Convergence rates for $\mu\{\hat{\mathcal{L}}(\hat{c}_\alpha)\triangle\mathcal{L}(c_\alpha)\}$ were provided in Baíllo (2003), where they were shown to depend on the flatness of $f$ at $c$. Cadre (2006) showed that the random variable $(nh^d)^{1/2}\mu\{\hat{\mathcal{L}}(\hat{c})\triangle\mathcal{L}(c)\}$ converges to a degenerate deterministic distribution taking only a single value, which implies that $\mu\{\hat{\mathcal{L}}(c)\Delta\mathcal{L}(c)\}$ converges in probability to 0 at rate $(nh^d)^{-1/2}$. We conjecture that this rate is $n^{-1/2}|\mathbf{H}|^{-1/4}$ for unconstrained matrices.

Samworth & Wand (2010) obtained an asymptotic expansion for the expected distance-in-measure of kernel level set estimators in the univariate case, and used it to derive an automatic bandwidth selection procedure for level set estimation. On the other hand, the simulation study of Saavedra-Nieves et al. (2014) revealed that bandwidth selectors designed for density estimation appeared to be as competitive as the bandwidth selectors specifically aimed for level set estimation, so the bandwidth selectors introduced in Chapter 3 are also useful for multivariate level set estimation.

### *6.1.1 Modal region and bump estimation*

A local mode of a (density) function $f$ is the point whose value is greater than all functional values in a local neighbourhood. A local mode belongs to the class of fixed points of a function which are crucial in characterising the behaviour of the function. As an isolated mode has a zero probability mass, we extend it to cover a modal region so that the latter has positive mass. A bump is the graph of a modal region. Thus modal regions and bumps are more suitable for visualisation than single modal points. Furthermore, modal regions and bumps can be identified with the important data-rich regions.

Univariate bump-hunting (as coined by Good & Gaskins, 1980) is greatly assisted by the unambiguously defined upwards and downward tangent slope for establishing local convexity and determining when a modal region/bump is separated by regions of lower density. With multivariate data, there is a richer class of behaviour at the tangent plane so an unambiguous direction of the slope is not well-defined. Locally following a single direction, e.g., steepest descent, is insufficient to establish the location and extent of a $d$-dimensional modal region. So we seek an alternative characterisation of a modal region.

Considering that a mode at $\boldsymbol{x}$ in a density function $f$ requires the inequality $f(\boldsymbol{x}) \geq f(\boldsymbol{y})$ to hold for all $\boldsymbol{y}$ in a local neighbourhood, then this recalls the form of a level set. Now, denote as $f_\tau$ the $\tau$-th quantile of $f(\boldsymbol{X})$ for $\boldsymbol{X} \sim f$, so that $\mathbb{P}(\boldsymbol{X} \in \mathcal{L}(f_\tau)) = 1 - \tau$. The modal regions of $f$ can be recovered by inspecting the level set $\mathcal{L}(f_\tau)$ for suitable values of $\tau$, as this is the result of a global thresholding procedure and so it is not restricted to producing a single simply connected region. Rather than being a disadvantage, this approach avoids a two-stage local approach consisting of searching first for the local modes individually and then subsequently expanding them into their respective modal regions.

Alternatively, the data-rich regions can be also defined as the modal regions of the summary density curvature $s(\boldsymbol{x}) = -\mathbf{1}\{\mathsf{H}f(\boldsymbol{x}) < 0\}\,\mathrm{abs}(|\mathsf{H}f(\boldsymbol{x})|)$ from Equation (5.4), namely $\mathcal{L}(s_\tau) = \{\boldsymbol{x} : -s(\boldsymbol{x}) \geq -s_\tau\}$ where $s_\tau$ is the $\tau$-th quantile of $s(\boldsymbol{X})|\{s(\boldsymbol{X}) < 0\}$, since $s$ is a non-positive function.

**Example 6.1** For the daily temperature data, these two modal regions as the level sets of the density and the summary density curvature estimates are displayed in Figure 6.1(a) in the purple regions and (b) in the orange regions. The selected threshold for these level sets is $\tau = 0.25$ and the levels $\hat{f}_{0.25}$ and $\hat{s}_{0.25}$ have been obtained from the corresponding kernel estimates of $f$ and $s$. The $\tau = 0.25$ threshold may not appear to be a priori sufficiently low to define the modal regions, if we infer that this corresponds to a univariate upper quartile. This intuition from univariate analysis does not carry over well as the rectangle subtended by the data range $\Omega = [-7.2, 26.0] \times [1.4, 44.8]$ has an area of 1440, whereas the area of the 25% contour region $\hat{\mathcal{L}}(\hat{f}_{0.25})$ is 3.59% of the area of $\Omega$, whilst it contains 25% of the data points. The 25% contour regions for the summary density curvature contain 10.48% of the data points and comprise an area which is 1.28% of the area of $\Omega$. In this case, $\hat{\mathcal{L}}(\hat{s}_{0.25}) \subset \hat{\mathcal{L}}(\hat{f}_{0.25})$ though this is not always guaranteed. That small regions contain considerable proportions of the data indicates that for this data set, the 25% contours are a reasonable visualisation of the data-rich modal regions.

□

**Example 6.2** For the stem cell data, we use the 50% modal regions, as $\tau = 0.25$ results in insufficiently sizeable regions, as illustrated in Figure 6.2. Whilst 50% appears to be too high a threshold to define the modal regions, for these data, due the highly unbalanced sample sizes within each of these regions, only the two highest modes are identified using global thresholds with the level sets $\hat{\mathcal{L}}(\hat{f}_{0.5}), \hat{\mathcal{L}}(\hat{s}_{0.5})$. □

As Figures 6.1 and 6.2 give a graphical intuition of modal regions, we proceed to a more formal description of how they are estimated from the data sample. Recall again that the estimator of the probability contour level $f_\tau$ is the $\tau$-th quantile of $\hat{f}(\boldsymbol{X}_i; \mathbf{H}), i = 1, \ldots, n$. This leads to the plug-in es-

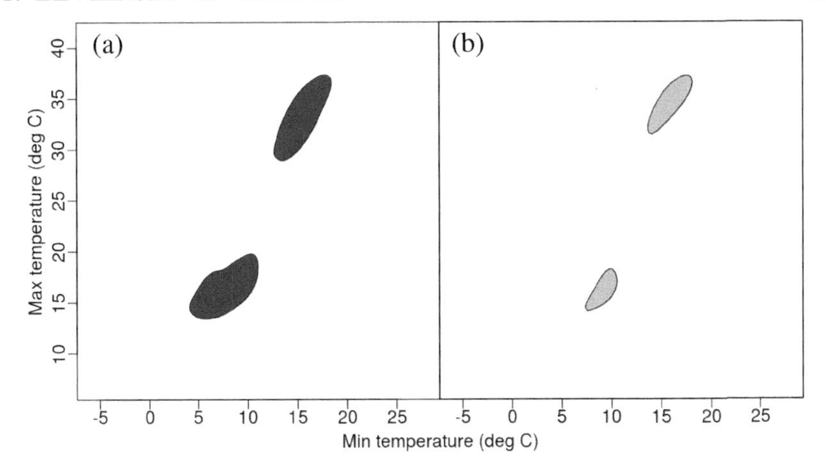


Figure 6.1 *Modal regions of the kernel estimates for the daily temperature data. (a)–(b) Modal regions. (a) 25% contour of density estimate* $\hat{f}$ *(purple). (b) Quasi-25% contour of summary density curvature estimate* $\hat{s}$ *(orange).*

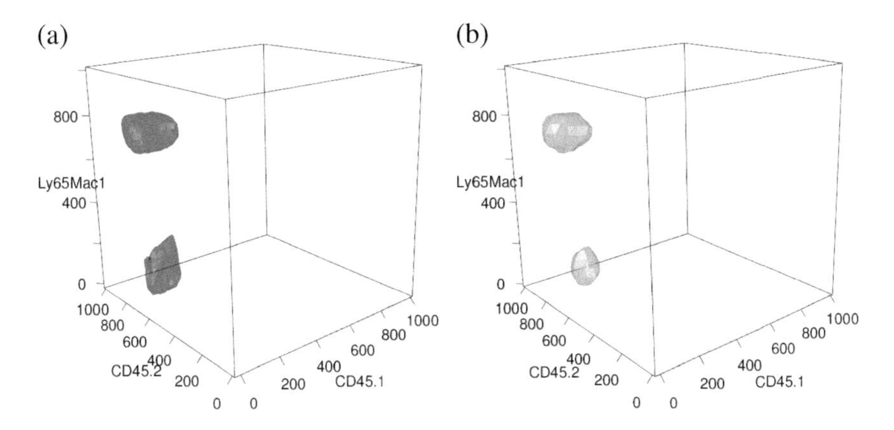


Figure 6.2 *Modal regions of the kernel estimates for the stem cell data, for subject #12. (a) 50% contour of density estimate* $\hat{f}$ *(purple). (b) Quasi-50% contour of summary density curvature estimate* $\hat{s}$ *(orange).*

timator of the probability contour region $\hat{\mathcal{L}}(\hat{f}_\tau) = \{\boldsymbol{x} : \hat{f}(\boldsymbol{x};\mathbf{H}) \geq \hat{f}_\tau\}$. By construction $\int_{\hat{\mathcal{L}}(\hat{f}_\tau)} \hat{f}(\boldsymbol{x};\mathbf{H})\, d\lambda(\boldsymbol{x}) = 1 - \tau$, and Cadre (2006) showed that $\int_{\hat{\mathcal{L}}(\hat{f}_\tau)} f(\boldsymbol{x})\, d\lambda(\boldsymbol{x}) = (1-\tau)\{1 + o_p(1)\}$. Due to the minimal hypervolume property of $\hat{\mathcal{L}}$ then as $\tau$ increases to 1, the $\{\hat{\mathcal{L}}(\hat{f}_\tau)\}$ form a sequence of sets with monotonically decreasing hypervolumes. In the limit as $\tau \to 1$, $\hat{\mathcal{L}}(\hat{f}_\tau)$ converges to the isolated mode(s) of the density estimate. For a suitable value of $\tau$, $\hat{\mathcal{L}}(\hat{f}_\tau)$ is an estimator of the modal region(s), and $\{(\boldsymbol{x}, \hat{f}(\boldsymbol{x};\mathbf{H})) : \boldsymbol{x} \in \hat{\mathcal{L}}(\hat{f}_\tau)\}$ of the bumps of $f$.

The estimation of modal regions as the level sets of the summary density curvature follows analogously. For the Hessian $\mathsf{H}\hat{f}(\boldsymbol{x};\mathbf{H})$ we have $\hat{s}(\boldsymbol{x};\mathbf{H}) = -\mathbf{1}\{\mathsf{H}\hat{f}(\boldsymbol{x};\mathbf{H}) < 0\}\,\mathrm{abs}(|\mathsf{H}\hat{f}(\boldsymbol{x};\mathbf{H})|)$. The threshold estimate is the $\tau$-quantile of the reduced sample $\{\hat{s}(\boldsymbol{X}_i;\mathbf{H}) : \hat{s}(\boldsymbol{X}_i;\mathbf{H}) < 0\}$ which consists of only those data points whose Hessian matrix is negative definite. Despite that $\hat{\mathcal{L}}(\hat{s}_\tau) = \{\boldsymbol{x} : -\hat{s}(\boldsymbol{x};\mathbf{H}) \geq -\hat{s}_\tau\}$ does not have the same probabilistic interpretation as $\hat{\mathcal{L}}(\hat{f}_\tau)$ as $\hat{s}$ is not a true density function, it remains that $\hat{\mathcal{L}}(\hat{s}_\tau)$ and $\{(\boldsymbol{x}, \hat{s}(\boldsymbol{x};\mathbf{H})) : \boldsymbol{x} \in \hat{\mathcal{L}}(\hat{s}_\tau)\}$ also yield suitable visualisations of the modal regions and bumps.

### 6.1.2 *Density support estimation*

Whilst modal regions are level sets with a high threshold, at the other end is the support of $f$ which is the set of points with strictly positive density value:

$$S = \overline{\mathcal{L}^*(0)} = \overline{\mathcal{L}^*(f;0)} = \overline{\{\boldsymbol{x} : f(\boldsymbol{x}) > 0\}}$$

where $\mathcal{L}^*$ is the level set $\mathcal{L}$ except that the greater than or equal to inequality is replaced with a strict inequality, and $\overline{A}$ is the closure of a set $A$.

The first density support estimator was introduced by Devroye & Wise (1980), along with applications to abnormal behaviour detection in statistical quality control (see also Baíllo et al., 2000). They proposed to estimate $S$ by the union of closed balls of radius $\varepsilon > 0$, centred at each of the sample points. We recognise this as the support of a kernel density estimator whose kernel is supported on the unit closed ball and whose bandwidth is $\varepsilon^2\mathbf{I}_d$. Elaborating on this idea, Dobrow (1992) suggested the plug-in estimator $\hat{S}(\mathbf{H}) = \overline{\hat{\mathcal{L}}^*(0)} = \overline{\mathcal{L}^*(\hat{f};0)} = \overline{\{\boldsymbol{x} : \hat{f}(\boldsymbol{x};\mathbf{H}) > 0\}}$ using a kernel density estimator $\hat{f}$. In this case then, $\hat{S}(\mathbf{H})$ resembles the structure of the Devroye-Wise estimator, since it can be written as $\hat{S}(\mathbf{H}) = \bigcup_{i=1}^n \{\boldsymbol{X}_i + \mathbf{H}^{1/2}\boldsymbol{y} \colon \boldsymbol{y} \in S_K\}$, where $S_K$ denotes the support of $K$. In practise, a kernel with compact support must be used, for otherwise the degenerate estimator $\hat{S}(\mathbf{H}) = \mathbb{R}^d$ is obtained.

An alternative, suggested by Cuevas & Fraiman (1997), is to replace the zero threshold with a small non-zero one, which acts as a new tuning parameter. Whenever the threshold is non-zero, then we can consider the usual level sets $\mathcal{L}$ rather than $\mathcal{L}^*$, since for any $c > 0$ the symmetric difference $\hat{\mathcal{L}}(c)\Delta\hat{\mathcal{L}}^*(c) = \{\boldsymbol{x} : \hat{f}(\boldsymbol{x};\mathbf{H}) = c\} = \partial\hat{\mathcal{L}}(c)$ has zero Lebesgue measure in most cases. A suitable data-based threshold would be the probability contour threshold $\hat{f}_\tau$ for a small value of $\tau$ close to zero, which leads to the support estimator $\hat{\mathcal{L}}(\hat{f}_\tau)$. A side benefit of a non-zero threshold is that it typically results in a differentiable boundary $\partial\hat{\mathcal{L}}(\hat{f}_\tau)$ whereas $\partial\hat{\mathcal{L}}^*(0)$ typically does not.

**Example 6.3** Two density support estimates for the daily temperature data are displayed in Figure 6.3. In Figure 6.3(a) is the Devroye-Wise estimate $\hat{S}$ with a truncated normal kernel with $S_K = [-3.7, 3.7] \times [-3.7, 3.7]$. By construction, it covers all the data points, though we observe that the boundary of $\hat{S}$ is highly non-smooth as it is composed of intersections of rectangles. Its convex hull is the grey dotted curve. In Figure 6.3(b), we have $\hat{\mathcal{L}}(\hat{f}_\tau)$ for $\tau = 0.05, 0.005, 0.0005$. Despite their under-estimation of the target support, they arguably provide a more visually informative estimate than the previous ones for these data. This under-estimation can be attenuated by considering the convex hull of $\hat{\mathcal{L}}(\hat{f}_{0.0005})$ (the dashed purple curve). □

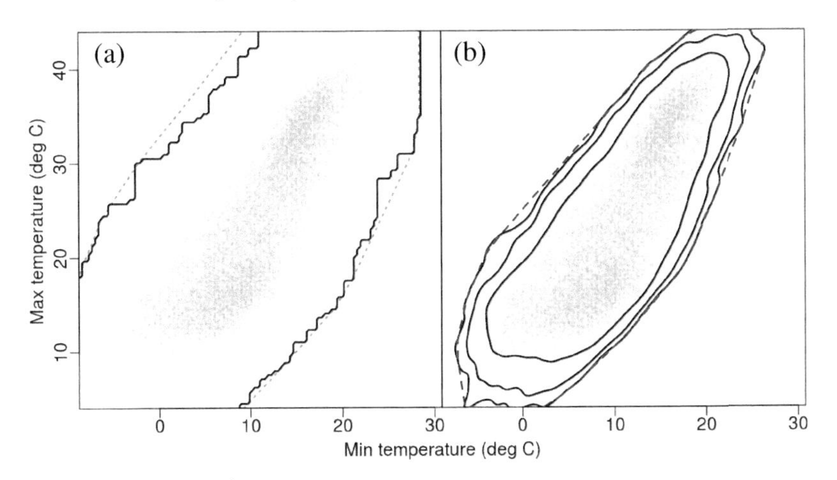


Figure 6.3 *Density support estimates for the daily temperature data. The temperature measurements are the green points. The support estimates are the solid black curves. (a) Devroye-Wise estimate $\hat{S}$. The convex hull of $\hat{S}$ is the dotted grey curve. (b) Probability contour thresholds $\hat{\mathcal{L}}(\hat{f}_\tau), \tau = 0.05, 0.005, 0.0005$. The convex hull of $\hat{\mathcal{L}}(\hat{f}_{0.0005})$ is the dashed purple curve.*

The under-estimation of the density support induced by $\hat{\mathcal{L}}(\hat{f}_\tau)$ is not always a disadvantage. In biogeographic studies, the home range is 'that area traversed by the individual in its normal activities of food gathering, mating and caring for young. Occasional sallies outside the area, perhaps exploratory in nature, should not be considered as in part of the home range' (Burt, 1943). This explains why the convex hull of all the animal locations (and other related deterministic geometric methods) tend to over-estimate the home range as they are unable to exclude these 'occasional sallies'. To resolve this, Jennrich & Turner (1969) took as their starting point the utilisation distribution, which is the term for the density function of the animal locations in these types of studies, to define the home range as the smallest region which accounts for a specified proportion of its total utilisation. This corresponds exactly to the level set of the density with probability contour threshold $\mathcal{L}(f_\tau)$, with $\tau = 0.01, 0.05$ as common choices. Since Worton (1989) demonstrated the usefulness of kernel estimators of the utilisation distribution and the home range, they have been widely adopted for analysing these animal movement data. See Baíllo & Chacón (2018) for a recent detailed survey of home range estimation techniques.

**Example 6.4** The concepts of home range and the utilisation distribution can be adapted to other organisms like plants, as illustrated in Figure 6.4 for the grevillea locations (green points). The 1% probability contour $\hat{\mathcal{L}}(\hat{f}_{0.01})$ is the solid black curve and its convex hull is the dashed purple curve. In comparison, the convex hull of the locations is the grey dotted curve. □

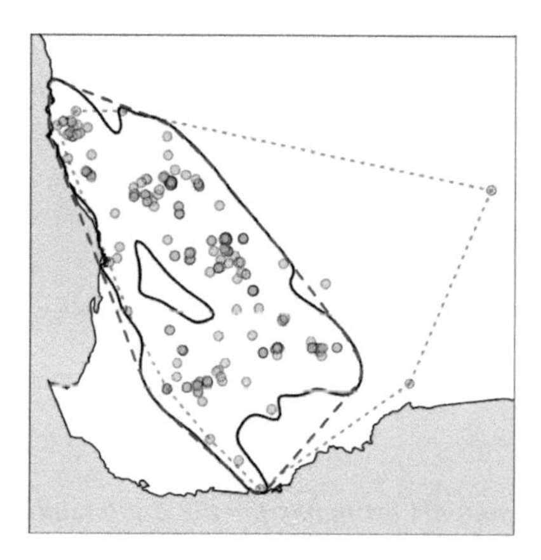

Figure 6.4 *Density support estimates for the* Grevillea *data. The* Grevillea *locations are the green points. The convex hull of these locations is the grey dotted curve. The 1% probability contour* $\hat{\mathcal{L}}(\hat{f}_{0.01})$ *is the solid black curve and its convex hull is the dashed purple curve.*

Implicit in this treatment of support estimation so far is that the density support is finite. Nonetheless for infinitely supported densities, it is still of interest to compute $\mathcal{L}(f_\tau)$ as an 'effective' support. Moreover, as $\mathcal{L}(f_\tau) \subseteq \mathcal{L}^*(0)$, the under-coverage of this effective support can be controlled since $\mathbb{P}(\mathcal{L}^*(0) \Delta \mathcal{L}(f_\tau)) = \mathbb{P}(\mathcal{L}(f_\tau)) = \tau$.

If additional geometric assumptions on the support $S$ are imposed, then further alternatives not requiring smoothing parameters are available. For example, if $S$ is assumed to be convex then a natural estimator of $S$ is the convex hull of the sample points (Rényi & Sulanke, 1963). Korostelev & Tsybakov (1993) showed that this is the maximum likelihood estimator in the family of all closed convex sets, and Dümbgen & Walther (1996) obtained its convergence rates. A related but less restrictive geometric assumption on $S$ is $\alpha$-convexity, which allows for the corresponding $\alpha$-convex hull estimator of $S$, introduced in Rodríguez-Casal (2007).

## 6.2 Density-based clustering

As a data analysis problem, the goal of cluster analysis is to discover homogeneous subgroups within a data set. We intuitively understand the construction of data clusters in a trade-off of similarity/dissimilarity: members of the same cluster are similar to each other and members of different clusters are dissimilar to each other. Many clustering techniques and algorithms have been proposed in the literature: Everitt et al. (2011) provide a complete reference for the classical approaches. Here we will adopt a modern nonparametric setup, in which clusters are naturally associated with the density modes.

From a nonparametric point of view, the task of placing the classical heuristic definition within a formal statistical framework, say, analogous to the squared error framework for density estimation, has only recently been addressed (Chacón, 2015). The first step towards this is to identify a cluster as a data-rich region (high density values) which is separated from another data-rich region by a data-poor region (low density values), as introduced by Hartigan (1975). From this arises the class of nonparametric density-based or modal clustering approaches (Stuetzle, 2003; Li, Ray & Lindsay, 2007).

The population equivalent to a partition of a finite data set is to consider a clustering $\mathscr{C} = \{C_1, \dots, C_q\}$ as a partition of the entire data space $\mathbb{R}^d$. Equivalently, such a whole-space clustering can be specified in terms of a cluster labelling function $\gamma: \mathbb{R}^d \to \mathbb{N}$ by defining $\gamma(\boldsymbol{x}) = \ell$ whenever $\boldsymbol{x} \in C_\ell$. The goal of cluster analysis is to compute an estimate $\hat{\gamma}$ of this cluster labelling function (or an estimate $\hat{\mathscr{C}}$ of the partition $\mathscr{C}$) from the data. Historically many data-based clustering algorithms were satisfied to separate a given data set $\boldsymbol{X}_1, \dots, \boldsymbol{X}_n$ into clusters, i.e., to compute $\hat{\gamma}(\boldsymbol{X}_1), \dots, \hat{\gamma}(\boldsymbol{X}_n)$ without having to consider the closeness of $\hat{\gamma}$ to $\gamma$. Whilst this suffices from a purely empirical point of view as we have estimated the labels for all the data, it is theoretically insufficient, as the accuracy of the estimated cluster labelling function $\hat{\gamma}$ remains unquantified, and the convergence of $\hat{\gamma}$ to the target labelling function $\gamma$ is not guaranteed. Our first step is to define a population goal for nonparametric density-based clustering which does not involve the data sample at hand.

### 6.2.1 Stable/unstable manifolds

Suppose that $\boldsymbol{\xi}$ is a critical point of a function $f$. The basin of attraction of $\boldsymbol{\xi}$ is defined as the set of all points which converge to $\boldsymbol{\xi}$ by following the ascent paths determined by the gradient of $f$. 'Basin of attraction' is a term borrowed from dynamical systems analysis and it is more common to employ instead 'stable manifold' in the context of statistical data analysis. The definition of a stable manifold of a critical point $\boldsymbol{\xi}$ of $f$ is

$$W_+^s(\boldsymbol{\xi}) = \{\boldsymbol{x} : \lim_{t\to\infty} \boldsymbol{\varphi}_{\boldsymbol{x}}(t) = \boldsymbol{\xi}\} \tag{6.1}$$

where $\boldsymbol{\varphi}_{\boldsymbol{x}} : \mathbb{R} \to \mathbb{R}^d$ is the parametric form of the curve, indexed by $\boldsymbol{x}$, which is the solution to the initial value problem involving the positive density gradient flow: $(d/dt)\boldsymbol{\varphi}_{\boldsymbol{x}}(t) = \mathsf{D}f(\boldsymbol{\varphi}_{\boldsymbol{x}}(t)),\ \boldsymbol{\varphi}_{\boldsymbol{x}}(0) = \boldsymbol{x}$.

Chacón (2015) observed that this stable manifold of the positive gradient is equivalent to the unstable manifold of the negative gradient $W_-^u(\boldsymbol{\xi}) = \{\boldsymbol{x} : \lim_{t\to-\infty} \boldsymbol{b}_{\boldsymbol{x}}(t) = \boldsymbol{\xi}\}$ where $\boldsymbol{b}_{\boldsymbol{x}} : \mathbb{R} \to \mathbb{R}^d$ is a solution to the initial value problem $(d/dt)\boldsymbol{b}_{\boldsymbol{x}}(t) = -\mathsf{D}f(\boldsymbol{b}_{\boldsymbol{x}}(t)),\ \boldsymbol{b}_{\boldsymbol{x}}(0) = \boldsymbol{\xi}$; i.e., $W_+^s(\boldsymbol{\xi}) = W_-^u(\boldsymbol{\xi})$. The unstable manifold has a physical interpretation as the region that is covered by the trajectories under the action of gravity of a water droplets, which originate from a continuous source located directly above the local mode, over a surface that is described by $f$. For statistical clustering, the stable manifold parametrisation is preferred.

The population goal for nonparametric density-based clustering is defined in terms of the stable manifolds of the $q$ modes $\boldsymbol{\xi}_1, \dots, \boldsymbol{\xi}_q$ of the density $f$. The population goal of modal clustering is $\mathscr{C} = \{C_1, \dots, C_q\}$, where $C_\ell = W_+^s(\boldsymbol{\xi}_\ell)$ for $\ell = 1, \dots, q$, and the induced cluster labelling function is

$$\gamma(\boldsymbol{x}) = \sum_{\ell=1}^{q} \ell \mathbf{1}\{\boldsymbol{x} \in W_+^s(\boldsymbol{\xi}_\ell)\}. \tag{6.2}$$

That is, a candidate point $\boldsymbol{x}$ is assigned the label of the stable manifold in which it is located. Thom's theorem ensures that $\mathscr{C}$ indeed constitutes a well-defined clustering of almost all of the whole space, i.e., the set of points not contained in any of the clusters of $\mathscr{C}$ has null probability (Thom, 1949).

**Example 6.5** To appreciate that this implicit definition does correspond to an intuitive understanding of clusters, consider the Trimodal III normal mixture density from Section 3.8. Figure 6.5(a) shows its contour plot with the 10%, 30%, 50%, 70% and 90% contours. The three local modes $\boldsymbol{\xi}_1 = (-0.998, 0.002), \boldsymbol{\xi}_2 = (0.990, 1.153), \boldsymbol{\xi}_3 = (1.000, -1.120)$ are the orange points. They are very close but not exactly the same as the means of the individual mixture components. Figure 6.5(b) shows the three stable manifolds corresponding to these modes as the green, purple and blue regions. The density gradient ascent flows are represented in the quiver plot by the black arrows with the arrow head size indicating the relative magnitude of the gradient. The extent of each of the stable manifolds consists of all those points whose gradient ascent flow terminates at the same local mode. In turn, their boundaries (solid black lines) are made of those points whose destination through the flow lines are the two existing saddle points of this density (not shown). From this cluster partition $\{W_+^s(\boldsymbol{\xi}_1), W_+^s(\boldsymbol{\xi}_2), W_+^s(\boldsymbol{\xi}_3)\}$, it is straightforward to define the cluster label function $\gamma$. □

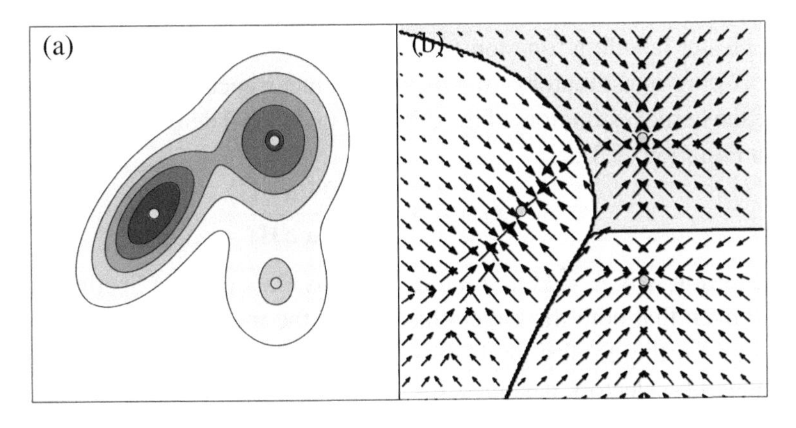


Figure 6.5 *Stable manifolds and population cluster partition for the trimodal normal mixture density. The local modes* $\boldsymbol{\xi}_1 = (-0.998, 0.002), \boldsymbol{\xi}_2 = (0.990, 1.153), \boldsymbol{\xi}_3 = (1.000, -1.120)$ *are the orange points. (a) Contour plot of the density function with 10, 30, 50, 70, 90% contours. (b) Stable manifolds. The black arrows are the gradient ascent flows. The stable manifolds* $\{W_+^s(\boldsymbol{\xi}_1), W_+^s(\boldsymbol{\xi}_2), W_+^s(\boldsymbol{\xi}_3)\}$ *are the green, purple and blue regions.*

### 6.2.2 *Mean shift clustering*

Mean shift clustering is a clustering algorithm, introduced by Fukunaga & Hostetler (1975), that is designed to estimate the population modal clustering defined through these stable manifolds. For a candidate point $\boldsymbol{x}$, the mean shift recurrence relation is

$$\boldsymbol{x}_{j+1} = \boldsymbol{x}_j + \frac{\mathbf{A}\mathsf{D}f(\boldsymbol{x}_j)}{f(\boldsymbol{x}_j)} \tag{6.3}$$

for a given positive-definite matrix $\mathbf{A}$, for $j \geq 1$ and $\boldsymbol{x}_0 = \boldsymbol{x}$, and the output from Equation (6.3) is the sequence $\{\boldsymbol{x}_j\}_{j\geq 0}$. Equation (6.3) can be easily recognized as a classical gradient ascent algorithm, usually employed to numerically find the local maxima of a function, though using the normalized gradient (i.e., the gradient divided by the density) to accelerate the convergence in low density regions.

Suppose that $\boldsymbol{\xi}_\ell$ is the local mode to which the flow line $\boldsymbol{\varphi}_{\boldsymbol{x}}(t)$ converges as $t \to \infty$; i.e., $\boldsymbol{x} \in W_+^s(\boldsymbol{\xi}_\ell)$. Then, Arias-Castro et al. (2016*b*) showed that $\boldsymbol{x}_j \to \boldsymbol{\xi}_\ell$ as $j \to \infty$ for the unnormalised mean shift algorithm. These authors also asserted that the polygonal line defined by this sequence provides a uniformly convergent approximation to the flow line $\{\boldsymbol{\varphi}_{\boldsymbol{x}}(t) : t \geq 0\}$. So if we apply the mean shift iterations to all the points in the sample space, then by inspecting the limit values of these sequences, each point can be assigned to one of the target population clusters $\{W_+^s(\boldsymbol{\xi}_1), \ldots, W_+^s(\boldsymbol{\xi}_q)\}$ and we recover the cluster labelling function $\gamma$.

A plug-in estimator of the mean shift recurrence is obtained if we substitute the kernel estimators of the density $\hat{f}(\boldsymbol{x};\mathbf{H})$ and density gradient $\mathsf{D}\hat{f}(\boldsymbol{x};\mathbf{H})$ into Equation (6.3):

$$\boldsymbol{x}_{j+1} = \boldsymbol{x}_j + \frac{\mathbf{H}\mathsf{D}\hat{f}(\boldsymbol{x}_j;\mathbf{H})}{\hat{f}(\boldsymbol{x}_j;\mathbf{H})}. \tag{6.4}$$

The reason why $\mathbf{A} = \mathbf{H}$ is suitable choice is elaborated in Section 6.2.3.

Given that both the target clusters and the mean shift algorithm strongly depend on the density gradient, Chacón & Duong (2013) suggested to use bandwidth selectors designed for gradient density estimation for the bandwidth for mean shift clustering. The simulation study of Chacón & Monfort (2014) showed that this strategy yields good results in practice. This choice is also supported by the results in Vieu (1996) and Chen et al. (2016), where it was shown that the optimal bandwidth choice for estimating the mode of a density is closely related to the problem of density derivative estimation.

**Example 6.6** The action of the mean shift on a regular $9 \times 9$ grid of points is illustrated in Figure 6.6. The data sample $\boldsymbol{X}_1, \ldots, \boldsymbol{X}_{1000}$ is drawn from the Trimodal III normal mixture density, and are the green points. The estimated local modes are the orange points. In Figure 6.6(a) the initial regular grid points are the grey points. In (b), the displacement of this grid after a single iteration is shown, where the points farthest away from the local modes have shifted the most. In (c), after 10 iterations, we observe that initially, regularly spaced grid points are migrating along the gradient ascent paths. In (d), after 45 iterations, even the initially distant grid points have converged to a local mode. The three estimated modes $\hat{\boldsymbol{\xi}}_1 = (-0.87, 0.04), \hat{\boldsymbol{\xi}}_2 = (0.91, 1.04), \hat{\boldsymbol{\xi}}_2 = (1.12, -1.30)$

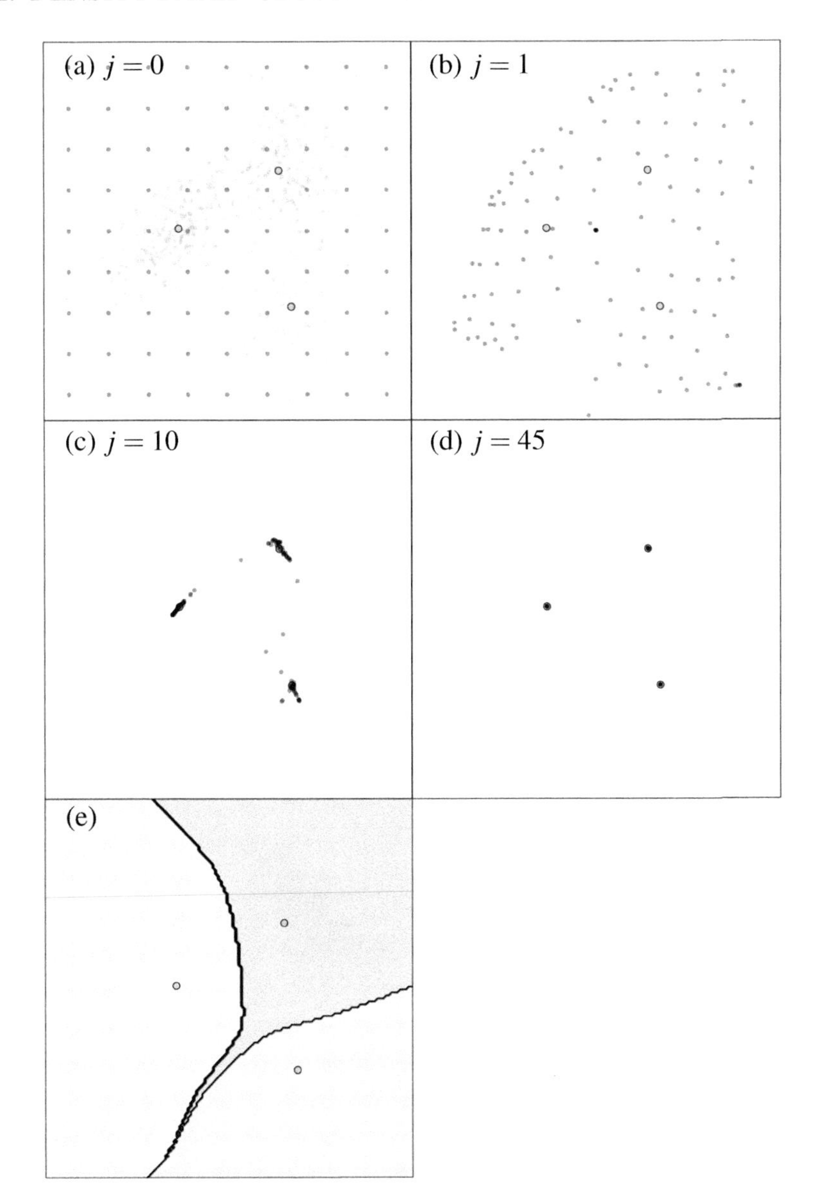


Figure 6.6 *Mean shift recurrence for the trimodal normal mixture density. The* $n = 1000$ *random data sample are the green points. The estimated local modes* $\hat{\boldsymbol{\xi}}_1 = (-0.87, 0.04), \hat{\boldsymbol{\xi}}_2 = (0.91, 1.04), \hat{\boldsymbol{\xi}}_2 = (1.12, -1.30)$ *are the orange points. (a) Initial regular* $9 \times 9$ *grid at* $j = 0$*. (b) Displacement of the grid after* $j = 1$ *iteration of the mean shift recurrence. (c) Displacement after* $j = 10$ *iterations. (d) Displacement after* $j = 45$ *iterations. (e) Estimated cluster partitions are the green, purple and blue regions.*

are the orange points. If we apply the mean shift recurrence to all the points in the sample space, then we obtain estimates of the population clusters/stable manifolds $\{W_+^s(\hat{\boldsymbol{\xi}}_1), W_+^s(\hat{\boldsymbol{\xi}}_2), W_+^s(\hat{\boldsymbol{\xi}}_3)\}$, displayed as the green, purple and blue regions in Figure 6.6(e). □

For many data analysis applications, it is not common to cluster the entire data space as in Figure 6.6(e), rather it is more usual to cluster a given data set, as in the following example.

**Example 6.7** We apply the mean shift clustering to the daily temperature data, with the plug-in bandwidth for the density gradient $\hat{\mathbf{H}}_{\mathrm{PI},1} = [1.04, 0.98; 0.98, 1.69]$, with an additional constraint that the minimum cluster size is $\lfloor 0.05n \rfloor = 1095$. In Figure 6.7 are the estimated cluster labels of the data $\boldsymbol{X}_1, \ldots, \boldsymbol{X}_n$, in green $\{\boldsymbol{X}_i : \hat{\gamma}(\boldsymbol{X}_i) = 1\}$ and purple $\{\boldsymbol{X}_i : \hat{\gamma}(\boldsymbol{X}_i) = 2\}$. The estimated local modes $\hat{\boldsymbol{\xi}}_1 = (8.7°\mathrm{C}, 16.4°\mathrm{C})$, $\hat{\boldsymbol{\xi}}_2 = (15.5°\mathrm{C}, 33.7°\mathrm{C})$ are shown as the two orange points. The cluster sizes are $n_1 = 13367\,(61.0\%), n_2 = 8541\,(39.0\%)$. The members of the mean shift clusters do not form (intersections of) symmetric ellipsoids, demonstrating the flexibility of the mean shift to detect asymmetric, non-ellipsoidal clusters. Moreover, the mean shift does not require that the number of clusters to be specified before the cluster discovery can commence: the estimated number of clusters emerges naturally from the estimated local modes. □

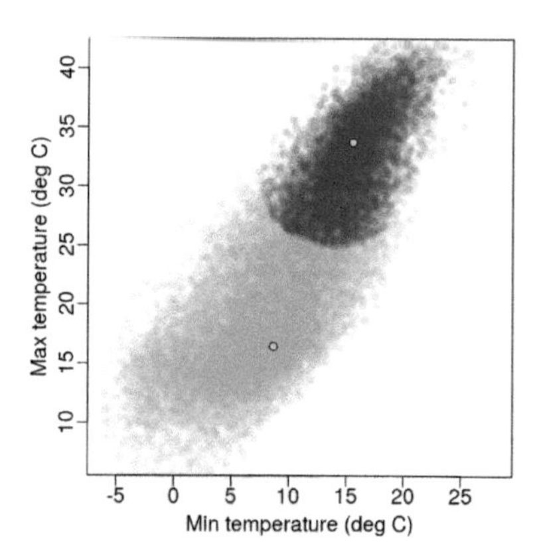


Figure 6.7 *Mean shift clusters for the daily temperature data. The estimated clusters are in green* ($n_1 = 13367$), *and purple* ($n_2 = 8541$). *The cluster modes are the orange points. The bandwidth is* $\hat{\mathbf{H}}_{\mathrm{PI},1} = [1.04, 0.98; 0.98, 1.69]$.

**Example 6.8** For the stem cell data which consists of mixed cell populations from the donor and the recipients, the goal is to identify the presence of monocytes/granulocytes in the recipient which indicates a successful graft. An elevated level of the CD45.1 (denoted CD45.1$^+$) indicates a donor cell and CD45.2$^+$ a recipient cell, and Ly65$^+$Mac1$^+$ indicates a monocyte/granulocyte. Since all three fluorochromes are simultaneously required to identify recipient monocytes/granulocytes, a complete 3-dimensional analysis is required.

Whilst a 3-dimensional scatter plot can already give visual cues on the point sub-cloud that corresponds to the recipient monocytes/granulocytes, this task is clarified, and more importantly automated, using the mean shift cluster labels displayed in Figure 6.8 for mouse subject #12. Our mean shift analysis was carried out with the bandwidth $\hat{\mathbf{H}}_{\mathrm{PI},1} = [2323, -1397, -211; -1397, 1849, 69.7, -211, 69.7, 3701]$. There are 5 clusters, coloured in green, blue, purple, red and orange, which correspond well to the 5 different cell types as identified by biological experts (Aghaeepour et al., 2013, Figure 3(a–b)). The cluster counts are $n_1 = 2478, n_2 = 2591, n_3 = 105, n_4 = 555, n_5 = 507$. Of the most interest are blue and green clusters which are comprised of CD45.2$^+$ recipient cells. Of these two, the blue cluster represents monocytes/granulocytes as it is also Ly65$^+$Mac1$^+$. The presence of this CD45.2$^+$/Ly65$^+$Mac1$^+$ blue cluster indicates that the graft operation was successful for this mouse.

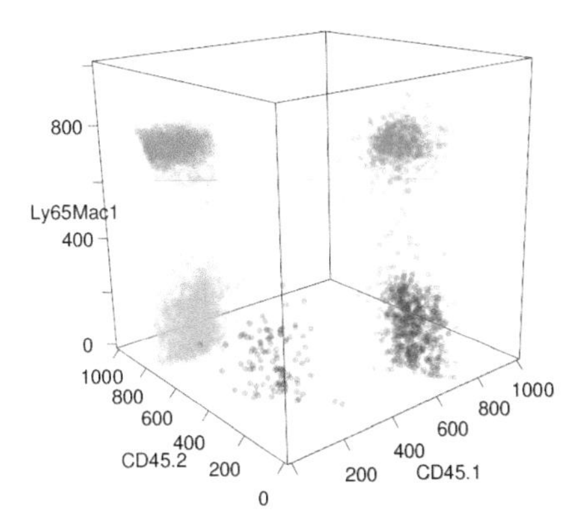


Figure 6.8 *Mean shift clusters for the stem cell data. The 5 estimated cluster labels of mouse subject #12 are in green ($n_1 = 2478$), blue ($n_2 = 2591$), purple ($n_3 = 105$), red ($n_4 = 555$) and orange ($n = 507$), corresponding to the 5 different cell types. The bandwidth is* $\hat{\mathbf{H}}_{\mathrm{PI},1} = [2323, -1397, -211; -1397, 1849, 69.7, -211, 69.7, 3701]$.

Again, the unbalanced cluster sizes, as the smallest cluster $n_3 = 105$ is approximately 25 times smaller than the two largest clusters $n_1 = 2478, n_2 = 2591$, do not pose difficulties, for the mean shift to simultaneously find the different clusters. It relies on local density gradients which are robust to unbalanced cluster sizes. □

The mean shift clustering algorithm is described in Algorithm 10. The inputs are the data sample $\boldsymbol{X}_1, \dots, \boldsymbol{X}_n$ and the candidate points $\boldsymbol{y}_1, \dots, \boldsymbol{y}_m$ which we wish to cluster (these can be the same as $\boldsymbol{X}_1, \dots, \boldsymbol{X}_n$, but this is not required). The tuning parameters are: the bandwidth matrix $\mathbf{H}$, the tolerance $\varepsilon_1$ under which subsequent iterations in the mean shift update are considered convergent, the maximum number of iterations $j_{\max}$, and the tolerance $\varepsilon_2$ under which two close cluster centres are considered to form a single cluster. The output is the cluster labels of the candidate points $\{\hat{\gamma}(\boldsymbol{y}_1), \dots, \hat{\gamma}(\boldsymbol{y}_m)\}$. Lines 1–5 correspond to gradient ascent paths in Equation (6.4), with $\boldsymbol{\eta}(\boldsymbol{y};\mathbf{H}) = \mathbf{H}\mathsf{D}\hat{f}(\boldsymbol{y};\mathbf{H})/\hat{f}(\boldsymbol{y};\mathbf{H})$, which are iterated until subsequent iterates are less than $\varepsilon_1$ apart or the maximum number of iterations $j_{\max}$ is reached. The output from these lines is the final iterates $\boldsymbol{y}_1^*, \dots, \boldsymbol{y}_m^*$. Lines 6–8 concern merging the final iterates within $\varepsilon_2$ distance of each other into a single cluster, thus creating a clustering of $\boldsymbol{y}_1^*, \dots, \boldsymbol{y}_m^*$, which is then transferred to the original data $\boldsymbol{y}_1, \dots, \boldsymbol{y}_m$ in Line 9.

**Algorithm 10** Mean shift clustering

**Input:** $\{\boldsymbol{X}_1, \dots, \boldsymbol{X}_n\}, \{\boldsymbol{y}_1, \dots, \boldsymbol{y}_m\}, \mathbf{H}, \varepsilon_1, \varepsilon_2, j_{\max}$
**Output:** Cluster labels $\{\hat{\gamma}(\boldsymbol{y}_1), \dots, \hat{\gamma}(\boldsymbol{y}_m)\}$
Compute gradient ascent paths
1: **for** $\ell := 1$ **to** $m$ **do**
2: **repeat** Iterate mean shift $\boldsymbol{y}_{\ell,j+1} := \boldsymbol{y}_{\ell,j} + \boldsymbol{\eta}(\boldsymbol{y}_{\ell,j};\mathbf{H})$
3: **until** $\|\boldsymbol{y}_{\ell,j+1} - \boldsymbol{y}_{\ell,j}\| \le \varepsilon_1$ **or** $j > j_{\max}$
4: Store final iterates $\boldsymbol{y}_\ell^* := \boldsymbol{y}_{\ell,j+1}$
5: **end for**
Assign same label to close final iterates
6: **for** $\ell_1, \ell_2 := 1$ **to** $m$ **do**
7: **if** $\|\boldsymbol{y}_{\ell_1}^* - \boldsymbol{y}_{\ell_2}^*\| \le \varepsilon_2$ **then** $\hat{\gamma}(\boldsymbol{y}_{\ell_1}^*) := \hat{\gamma}(\boldsymbol{y}_{\ell_2}^*)$ **end if**
8: **end for**
Assign labels to original data
9: **for** $\ell := 1$ **to** $m$ **do** $\hat{\gamma}(\boldsymbol{y}_\ell) := \hat{\gamma}(\boldsymbol{y}_\ell^*)$ **end for**

Reasonable choices, from simulation experiments, for the tuning parameters are: the iteration convergence threshold $\varepsilon_1 = 0.001 \min_{1 \le j \le d}\{\mathrm{IQR}_j\}$ where $\mathrm{IQR}_j$ is the $j$-th marginal sample interquartile range, the threshold for merging two cluster centres into a single cluster $\varepsilon_2 = 0.01 \max_{1 \le j \le d}\{\mathrm{IQR}_j\}$, and the maximum number of iterations $j_{\max} = 400$. A range of these tuning parameters yield similar clustering results for a given the bandwidth $\mathbf{H}$, and it is this latter choice which requires more careful consideration.

The convergence of the mean-shift recurrence $\boldsymbol{x}_{j+1} = \boldsymbol{x}_j + \boldsymbol{\eta}(\boldsymbol{x}_j;\mathbf{H})$ in Equation (6.4) has been studied by many authors, including Comaniciu & Meer (2002), though it has been subsequently discovered that these incorrectly inferred convergence of the sequence $\{\boldsymbol{x}_j\}_{j\geq 0}$. Correct results are provided, for example, by Li, Hu & Wu (2007), Aliyari Ghassabeh (2013) and Arias-Castro et al. (2016*a*) in different setups.

### 6.2.3 *Choice of the normalising matrix in the mean shift*

We provide the rationale for the choice for the normalising matrix $\mathbf{A} = \mathbf{H}$ in the kernel mean shift recurrence $\boldsymbol{x}_{j+1} = \boldsymbol{x}_j + \mathbf{A}\mathsf{D}\hat{f}(\boldsymbol{x}_j;\mathbf{H})/\hat{f}(\boldsymbol{x}_j;\mathbf{H})$. Since the kernel $K$ is spherically symmetric, it can be written as $K(\boldsymbol{x}) = \frac{1}{2}k(\|\boldsymbol{x}\|^2)$, where the function $k\colon \mathbb{R}_+ \to \mathbb{R}$ is known as the profile of $K$. Under the usual conditions this profile is a decreasing function, so that $g(x) = -k'(x) \geq 0$ for all $x$. Observing that $\mathsf{D}K(\boldsymbol{x}) = -\boldsymbol{x}g(\|\boldsymbol{x}\|^2)$, the kernel density gradient estimator can be written as

$$\begin{aligned}\mathsf{D}\hat{f}(\boldsymbol{x};\mathbf{H}) &= n^{-1}\sum_{i=1}^{n}|\mathbf{H}|^{-1/2}\mathbf{H}^{-1}(\boldsymbol{X}_i-\boldsymbol{x})g((\boldsymbol{x}-\boldsymbol{X}_i)^\top\mathbf{H}^{-1}(\boldsymbol{x}-\boldsymbol{X}_i))\\ &= \mathbf{H}^{-1}\tilde{f}(\boldsymbol{x};\mathbf{H})\boldsymbol{\eta}(\boldsymbol{x};\mathbf{H})\end{aligned}$$

where $\tilde{f}(\boldsymbol{x};\mathbf{H}) = n^{-1}|\mathbf{H}|^{-1/2}\sum_{i=1}^n g((\boldsymbol{x}-\boldsymbol{X}_i)^\top\mathbf{H}^{-1}(\boldsymbol{x}-\boldsymbol{X}_i))$, and $\boldsymbol{\eta}(\boldsymbol{x};\mathbf{H}) = [\sum_{i=1}^n \boldsymbol{X}_i g((\boldsymbol{x}-\boldsymbol{X}_i)^\top\mathbf{H}^{-1}(\boldsymbol{x}-\boldsymbol{X}_i))/\sum_{i=1}^n g((\boldsymbol{x}-\boldsymbol{X}_i)^\top\mathbf{H}^{-1}(\boldsymbol{x}-\boldsymbol{X}_i))] - \boldsymbol{x}$. As $\tilde{f}(\boldsymbol{x};\mathbf{H})$ can be considered to be an unnormalised density estimator, the ratio of the density gradient and the density is $\mathsf{D}\hat{f}(\boldsymbol{x};\mathbf{H})/\hat{f}(\boldsymbol{x};\mathbf{H}) = c\mathbf{H}^{-1}\boldsymbol{\eta}(\boldsymbol{x};\mathbf{H})$, i.e., $\mathbf{H}^{-1}\boldsymbol{\eta}(\boldsymbol{x};\mathbf{H})$ estimates (up to the constant $c$) the normalised density gradient. If we set $\mathbf{A} = c^{-1}\mathbf{H}$, the mean shift recurrence becomes $\boldsymbol{x}_{j+1} = \boldsymbol{x}_j + c^{-1}\mathbf{H}\mathsf{D}\hat{f}(\boldsymbol{x}_j;\mathbf{H})/\hat{f}(\boldsymbol{x}_j;\mathbf{H}) = \boldsymbol{x}_j + \boldsymbol{\eta}(\boldsymbol{x}_j)$. Observe that we are not required to compute this constant $c$ in order to use the mean shift recurrence relation, though for the normal kernel we have $c = 1$.

## 6.3 Density ridge estimation

For data analysis, the principal components can be considered as the directions with the largest variations in the data (Mardia et al., 1979). They are based on the singular value decomposition (SVD) of the variance of the underlying random variable $\boldsymbol{X}$, i.e., $\operatorname{Var}\boldsymbol{X} = \mathbf{U}\boldsymbol{\Lambda}\mathbf{U}^\top$ with $\mathbf{U} = [\boldsymbol{u}_1|\cdots|\boldsymbol{u}_d] \in \mathcal{M}_{d\times d}$ and $\boldsymbol{\Lambda} = \operatorname{diag}(\lambda_1,\ldots,\lambda_d) \in \mathcal{M}_{d\times d}$, where the $\lambda_1 \geq \cdots \geq \lambda_d$ are the ordered eigenvalues, and $\boldsymbol{u}_1, \dots, \boldsymbol{u}_d$ the corresponding orthonormal eigenvectors.

The $p$ principal components become a dimension reduction technique whenever $p < d$. Common choices are $p = 1, 2, 3$ as these render the $d$-dimensional data susceptible to graphical exploratory analyses. We collate the $p$ eigenvectors associated with the $p$ largest eigenvalues, and transform the data point $\boldsymbol{X}_i \in \mathbb{R}^d$ to $\boldsymbol{X}_i^{(p)} = \mathbf{U}^{(p)\top}\boldsymbol{X}_i \in \mathbb{R}^p$ where $\mathbf{U}^{(p)} = [\boldsymbol{u}_1|\cdots|\boldsymbol{u}_p] \in \mathcal{M}_{d\times p}$. Whilst principal components is a neat mathematical technique for reducing the complexity of the data $\boldsymbol{X}_i$, the difficulty is the interpretation of the resulting $\boldsymbol{X}_i^{(p)}$, as it is not straightforward to back-transform the visualisations and conclusions based on $p$ principal component space to the original data space.

**Example 6.9** Whilst the first principal component offers a convenient univariate analysis, it can suffer from a drastic reduction in the information of the structure of the original $d$-dimensional data, even for $d = 2$, as demonstrated in Figure 6.9 for the severe earthquake locations. The principal component

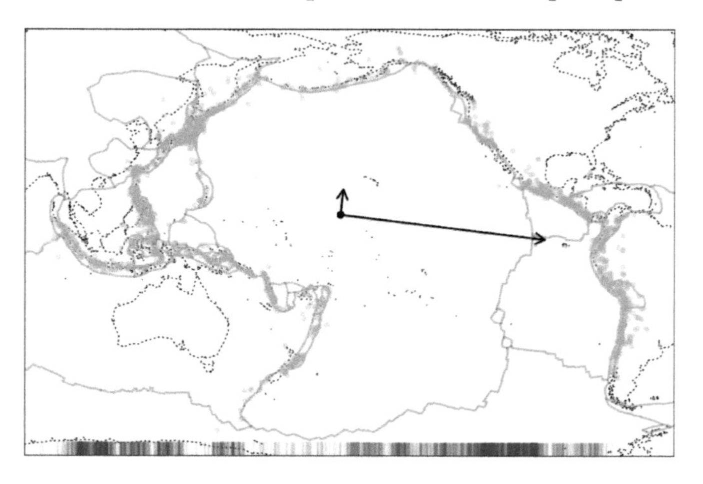

Figure 6.9 *First principal components of the earthquake data. The (longitude, latitude) locations of the $n = 2646$ major earthquakes in the Circum-Pacific belt are the green points. The boundaries of the tectonic plates are the solid blue curves. The principal component directions are the solid black arrows, with lengths proportional to their eigenvalues. The first principal components are the purple rug plot on the horizontal axis.*

directions are given by the solid black arrows, whose lengths are proportional to their eigenvalues. The first principal components are the projections of the locations onto the line parallel to the longer arrow. After being suitably transformed, they form the purple rug plot on the horizontal axis. In the process of reducing the bivariate data to univariate data, important spatio-geographical information has been lost, e.g., their 2-dimensional spatial distribution and their relative positions with respect to the tectonic plate boundaries (the solid blue curves). □

Principal curves, as 1-dimensional curves embedded in the $d$-dimensional data space, aim to preserve more of the original data structure than the 1-dimensional orthogonal projections that are the first principal components. Hastie & Stuetzle (1989) proposed that the defining characteristic of a principal curve be self-consistency, i.e., for any point $\boldsymbol{x}$ on a principal curve, if we collate all the points which project onto $\boldsymbol{x}$ then the mean of all these points coincides with $\boldsymbol{x}$. These authors introduced an algorithm to construct self-consistent principal curves. Initialising with the first principal components, then a procedure of projection and averaging is iterated until self-consistency is achieved. To avoid overfitting in this projection/averaging procedure, a regularisation penalty is introduced to impose sufficient smoothness in the resulting principal curve. The intuitiveness of self-consistent principal curves has ensured that it has formed the basis of many of the subsequent refinements in principal curve estimation, despite, as Hastie & Stuetzle (1989) themselves admit, that their procedure is not guaranteed to be convergent as it is not informed by a sufficiently rigorous statistical framework.

More recently, Ozertem & Erdogmus (2011) proposed an alternative characterisation of principal curves as the ridges of the density function. A ridge is a generalisation of a mode for filament structures like those found in the earthquake data. Whilst Ozertem & Erdogmus (2011) and subsequent authors replace the self-consistent curves in the definition of principal curves with density ridges, in order to avoid adding to the confusion of these two distinct mathematical quantities, we denote the latter only as density ridges.

Suppose that at an estimation point $\boldsymbol{x}$ the Hessian matrix of the density $\mathsf{H}f(\boldsymbol{x})$ is of full rank $d$ such that there are no repeated or zero eigenvalues. Let $\lambda_1(\boldsymbol{x}) > \cdots > \lambda_d(\boldsymbol{x})$ be the ordered eigenvalues of $\mathsf{H}f(\boldsymbol{x})$, and $\boldsymbol{u}_1(\boldsymbol{x}), \dots, \boldsymbol{u}_d(\boldsymbol{x})$ be the corresponding orthonormal eigenvectors. As the density Hessian matrix is negative definite (all eigenvalues are negative) near a ridge, then we focus on the smallest eigenvalues for the density ridge in contrast to the largest eigenvalues for the principal components. The singular value decomposition of the density Hessian is $\mathsf{H}f(\boldsymbol{x}) = \mathbf{U}(\boldsymbol{x})\boldsymbol{\Lambda}(\boldsymbol{x})\mathbf{U}(\boldsymbol{x})^\top$, and the eigenvectors matrix is partitioned as $\mathbf{U}(\boldsymbol{x}) = [\mathbf{U}^{(1)}|\mathbf{U}_{(d-1)}]$ where $\mathbf{U}^{(1)}(\boldsymbol{x}) = \boldsymbol{u}_1(\boldsymbol{x}) \in \mathbb{R}^d$ and $\mathbf{U}_{(d-1)}(\boldsymbol{x}) = [\boldsymbol{u}_2(\boldsymbol{x})|\cdots|\boldsymbol{u}_d(\boldsymbol{x})]\mathcal{M}_{d\times(d-1)}$. Let the 1-dimensional projected density gradient $\mathsf{D}f_{(d-1)} : \mathbb{R}^d \to \mathbb{R}$ be

$$\mathsf{D}f_{(d-1)}(\boldsymbol{x}) = \mathbf{U}_{(d-1)}(\boldsymbol{x})\mathbf{U}_{(d-1)}^\top(\boldsymbol{x})\mathsf{D}f(\boldsymbol{x}).$$

Genovese et al. (2014) define the density ridge $\mathcal{P}$ as consisting of all the points with zero projected gradient and $(d-1)$ negative eigenvalues

$$\mathcal{P} \equiv \mathcal{P}(f) = \{\boldsymbol{x} : \|\mathsf{D}f_{(d-1)}(\boldsymbol{x})\| = 0, \lambda_2(\boldsymbol{x}), \ldots, \lambda_d(\boldsymbol{x}) < 0\}.$$

Therefore $\boldsymbol{x}$ lies on the density ridge $\mathcal{P}$ if its density gradient $\mathsf{D}f(\boldsymbol{x})$ is parallel to the first eigenvector $\boldsymbol{u}_1(\boldsymbol{x})$, and is orthogonal to the other $(d-1)$ eigenvectors $\boldsymbol{u}_2(\boldsymbol{x}), \ldots, \boldsymbol{u}_d(\boldsymbol{x})$ of its density Hessian $\mathsf{H}f(\boldsymbol{x})$. Equivalently the density ridge is $\mathcal{P} = \{\boldsymbol{x} : \lim_{t\to\infty} \boldsymbol{a}(t) = \boldsymbol{x}\}$ where $\boldsymbol{a} : \mathbb{R} \to \mathbb{R}^d$ is an integral curve which is a solution of the differential equation $(d/dt)\boldsymbol{a}(t) = \mathsf{D}f_{(d-1)}(\boldsymbol{a}(t))$. This recalls the integral curve parametrisation of the stable manifold of a local mode $\boldsymbol{\xi}$ in Equation (6.1): $W_+^s(\boldsymbol{\xi}) = \{\boldsymbol{x} : \lim_{t\to\infty} \boldsymbol{a}(t) = \boldsymbol{\xi}\}$ where $\boldsymbol{a}$ is a solution of the initial value problem $(d/dt)\boldsymbol{a}(t) = \mathsf{D}f(\boldsymbol{a}(t))$, $\boldsymbol{a}(0) = \boldsymbol{x}$.

Whilst the change from self-consistent curves to density ridges may appear to be innocuous, the latter lead to a more rigorous estimation framework. Ozertem & Erdogmus (2011) observed that since the mean shift is the basis for the estimating the gradient ascent paths of the stable manifold, then it could be adapted for estimating the density ridge. These authors proposed to replace the density gradient $\mathsf{D}\hat{f}$ in the mean shift recurrence by the projected density gradient $\mathsf{D}\hat{f}_{(d-1)}(\boldsymbol{x};\mathbf{H}) = \mathbf{U}_{(d-1)}(\boldsymbol{x};\mathbf{H})\mathbf{U}_{(d-1)}(\boldsymbol{x};\mathbf{H})^\top \mathsf{D}\hat{f}(\boldsymbol{x};\mathbf{H})$ where $\mathbf{U}_{(d-1)}(\boldsymbol{x};\mathbf{H})$ is derived from the singular value decomposition of the density Hessian estimator $\mathsf{H}\hat{f}(\boldsymbol{x};\mathbf{H}) = \mathbf{U}(\boldsymbol{x};\mathbf{H})\boldsymbol{\Lambda}(\boldsymbol{x};\mathbf{H})\mathbf{U}(\boldsymbol{x};\mathbf{H})^\top$. The mean shift recurrence for the projected gradient is therefore

$$\boldsymbol{x}_{j+1} = \boldsymbol{x}_j + \mathbf{U}_{(d-1)}(\boldsymbol{x}_j;\mathbf{H})\mathbf{U}_{(d-1)}(\boldsymbol{x}_j;\mathbf{H})^\top \boldsymbol{\eta}(\boldsymbol{x}_j;\mathbf{H}) \tag{6.5}$$

where $\boldsymbol{\eta}(\boldsymbol{x};\mathbf{H}) = \dfrac{\sum_{i=1}^n \boldsymbol{X}_i g((\boldsymbol{x}-\boldsymbol{X}_i)^\top \mathbf{H}^{-1}(\boldsymbol{x}-\boldsymbol{X}_i))}{\sum_{i=1}^n g((\boldsymbol{x}-\boldsymbol{X}_i)^\top \mathbf{H}^{-1}(\boldsymbol{x}-\boldsymbol{X}_i))} - \boldsymbol{x}$ is the non-projected mean shift from Equation (6.4).

In comparison to self-consistent principal curves, density ridges have two main statistical advantages. First, the latter inherit the smoothness of the underlying density, and so no additional regularisation penalty is required as for the former to avoid overfitting. Second, as the mean shift is a convergent algorithm, then the estimated density ridge $\hat{\mathcal{P}}$ converges to its target $\mathcal{P}$, whereas convergence is not guaranteed for the estimated self-consistent curves.

**Example 6.10** The density ridge estimate for the earthquake data computed from the mean shift of the projected density gradient in Equation (6.5) is illustrated in Figure 6.10. The density ridge estimates are the solid purple curves. In contrast to the rug plot of the first principal components in Figure 6.9, the density ridges retain more spatial information of the structure of the data. Whilst the scatter plot already indicates that the earthquake locations follow the tectonic plate boundaries, the density ridges offer a more compelling, less cluttered visualisation of this spatio-geographic relationship. The bandwidth utilised is the plug-in matrix for the second density derivative $\hat{\mathbf{H}}_{\text{PI},2} = [74.6, -8.53; -8.53, 13.3]$ since the density Hessian should be optimally estimated. □

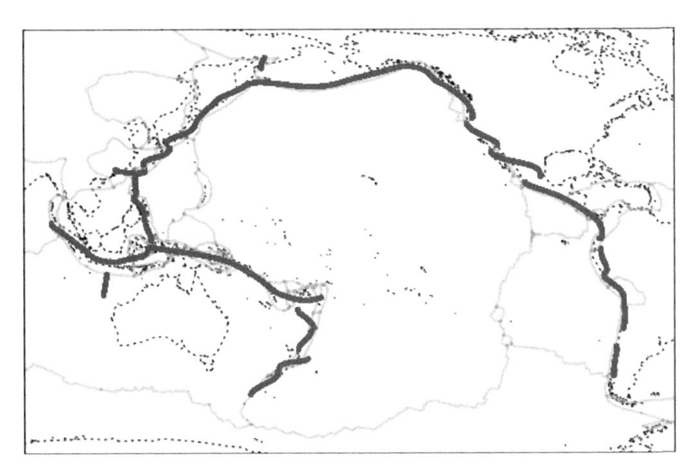

Figure 6.10 *Density ridge estimate for the earthquake data. The density ridges are the solid purple curves, based on the plug-in bandwidth* $\hat{\mathbf{H}}_{\text{PI},2} = [74.6, -8.53; -8.53, 13.3]$.

Algorithm 11 details the construction of the density ridge estimators in Figure 6.10, as introduced by Genovese et al. (2014). The inputs are the data sample $\boldsymbol{X}_1, \dots, \boldsymbol{X}_n$ and the grid of estimation points $\boldsymbol{x}_1, \dots, \boldsymbol{x}_m$ which cover the data range. The tuning parameters are: the bandwidth matrix $\mathbf{H}$, the tolerance $\varepsilon_1$, under which subsequent iterations in the mean shift update are considered convergent, the maximum number of iterations $j_{\max}$, and the probability contour threshold $\tau$ under which the iterations are not carried out. The output is the estimated density ridge set $\hat{\mathcal{P}}$. Line 2 verifies that the estimated density $\hat{f}$ at an estimation grid point $\boldsymbol{x}_\ell$ is higher than the $100\tau\%$ probability contour level $\hat{f}_\tau$. If so, then Lines 3–8 iterate the projected density gradient mean shift recurrence until subsequent iterates are less than $\varepsilon_1$ apart or the maximum number of iterations $j_{\max}$ is reached. Line 9 then increments the density ridge set $\hat{\mathcal{P}}$ by adding the final iterate $\boldsymbol{x}_{\ell,j+1}$ to it.

For 2-dimensional data, an $m = 151^2$ grid of estimation points is commonly used. The probability threshold $\tau$ can be set to a high value to avoid discarding too many grid points, whilst still reducing the time complexity, e.g., for the earthquake data, only 6935 of the $m = 22801$ grid points

**Algorithm 11** Density ridge estimator

**Input:** $\{\boldsymbol{X}_1,\ldots,\boldsymbol{X}_n\},\{\boldsymbol{x}_1,\ldots,\boldsymbol{x}_m\},\mathbf{H},\varepsilon_1,j_{\max},\tau$
**Output:** Density ridge $\hat{\mathcal{P}}$
1: **for** $\ell := 1$ **to** $m$ **do**
2: **if** $\hat{f}(\boldsymbol{x}_\ell;\mathbf{H}) > \hat{f}_\tau$ **then**
3: **repeat**
4: Compute density Hessian estimate $\mathsf{H}\hat{f}(\boldsymbol{x}_{\ell,j})$
5: Perform SVD of $\mathsf{H}\hat{f}(\boldsymbol{x}_{\ell,j}) := \mathbf{U}(\boldsymbol{x}_{\ell,j})\boldsymbol{\Lambda}(\boldsymbol{x}_{\ell,j})\mathbf{U}(\boldsymbol{x}_{\ell,j})^\top$
6: Collate $(d-1)$ eigenvectors with smallest eigenvalues
$\mathbf{U}_{(d-1)}(\boldsymbol{x}_{\ell,j}) = [\boldsymbol{u}_{d-1}(\boldsymbol{x}_{\ell,j})|\cdots|\boldsymbol{u}_d(\boldsymbol{x}_{\ell,j})]$
7: Iterate projected mean shift
$\boldsymbol{x}_{\ell,j+1} := \boldsymbol{x}_{\ell,j} + \mathbf{U}_{(d-1)}(\boldsymbol{x}_{\ell,j})\mathbf{U}_{(d-1)}(\boldsymbol{x}_{\ell,j})^\top\boldsymbol{\eta}(\boldsymbol{x}_{\ell,j};\mathbf{H})$
8: **until** $\|\boldsymbol{x}_{\ell,j+1} - \boldsymbol{x}_{\ell,j}\| \leq \varepsilon_1$ **or** $j > j_{\max}$
9: Increment $\hat{\mathcal{P}} := \hat{\mathcal{P}} \cup \{\boldsymbol{x}_{\ell,j+1}\}$
10: **end if**
11: **end for**

on $[70,310]\times[-70,80]$ exceed $\hat{f}_{0.99}$. The iteration convergence threshold $\varepsilon_1 = 0.001\min_{1\leq j\leq d}\{\mathrm{IQR}_j\}$ where $\mathrm{IQR}_j$ is the $j$-th marginal sample interquartile range, and the maximum number of iterations $j_{\max} = 400$, are the same as for the mean shift clustering in Algorithm 10.

For the selection of the bandwidth $\mathbf{H}$, Ozertem & Erdogmus (2011) and Genovese et al. (2014) utilised scalar bandwidths of class $\mathcal{A}$ optimally designed for $\hat{f}$. Using the unconstrained MISE-optimal bandwidth for the $r$-th density derivative $\mathbf{H} = O(n^{-2/(d+2r+4)})$ from Section 5.6, the Hausdorff distance between the target density ridge $\mathcal{P}$ and its kernel estimate $\hat{\mathcal{P}}$ is $\mathrm{Haus}(\mathcal{P},\hat{\mathcal{P}}) = O(n^{-2/(d+2r+4)}) + O_P((\log n)^{1/2}n^{-r/(d+2r+4)})$, as asserted by Genovese et al. (2014, Theorems 4–5). Using a density MISE optimal bandwidth as above (i.e., $r=0$), then $\mathrm{Haus}(\mathcal{P},\hat{\mathcal{P}}) = O_P((\log n)^{1/2})$ which is not convergent as $n\to\infty$. To ensure that the latter order in probability is the dominant term but still converges to 0, we require $r=2$, which yields $\mathrm{Haus}(\mathcal{P},\hat{\mathcal{P}}) = O_P((\log n)^{1/2}n^{-2/(d+8)})$. This justifies our proposal to employ an optimal selector for the density second derivative estimator $\mathsf{D}^{\otimes 2}\hat{f}$, e.g., $\hat{\mathbf{H}}_{\mathrm{PI},2}$.

A further advantage of density ridges is that their extension as $p$-dimensional hyper-ridges is straightforward, unlike for self-consistent principal curves. Given the singular decomposition $\mathsf{H}f(\boldsymbol{x}) = \mathbf{U}(\boldsymbol{x})\boldsymbol{\Lambda}(\boldsymbol{x})\mathbf{U}(\boldsymbol{x})^\top$, the eigenvectors matrix is partitioned as $\mathbf{U}(\boldsymbol{x}) = [\mathbf{U}^{(p)}|\mathbf{U}_{(d-p)}]$ where $\mathbf{U}^{(p)}(\boldsymbol{x}) = [\boldsymbol{u}_1(\boldsymbol{x})|\cdots|\boldsymbol{u}_p(\boldsymbol{x})] \in \mathcal{M}_{d\times p}$ and $\mathbf{U}_{(d-p)}(\boldsymbol{x}) = [\boldsymbol{u}_{p+1}(\boldsymbol{x})|\cdots|\boldsymbol{u}_d(\boldsymbol{x})] \in \mathcal{M}_{d\times(d-p)}$.

The projected density gradient is $\mathsf{D}f_{(d-p)}(\boldsymbol{x}) = \mathbf{U}_{(d-p)}(\boldsymbol{x})\mathbf{U}_{(d-p)}(\boldsymbol{x})^\top \mathsf{D}f(\boldsymbol{x}) \in \mathbb{R}^{d-p}$. Then the $p$-dimensional density ridge $\mathcal{P}_p$ consists of all the points with zero projected gradient and $(d-p)$ negative eigenvalues: $\mathcal{P}_p = \{\boldsymbol{x} : \|\mathsf{D}f_{(d-p)}(\boldsymbol{x})\| = 0, \lambda_{p+1}(\boldsymbol{x}), \dots, \lambda_d(\boldsymbol{x}) < 0\}$. This density hyper-ridge $\mathcal{P}_p$ has an intrinsic dimension $p$ and cannot be further reduced in a lower dimensional set without losing important information (Ozertem & Erdogmus, 2011).

## 6.4 Feature significance

Feature significance denotes the suite of formal inferential methods utilised in conjunction with exploratory data analytic methods, as introduced by Chaudhuri & Marron (1999) and Godtliebsen et al. (2002). In this context, a 'feature' refers to an important characteristic of the density function $f$, such as a local extremum. We focus on modal regions as these are data-rich regions of interest, and which were investigated in Section 6.1 as globally thresholded level sets of the density or the summary density curvature function. Here we characterise the modal regions in terms of the local significance tests for the density curvature function. At each estimation point $\boldsymbol{x}$, let the local null hypothesis be

$$H_0(\boldsymbol{x}) : \|\mathsf{D}^{\otimes 2} f(\boldsymbol{x})\| = 0.$$

The significant modal region is the rejection region, i.e., the zone where the density curvature is significantly non-zero, with the additional condition of negative definiteness of the density Hessian to ensure a local mode rather than another type of local extremum:

$$\mathcal{M} \equiv \mathcal{M}(f) = \{\boldsymbol{x} : \text{reject } H_0(\boldsymbol{x}), \mathsf{H}f(\boldsymbol{x}) < 0\}. \quad (6.6)$$

We focus on this characterisation of a density mode via the Hessian $\mathsf{H}f$ rather than the gradient $\mathsf{D}f$. At first glance, the latter characterisation appears to be simpler mathematically and computationally. A more thorough reasoning reveals that the rejection region for the gradient-based hypothesis test $\|\mathsf{D}f(\boldsymbol{x})\| = 0$ tends to increase in size to cover the entire sample space, and hence suffers from low discriminatory power (Genovese et al., 2016).

A natural test statistic is obtained by replacing the target density curvature $\mathsf{D}^{\otimes 2} f$ by its kernel estimator $\mathsf{D}^{\otimes 2}\hat{f}$. Under the hull hypothesis, the expected value is $\mathbb{E}\{\mathsf{D}^{\otimes 2}\hat{f}(\boldsymbol{x};\mathbf{H})\} = \mathsf{D}^{\otimes 2} f(\boldsymbol{x})\{1 + o(1)\} = \mathbf{0}_d + o(1)\mathbf{1}_d$ from Equation (5.12). From Equation (5.14), the null variance is $\operatorname{Var}\{\mathsf{D}^{\otimes 2}\hat{f}(\boldsymbol{x};\mathbf{H})\} = n^{-1}|\mathbf{H}|^{-1/2}(\mathbf{H}^{-1/2})^{\otimes 2}\mathbf{R}(\mathsf{D}^{\otimes 2}K)(\mathbf{H}^{-1/2})^{\otimes 2} f(\boldsymbol{x})\{1 + o(1)\}$. As $\mathbf{R}(\mathsf{D}^{\otimes 2}K) = \mathbf{R}(\operatorname{vec}\mathsf{H}K)$ is not invertible since it contains repeated rows, we replace it with its vector half form. Thus the null variance of $\operatorname{vech}\mathsf{H}\hat{f}(\boldsymbol{x};\mathbf{H})$ is

$n^{-1}|\mathbf{H}|^{-1/2}(\mathbf{H}^{-1/2})^{\otimes 2}\mathbf{R}(\operatorname{vech}\mathsf{H}K)(\mathbf{H}^{-1/2})^{\otimes 2}f(\boldsymbol{x})$. A suitable Wald test statistic for $H_0(\boldsymbol{x})$ is

$$W(\boldsymbol{x}) = \|\mathbf{S}(\boldsymbol{x})^{-1/2}\operatorname{vech}\mathsf{H}\hat{f}(\boldsymbol{x};\mathbf{H})\|^2$$

where $\mathbf{S}(\boldsymbol{x}) = n^{-1}|\mathbf{H}|^{-1/2}(\mathbf{H}^{-1/2})^{\otimes 2}\mathbf{R}(\operatorname{vech}\mathsf{H}K)(\mathbf{H}^{-1/2})^{\otimes 2}\hat{f}(\boldsymbol{x};\mathbf{H})$ is a plug-in estimator of the null variance. Duong et al. (2008) asserted that kernel density derivative estimators are pointwise asymptotically normal, so $\mathbf{S}(\boldsymbol{x})^{-1/2}\operatorname{vech}\mathsf{H}\hat{f}(\boldsymbol{x};\mathbf{H})$ is approximately standard normal $N(\mathbf{0}_{d^*},\mathbf{I}_{d^*})$, since the null expected value is $\mathbb{E}\{\operatorname{vech}\mathsf{H}\hat{f}(\boldsymbol{x};\mathbf{H})\} = \mathbf{0}_{d^*} + o(1)\mathbf{1}_{d^*}$, where $d^* = \frac{1}{2}d(d+1)$ is the length of the vector half of a $d \times d$ matrix. The asymptotic null distribution of $W(\boldsymbol{x})$ follows immediately as approximately chi-squared with $d^*$ degrees of freedom,

As the estimation points form a grid, so this sequence of local hypothesis tests are highly serially correlated. To adjust for this serial correlation, let the $p$-value from the local hypothesis test $H_0(\boldsymbol{x})$ at significance level $\alpha$ be $\mathbb{P}(W(\boldsymbol{x}) > \chi^2_{d^*}(1-\alpha))$. Let $p_{(1)} \leq \cdots \leq p_{(m)}$ be the order statistics of these $p$-values for the $m$ estimation points, with their corresponding hypotheses $H_{0,(1)},\ldots H_{0,(m)}$. The decision rule is to reject all the hypotheses $H_{0,(1)},\ldots H_{0,(j^*)}$ where $j^* = \operatorname{argmax}_{1\leq j\leq m}\{p_{(j)} \leq \alpha/(m-j+1)\}$. Hochberg (1988) demonstrated that the overall level of significance of this decision rule is $\alpha$. This testing procedure detects all departures from the null hypothesis of a zero second density derivative, so the estimator of the significant modal regions $\mathcal{M}$ is $\hat{\mathcal{M}} \equiv \hat{\mathcal{M}}(\hat{f}) = \{\boldsymbol{x} : \text{reject } H_0(\boldsymbol{x}), \mathsf{H}\hat{f}(\boldsymbol{x};\mathbf{H}) < 0\}$.

**Example 6.11** For the daily temperature data, the significant modal regions, at level of significance $\alpha = 0.05$, are displayed as the solid orange regions in Figure 6.11. The plug-in bandwidth matrix is $\hat{\mathbf{H}}_{\text{PI},2} = [1.44, 1.42; 1.42, 2.46]$. The boundaries of these significant modal regions $\hat{\mathcal{M}}$ are less smooth than those of the modal regions $\hat{\mathcal{L}}(\hat{s}_{0.25})$ in Figure 6.1(b), which are the superimposed purple dashed lines. The unsmooth appearance of the former results from their construction from the local properties of the density curvature in contrast to the global construction of the latter as a level set. Despite these differently motivated constructions (local significant modal regions and global level sets), they mostly agree with each other for this data set. □

**Example 6.12** For the stem cell data, the plug-in bandwidth matrix $\hat{\mathbf{H}}_{\text{PI},2} = [3497, -2258, -354; -2258, 2930, 75.4; -354, 75.4, 5894]$ produces significant modal regions at the $\alpha = 0.05$ level of significance in Figure 6.12. In contrast to the two modal regions of the level set $\hat{\mathcal{L}}(\hat{s}_{0.5})$ in Figure 6.2(b) which correspond only to the two most common cell types ($n_1 = 2478, n_2 = 2591$), the significant modal regions $\hat{\mathcal{M}}$ correspond to all five different cell types,

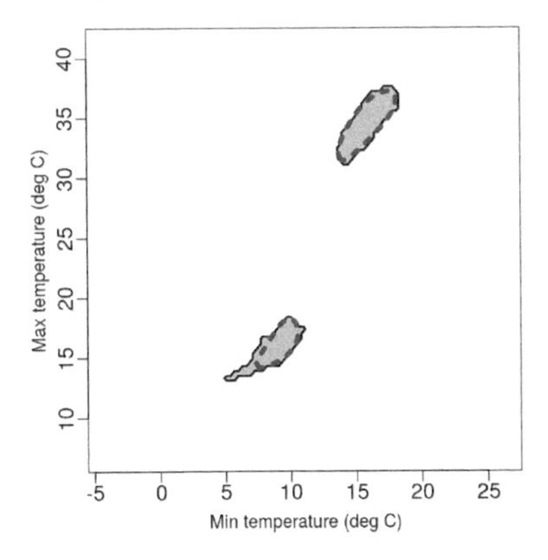


Figure 6.11 *Significant modal region estimates for the daily temperature data at significance level* $\alpha = 0.05$ *(solid orange regions). The bandwidth matrix* $\hat{\mathbf{H}}_{\text{PI},2} = [1.44, 1.42; 1.42, 2.46]$. *The 25% level set of the summary curvature* $\hat{\mathcal{L}}(\hat{s}_{0.25})$ *is delimited by the purple dashed lines.*

including the rarer cell types ($n_3 = 105, n_4 = 555, n_5 = 507$). Whilst these unbalanced subgroup sizes pose difficulties for the globally defined $\hat{\mathcal{L}}(\hat{s}_{0.5})$ to detect the lower modal regions, the locally defined significant modal region estimates $\hat{\mathcal{M}}$ do not encounter similar difficulties.

Superimposed on the significant modal regions in Figure 6.12 are the point clouds whose colours correspond to the mean shift clusters from Figure 6.8. There are seven significant modal regions, which is more than the five mean shift clusters corresponding to the five cell types. The blue, orange and red sub-point clouds correspond to a single modal region, whereas the green and purple ones each correspond to two modal sub-regions. Even if the significant modal regions are based on a sequential inferential procedure with the density curvature, and the mean shift clusters are based on a recursive estimation procedure with the density gradient, their statistical conclusions from these data mostly agree with each other. □

The significant modal region algorithm is detailed in Algorithm 12. The inputs are the data $\{\boldsymbol{X}_1, \dots, \boldsymbol{X}_n\}$ and the estimation grid points $\{\boldsymbol{x}_1, \dots, \boldsymbol{x}_m\}$. The tuning parameters are the bandwidth matrix $\mathbf{H}$, the level of significance $\alpha$ and the probability contour threshold $\tau$ under which the hypothesis tests are not carried out. Lines 1–7 loop over the estimation grid points to compute the individual Wald test statistic and $p$-values, for any estimation point that is inside the level set $\mathcal{L}(\hat{f}_\tau)$. Otherwise, there is insufficient data, so it does not contribute to $\hat{\mathcal{M}}$. The ordered statistics of the $p$-values are calculated in Line 8. In Lines 9–12, the Hochberg decision rule is applied to these ordered

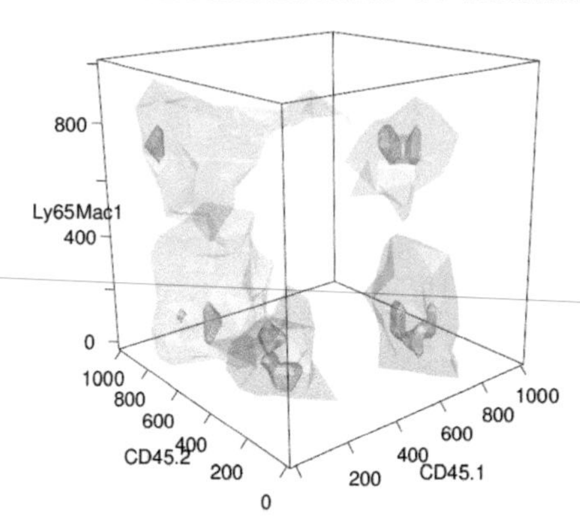


Figure 6.12 *Significant modal region estimates for the stem cell data, at significance level* $\alpha = 0.05$*, for subject #12 (opaque orange regions). The bandwidth matrix is* $\hat{\mathbf{H}}_{\mathrm{PI},2} = [3497, -2258, -354; -2258, 2930, 75.4; -354, 75.4, 5894]$. *The translucent shells (green, blue, purple, red and orange) are the convex hulls of the 5 mean shift clusters.*

**Algorithm 12** Significant modal region estimator

**Input:** $\{\boldsymbol{X}_1, \ldots, \boldsymbol{X}_n\}, \{\boldsymbol{x}_1, \ldots, \boldsymbol{x}_m\}, \mathbf{H}, \alpha, \tau$
**Output:** Significant modal regions $\hat{\mathcal{M}}$
1: **for** $\ell := 1$ **to** $m$ **do**
2: **if** $\hat{f}(\boldsymbol{x}_\ell; \mathbf{H}) > \hat{f}_\tau$ **then**
3: Compute density Hessian estimate $\mathsf{H}\hat{f}(\boldsymbol{x}_\ell)$
4: Compute Wald test statistic $W(\boldsymbol{x}_\ell)$
5: Compute $p$-value $\mathbb{P}(W(\boldsymbol{x}_\ell) \geq \chi^2_{d^*}(1-\alpha))$
6: **end if**
7: **end for**
8: Compute order statistics of $p$-values
9: **for** $\ell := 1$ **to** $m$ **do**
10: Reject hypotheses $H_0(\boldsymbol{x}_\ell)$ using Hochberg decision rule
11: Increment $\hat{\mathcal{M}} := \hat{\mathcal{M}} \cup \{\boldsymbol{x}_\ell : H_0(\boldsymbol{x}_\ell) \text{ is rejected}, \mathsf{H}\hat{f}(\boldsymbol{x}_\ell) < 0\}$
12: **end for**

$p$-values. Those estimation points where the local null hypothesis is rejected and all eigenvalues of $\mathsf{H}\hat{f}$ are negative are then added to $\hat{\mathcal{M}}$.

Reasonable choices for the tuning parameters are the level of significance $\alpha = 0.05$, and the probability threshold $\tau = 0.99$ to avoid discarding too many grid points. It is the choice of the bandwidth which requires more careful consideration. As the Wald test statistic involves $\mathsf{D}^{\otimes 2}\hat{f}$, an appropriate data-based bandwidth selector is $\hat{\mathbf{H}}_{\mathrm{PI},2}$ or any other of the cross validation selectors for the second density derivative from Section 5.6.

Chapter 7

# Supplementary topics in data analysis

The chapters so far in this monograph have focused on one-sample problems, that is, on analysing data samples drawn from a single probability distribution. In this chapter, we introduce kernel smoothing techniques in the scenarios where two or more distributions are involved. Section 7.1 introduces local significance testing for the difference between two density estimates. Section 7.2 extends to the comparison of multiple densities in the framework of classification/supervised learning. Section 7.3 examines deconvolution density estimation for smoothing data measured with error. Section 7.4 highlights the role that kernel estimation can play in learning about other non-parametric smoothing techniques, in this case, nearest neighbour estimation. Section 7.5 fills in the previously omitted mathematical details of the considered topics.

## 7.1 Density difference estimation and significance testing

A suitable way to compare two distributions, or the distribution of a variable of interest between two different populations, which we denote by $\boldsymbol{X}_1$ and $\boldsymbol{X}_2$, is by inspecting the difference of the density functions $f_{\boldsymbol{X}_1} - f_{\boldsymbol{X}_2}$. For an estimation point $\boldsymbol{x}$, the local null hypothesis is $H_0(\boldsymbol{x}) : f_{\boldsymbol{X}_1}(\boldsymbol{x}) \equiv f_{\boldsymbol{X}_2}(\boldsymbol{x})$. This means that, under this null hypothesis, the two density values and all their derivatives at $\boldsymbol{x}$ are equal. This more stringent null hypothesis than the hypothesis of equality of functional values in Duong (2013) is required in the sequel. We search for the regions of significant departures from a zero density difference:

$$\mathcal{U}^+ \equiv \mathcal{U}^+(f_{\boldsymbol{X}_1}, f_{\boldsymbol{X}_2}) = \{\boldsymbol{x} : \text{reject } H_0(\boldsymbol{x}) \text{ and } f_{\boldsymbol{X}_1}(\boldsymbol{x}) > f_{\boldsymbol{X}_2}(\boldsymbol{x})\}$$
$$\mathcal{U}^- \equiv \mathcal{U}^-(f_{\boldsymbol{X}_1}, f_{\boldsymbol{X}_2}) = \{\boldsymbol{x} : \text{reject } H_0(\boldsymbol{x}) \text{ and } f_{\boldsymbol{X}_1}(\boldsymbol{x}) \le f_{\boldsymbol{X}_2}(\boldsymbol{x})\}. \quad (7.1)$$

These are analogous to the modal regions $\mathcal{M}$ as significant departures from zero density curvature in the 1-sample feature significance in Section 6.4.

Let the density difference be $u(\boldsymbol{x}) = f_{\boldsymbol{X}_1}(\boldsymbol{x}) - f_{\boldsymbol{X}_2}(\boldsymbol{x})$. For the data samples

$\{\boldsymbol{X}_{1,1},\ldots,\boldsymbol{X}_{1,n_1}\}$ and $\{\boldsymbol{X}_{2,1},\ldots,\boldsymbol{X}_{2,n_2}\}$, a suitable test statistic is

$$W(\boldsymbol{x}) = \hat{u}(\boldsymbol{x};\mathbf{H}_1,\mathbf{H}_2)^2/S(\boldsymbol{x})^2$$

where $\hat{u}(\boldsymbol{x};\mathbf{H}_1,\mathbf{H}_2) = \hat{f}_{\boldsymbol{X}_1}(\boldsymbol{x};\mathbf{H}_1) - \hat{f}_{\boldsymbol{X}_2}(\boldsymbol{x};\mathbf{H}_2)$ is the plug-in estimator of $u$ and $S(\boldsymbol{x})^2 - R(K)[n_1^{-1}|\mathbf{H}_1|^{-1/2}\hat{f}_{\boldsymbol{X}_1}(\boldsymbol{x};\mathbf{H}_1) + n_2^{-1}|\mathbf{H}_2|^{-1/2}\hat{f}_{\boldsymbol{X}_2}(\boldsymbol{x};\mathbf{H}_2)]$ is the plug-in estimator of the variance of $\hat{u}$. Duong (2013) asserted that, since each of the kernel estimators $\hat{f}_{\boldsymbol{X}_1}, \hat{f}_{\boldsymbol{X}_2}$ are asymptotically normal, then their difference inherits this property. Moreover, the expected value is $\mathbb{E}\{\hat{u}(\boldsymbol{x};\mathbf{H}_1,\mathbf{H}_2)\} = u(\boldsymbol{x}) + \frac{1}{2}m_2(K)[\mathsf{D}^{\otimes 2}f_{\boldsymbol{X}_1}(\boldsymbol{x})^\top \operatorname{vec}\mathbf{H}_1 - \mathsf{D}^{\otimes 2}f_{\boldsymbol{X}_2}(\boldsymbol{x})^\top \operatorname{vec}\mathbf{H}_2] + o(\|\operatorname{vec}(\mathbf{H}_1 + \mathbf{H}_2)\|)$, and the variance is $\operatorname{Var}\{\hat{u}(\boldsymbol{x};\mathbf{H}_1,\mathbf{H}_2)\} = R(K)[n_1^{-1}|\mathbf{H}_1|^{-1/2}f_{\boldsymbol{X}_1}(\boldsymbol{x}) + n_2^{-1}|\mathbf{H}_2|^{-1/2}f_{\boldsymbol{X}_2}(\boldsymbol{x})] + o(n_1|\mathbf{H}_1|^{-1/2} + n_2^{-1}|\mathbf{H}_2|^{-1/2})$. Under the null hypothesis, $f_{\boldsymbol{X}_1}(\boldsymbol{x}) = f_{\boldsymbol{X}_2}(\boldsymbol{x}) = f(\boldsymbol{x})$ and $\mathsf{D}^{\otimes 2}f_{\boldsymbol{X}_1} = \mathsf{D}^{\otimes 2}f_{\boldsymbol{X}_2} = \mathsf{D}^{\otimes 2}f(\boldsymbol{x})$, so the null sampling distribution of $\hat{u}(\boldsymbol{x})$ is asymptotically normal with mean 0 and variance $[n_1^{-1}|\mathbf{H}_1|^{-1/2} + n_2^{-1}|\mathbf{H}_2|^{-1/2}]R(K)f(\boldsymbol{x})$, i.e., $W(\boldsymbol{x})$ is asymptotically $\chi^2$ with 1 degree of freedom.

The goal is to simultaneously test the hypotheses $H_0(\boldsymbol{x}_j)$, $j = 1,\ldots,m$, for a grid $\{\boldsymbol{x}_1,\ldots,\boldsymbol{x}_m\}$ of $m$ estimation points. We apply the Hochberg decision rule (Hochberg, 1988) introduced in Section 6.4 to adjust for the serially correlated local hypothesis tests. Let the $p$-value from the local hypothesis test $H_0(\boldsymbol{x}_j)$ at significance level $\alpha$ be $p_j = \mathbb{P}(W(\boldsymbol{x}_j) > \chi_1^2(1-\alpha))$. Let $p_{(1)} \leq \cdots \leq p_{(m)}$ be the order statistics of these $p$-values for the $m$ estimation points, with their corresponding hypotheses $H_{0,(1)},\ldots H_{0,(m)}$. The decision rule is to reject all the hypotheses $H_{0,(1)},\ldots H_{0,(j^*)}$ where $j^* = \operatorname{argmax}_{1\leq j\leq m}\{p_{(j)} \leq \alpha/(m-j+1)\}$. The estimators of the significant density difference regions $\mathcal{U}^+, \mathcal{U}^-$ are

$$\hat{\mathcal{U}}^+ = \{\boldsymbol{x} : \text{reject } H_0(\boldsymbol{x}) \text{ and } \hat{f}_{\boldsymbol{X}_1}(\boldsymbol{x};\mathbf{H}_1) > \hat{f}_{\boldsymbol{X}_2}(\boldsymbol{x};\mathbf{H}_2)\}$$
$$\hat{\mathcal{U}}^- = \{\boldsymbol{x} : \text{reject } H_0(\boldsymbol{x}) \text{ and } \hat{f}_{\boldsymbol{X}_1}(\boldsymbol{x};\mathbf{H}_1) \leq \hat{f}_{\boldsymbol{X}_2}(\boldsymbol{x};\mathbf{H}_2)\}.$$

**Example 7.1** Figure 7.1 illustrates these estimated density difference regions $\hat{\mathcal{U}}^+, \hat{\mathcal{U}}^-$ for the stem cell data from the control (subject #6) and treatment (subject #12) patients, for an $\alpha = 0.05$ level of significance. The purple regions are $\hat{\mathcal{U}}^+$ where the density of control cells is significantly greater than the treatment cells, and the orange regions $\hat{\mathcal{U}}^-$ where the density of control cells is significantly less than the treatment cells. The most crucial sub-region is the CD45.1$^-$/CD45.2$^+$/Ly65$^+$Mac1$^+$ orange region as it indicates that there are significantly more monocytes/granulocytes for the treatment subject than the control, indicating that the graft operation was successful.

The Hochberg decision rule controls the Type I error (false positive

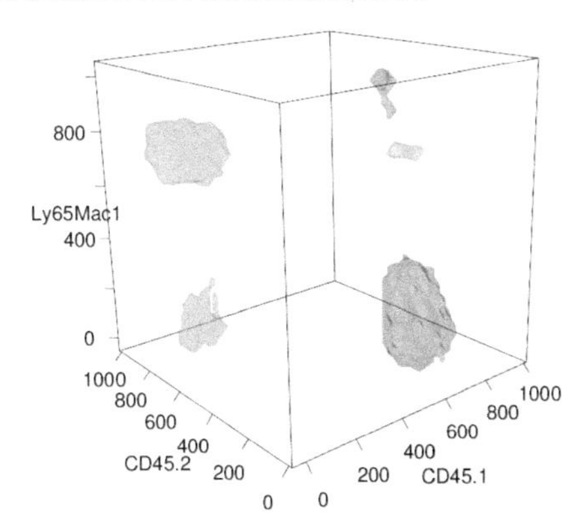


Figure 7.1 *Significant density difference region for the stem cell data, at significance level* $\alpha = 0.05$, *for subjects #6 (control) and #12 (treatment). The purple regions are* $\hat{\mathfrak{U}}^+$ *(control > treatment) and the orange regions* $\hat{\mathfrak{U}}^-$ *(control < treatment). The plug-in bandwidth matrices are* $\hat{\mathbf{H}}_1 = [117.4, -118.4, 31.9; -118.4, 158.7, -39.9; 31.9, -39.9, 1576.2]$ *and* $\hat{\mathbf{H}}_2 = [249.9, -223.3, -70.5; -223.3, 258.6, -1.3; -70.5, -1.3, 939.7]$.

rate) to be the specified level of significance $\alpha$. So it is instructive to examine the true negative rate and true positive rate (power) of these significant density difference regions. For the stem cell data, subject #5 is another control, and subject #9 is another treatment subject. The significant density difference regions for subjects #6 versus #5, and for subjects #9 versus #12 are shown in Figure 7.2. For the negative and positive controls in Figure 7.2(a)–(b), there are no significant density difference regions for CD45.1$^-$/CD45.2$^+$/Ly65$^+$Mac1$^+$, implying that there are no significant differences in the graft operation success between two control subjects or between two treatment subjects. □

Algorithm 13 details the steps to compute the significant density difference regions $\hat{\mathfrak{U}}^+, \hat{\mathfrak{U}}^-$. The inputs are the data $\{\boldsymbol{X}_{1,1}, \dots, \boldsymbol{X}_{1,n_1}\}$ (control), $\{\boldsymbol{X}_{2,1}, \dots, \boldsymbol{X}_{2,n_2}\}$ (treatment), and the estimation grid points $\{\boldsymbol{x}_1, \dots, \boldsymbol{x}_m\}$. The tuning parameters are the bandwidth matrices $\mathbf{H}_1, \mathbf{H}_2$, and the level of significance $\alpha$. Lines 1–5 loop over the estimation grid points to compute the individual Wald test statistics and $p$-values. In Lines 6–10, the Hochberg decision rule is applied to the ordered $p$-values. For those estimation points where the local null hypothesis is rejected: (a) they are added to $\hat{\mathfrak{U}}^+$ if the control density is greater than the treatment density, or (b) they are added to $\hat{\mathfrak{U}}^-$ if the control density is less than the treatment density.

Suitable bandwidth selectors for these density difference regions are the

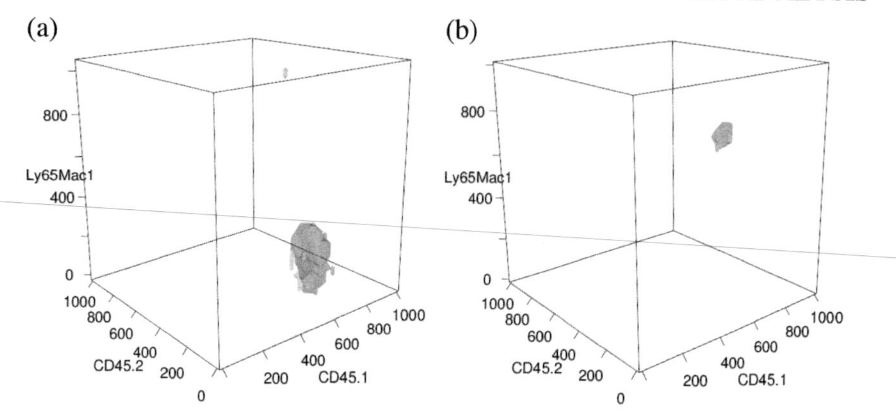


Figure 7.2 *Target negative and positive significant density difference regions for the stem cell data. The purple regions are* $\hat{\mathfrak{U}}^+(\hat{f}_{\mathbf{X}_1} > \hat{f}_{\mathbf{X}_2})$ *and the orange regions* $\hat{\mathfrak{U}}^-(\hat{f}_{\mathbf{X}_1} < \hat{f}_{\mathbf{X}_2})$. *(a) Negative control (true negative) between two control subjects #6 and #5. (b) Positive control (true positive) between two treatment subjects #9 and #12.*

**Algorithm 13** Significant density difference region estimator

**Input:** $\{\boldsymbol{X}_{1,1},\dots,\boldsymbol{X}_{1,n_1}\},\{\boldsymbol{X}_{2,1},\dots,\boldsymbol{X}_{2,n_2}\},\{\boldsymbol{x}_1,\dots,\boldsymbol{x}_m\},\mathbf{H}_1,\mathbf{H}_2,\alpha$
**Output:** Significant density difference regions $\hat{\mathfrak{U}}^+,\hat{\mathfrak{U}}^-$
1: **for** $\ell := 1$ **to** $m$ **do**
2: Compute density estimates $\hat{f}_{\mathbf{X}_1}(\boldsymbol{x}_\ell), \hat{f}_{\mathbf{X}_2}(\boldsymbol{x}_\ell)$
3: Compute Wald test statistic $W(\boldsymbol{x}_\ell)$
4: Compute $p$-value $\mathbb{P}(W(\boldsymbol{x}_\ell) \geq \chi_1^2(1-\alpha))$
5: **end for**
6: **for** $\ell := 1$ **to** $m$ **do**
7: Reject hypotheses $H_0(\boldsymbol{x}_\ell)$ using Hochberg decision rule
8: Increment $\hat{\mathfrak{U}}^+ := \hat{\mathfrak{U}}^+ \cup \{\boldsymbol{x}_\ell : H_0(\boldsymbol{x}_\ell) \text{ is rejected}, \hat{f}_{\mathbf{X}_1}(\boldsymbol{x}_\ell) > \hat{f}_{\mathbf{X}_2}(\boldsymbol{x}_\ell)\}$
9: Increment $\hat{\mathfrak{U}}^- := \hat{\mathfrak{U}}^- \cup \{\boldsymbol{x}_\ell : H_0(\boldsymbol{x}_\ell) \text{ is rejected}, \hat{f}_{\mathbf{X}_1}(\boldsymbol{x}_\ell) < \hat{f}_{\mathbf{X}_2}(\boldsymbol{x}_\ell)\}$
10: **end for**

bandwidths for the density function. These ensure that $\hat{u}(\boldsymbol{x};\mathbf{H}_1,\mathbf{H}_2)$ and $S(\boldsymbol{x})^2 = R(K)[n_1^{-1}|\mathbf{H}_1|^{-1/2}\hat{f}_{\mathbf{X}_1}(\boldsymbol{x};\mathbf{H}_1) + n_2^{-1}|\mathbf{H}_2|^{-1/2}\hat{f}_{\mathbf{X}_2}(\boldsymbol{x};\mathbf{H}_2)]$ are consistent estimators of $u(\boldsymbol{x})$ and $\mathrm{Var}\{\hat{u}(\boldsymbol{x};\mathbf{H}_1,\mathbf{H}_2)\}$. Hence the test statistic $W(\boldsymbol{x}) = \hat{u}(\boldsymbol{x};\mathbf{H}_1,\mathbf{H}_2)^2/S(\boldsymbol{x})^2$ asymptotically converges to its target $\chi_1^2$ distribution. We have employed the plug-in selectors $\hat{\mathbf{H}}_{\mathrm{PI}}$ to produce Figures 7.1–7.2, though any of those from Chapter 3 could also have been employed instead.

## 7.2 Classification

Density difference analysis is a useful visual tool for comparing two data samples, though it is difficult to extend to multiple sample comparisons as it relies heavily on the binary opposition of the relative density heights.

In the case of multiple sample comparisons, we know that our data come from $q$ different populations (or classes or groups), which are identified by the labels $\{1,2,\ldots,q\}$. We observe a $d$-dimensional random vector $\boldsymbol{X}$ that records some features of the individuals and we have a labelled sample $(\boldsymbol{X}_1,Y_1),\ldots,(\boldsymbol{X}_n,Y_n)$ from the random pair $(\boldsymbol{X},Y) \in \mathbb{R}^d \times \{1,\ldots,q\}$, with the label $Y_i$ meaning that $\boldsymbol{X}_i$ is associated with the population $Y_i$. The goal is to classify a new individual with observed features $\boldsymbol{x}$ to the most suitable group.

A function $\gamma$: $\mathbb{R}^d \to \{1,\ldots,q\}$ that determines how a feature vector $\boldsymbol{x}$ is assigned to a population $\gamma(\boldsymbol{x})$ is called a classifier. The performance of a given classifier $\gamma$ on the pair $(\boldsymbol{X},Y)$ is most commonly measured by the probability of making a classification error, or the misclassification rate $\mathrm{MR} = \mathbb{P}\left(\gamma(\boldsymbol{X}) \neq Y\right)$. Within this framework, it is well known that the optimal classifier is $\gamma_{\mathrm{Bayes}}$, the so-called Bayes rule, which assigns $\boldsymbol{x}$ to the group from which it is most likely to have been drawn, i.e., $\gamma_{\mathrm{Bayes}}(\boldsymbol{x}) = \mathrm{argmax}_{j\in\{1,\ldots,q\}} \mathbb{P}(Y = j|\boldsymbol{X} = \boldsymbol{x})$ (Devroye et al., 1996). This rule can be explicitly written in terms of the class-conditional densities and prior class probabilities. If we denote $f_j$ to be the density of the conditional random variable $\boldsymbol{X}|Y = j$ (or the density of $\boldsymbol{X}$ within the $j$-th class), and by $w_j = \mathbb{P}(Y = j)$ the prior probability of $\boldsymbol{X}$ belonging to the $j$-th class, then

$$\gamma_{\mathrm{Bayes}}(\boldsymbol{x}) = \underset{j\in\{1,\ldots,q\}}{\mathrm{argmax}}\ w_j f_j(\boldsymbol{x}). \tag{7.2}$$

The misclassification rate induced by the Bayes rule is known as the Bayes rate or error, as it is the minimal rate amongst all possible classifiers.

This data setup closely resembles that of clustering considered in Section 6.2, although with some notable differences. In classification, the number of populations is known in advance, while the number of groups is not known a priori in cluster analysis. Moreover, whilst the groups in modal clustering analysis are associated with a data-rich regions, the groups in classification are defined by the observed group labels and these latter are not necessarily based on these stable/unstable manifolds of local modes of the data density function. The data employed for classification provides the label indicating the corresponding population, which is not available for cluster analysis. In the machine learning community, classification is cast as a supervised learning problem and cluster analysis as unsupervised learning.

Since the Bayes rule depends on the unknown class probabilities and class-conditional densities, in practice these are estimated from the data $(\boldsymbol{X}_1, Y_1), \ldots, (\boldsymbol{X}_n, Y_n)$ to obtain a data-based classification rule $\hat{\gamma}(\boldsymbol{x}) = \text{argmax}_{j \in \{1,\ldots,q\}} \hat{w}_j \hat{f}_j(\boldsymbol{x})$. For this reason, such data are also called training data. This estimated rule can then be used to classify a new set of unlabelled points $\boldsymbol{x}_1, \ldots, \boldsymbol{x}_m$, which are known as the test data. These test data can be the same as the training data, though we emphasise that if the goal is to approximate the misclassification error of $\hat{\gamma}$, it is highly advisable to use test data which are independent of the training data, to avoid error under-estimation (Devroye et al., 1996).

The class probabilities are usually estimated from the sample proportions of data in each population; that is, $\hat{w}_j = n_j/n$, where $n_j = \sum_{i=1}^n \mathbf{1}\{Y_i = j\}$ is the number of observations belonging to the $j$-th population. It is the different methodologies employed to estimate the class-conditional densities which give rise to the different types of classifiers. The widely used parametric linear and quadratic classifiers replace the unknown densities in Equation (7.2) respectively by normal densities with a common variance matrix or with different variance matrices. In the context of kernel smoothing, the class-conditional densities are naturally estimated using kernel estimators. Thus the kernel classifier is

$$\hat{\gamma}(\boldsymbol{x}) \equiv \hat{\gamma}(\boldsymbol{x}; \mathbf{H}_1, \ldots, \mathbf{H}_q) = \underset{j \in \{1,\ldots,q\}}{\text{argmax}} \ \hat{w}_j \hat{f}_j(\boldsymbol{x}; \mathbf{H}_j) \tag{7.3}$$

where $\hat{f}_j$ is the kernel density estimator based on the data drawn from the $j$-th population.

**Example 7.2** An illustration of partitioning and classification using the kernel classifier in Equation (7.3) into three groups is given in Figure 7.3 for the cardiotocographic data. Three expert obstetricians classified the training data into three groups according the health status of the foetus: normal foetus ($n_1 = 412$, green), suspect ($n_2 = 83$, orange), pathological ($n_3 = 37$, purple) as illustrated in the scatter plot in Figure 1.5(a). From these training data, we construct the individual density estimates and the kernel classifier, whose induced partition is displayed in Figure 7.3(a). This classification rule is then applied to the test data set ($m = 1594$) to obtain the estimated group labels $\hat{\gamma}$, which are displayed in Figure 7.3(b). The frequencies of the estimated foetal states in the test data are: normal ($m_1 = 1339$, green points), suspect ($m_2 = 178$, orange), pathological ($m_3 = 77$, purple). □

The algorithm for kernel classification analysis is Algorithm 14. The inputs are the training data $\{(\boldsymbol{X}_1, Y_1), \ldots, (\boldsymbol{X}_n, Y_n)\}$, and the test data $\{\boldsymbol{x}_1, \ldots, \boldsymbol{x}_m\}$. The tuning parameters are the bandwidth matrices $\mathbf{H}_1, \ldots, \mathbf{H}_q$.

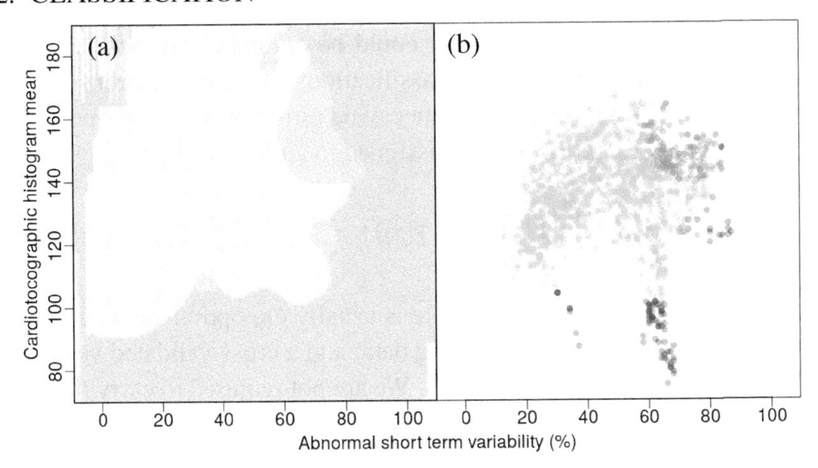


Figure 7.3 *Classification for the foetal cardiotocographic test data. The x-axis is the abnormal short-term variability (percentage), and the y-axis is the mean of the cardiotocographic histogram (0–255). (a) Partition induced by classification rule computed from training data ($n = 532$): normal (green), suspect (orange), pathological (purple). (b) Estimated test group labels ($m = 1592$): normal ($m_1 = 1339$, green), suspect ($m_2 = 178$, orange), pathological ($m_3 = 78$, purple).*

Lines 1–4 loop over each of the training data samples to compute the individual density estimates and weights. Line 5 assigns the group label to the test data according to the maximiser of the weighted density estimates.

---

**Algorithm 14** Classification

**Input:** $\{(\boldsymbol{X}_1, Y_1), \ldots, (\boldsymbol{X}_n, Y_n)\}, \{\boldsymbol{x}_1, \ldots, \boldsymbol{x}_m\}, \mathbf{H}_1, \ldots, \mathbf{H}_q$
**Output:** $\{\hat{\gamma}(\boldsymbol{x}_1), \ldots, \hat{\gamma}(\boldsymbol{x}_m)\}$
/* training data */
1: **for** $j := 1$ to $q$ **do**
2: Compute density estimate $\hat{f}_j(\boldsymbol{x}; \mathbf{H}_j)$
3: Compute sample proportion $\hat{w}_j$
4: **end for**
/* test data */
5: **for** $i := 1$ to $m$ **do** Assign group label $\hat{\gamma}(\boldsymbol{x}_i)$ **end for**

---

**Example 7.3** Figure 7.3(b) displays the estimated group labels for the test data, which can then be further scrutinised by the subject matter experts. To evaluate the accuracy of the classifier, suppose that the expert obstetricians had also classified these test data, providing the target group labels $\gamma(\boldsymbol{x}_1), \ldots, \gamma(\boldsymbol{x}_m)$. Then, the estimated group labels $\hat{\gamma}(\boldsymbol{x}_1), \ldots, \hat{\gamma}(\boldsymbol{x}_m)$ obtained automatically from the classifier could be compared to those expert-based labels. Assuming that the misclassifications in both directions are equally weighted, the misclassification rate can be estimated by the proportion of test data points $\boldsymbol{x}_1, \ldots, \boldsymbol{x}_m$ which are assigned to an incorrect group

$$\widehat{\text{MR}}(\hat{\gamma}) = m^{-1} \sum_{i=1}^{m} \mathbf{1}\{\hat{\gamma}(\boldsymbol{x}_i) \neq \gamma(\boldsymbol{x}_i)\} = 1 - m^{-1} \sum_{i=1}^{m} \mathbf{1}\{\hat{\gamma}(\boldsymbol{x}_i) = \gamma(\boldsymbol{x}_i)\}.$$

This misclassification rate estimate is usually too optimistic when this calculation is carried out on the training data, and a cross-validated version is preferred for more precise estimation. We are not required to carry out this cross validation for the foetal cardiotocographic example since the test data are independent from the training data. The obtained $\widehat{\text{MR}}(\hat{\gamma}) = 0.136$ corresponds to the more detailed cross classification in Table 7.1. From this table, the kernel classifier mostly identifies correctly the normal and pathological foetuses, though the intermediate group of suspect ones, which are less healthy than the normal foetuses but healthier than the pathological ones, is more difficult to identify correctly. □

| | | Estimated group labels | | | |
|---|---|---|---|---|---|
| | | Normal | Suspect | Patho. | Total |
| **Target** | Normal | 1195 | 45 | 3 | 1243 |
| **group** | Suspect | 106 | 106 | 0 | 212 |
| **labels** | Patho. | 38 | 27 | 74 | 139 |
| | Total | 1339 | 178 | 77 | 1594 |

Table 7.1 *Cross classification table for the foetal cardiotocographic test data. The rows are the target group labels and columns for the estimated group labels.*

It is widely acknowledged that it was precisely classification that originally motivated the development of kernel smoothing methods, as anticipated in the visionary technical report of Fix & Hodges (1951) (later republished as Fix & Hodges, 1989). Hand (1982) was the first monograph entirely devoted to kernel classification, and more general references on classification techniques include Devroye et al. (1996), Duda et al. (2000), or Webb & Copsey (2011).

Hand (1982) posited bandwidth selection for classification can proceed by selecting optimal bandwidths $\mathbf{H}_1, \ldots, \mathbf{H}_q$ (a) which are optimal for the individual kernel density estimates or (b) which directly optimise the misclassification rate, e.g., as Hall & Wand (1988) attempted for the two-class problem. Hand (1982) recommended the former approach as (a) accurate estimates of the individual density functions are useful in their own right and (b) are more likely to be useful if some accuracy measures other than the misclassification rate are considered, and as (c) direct optimisation with respect to a misclassification rate poses intractable mathematical obstacles, especially for more than two classes. Following these recommendations, we utilised the plug-in bandwidth matrices $\hat{\mathbf{H}}_{\rm PI}$, though any consistent bandwidth from Chapter 3 would also be valid choices in this sense.

Nevertheless, it is worth mentioning that more recent studies (Ghosh & Chaudhuri, 2004; Hall & Kang, 2005; Ghosh & Hall, 2008) suggested that, even if the optimal bandwidths that minimise the misclassification rate for kernel classification are of the same order as for density estimation, significant reductions in misclassification rate are possible over the MISE-optimal bandwidths for density estimation, especially for moderate-to-high dimensions. As for the problem of level set estimation, further investigation is required to demonstrate the gains in practice for bandwidth selectors specifically derived from (asymptotic) misclassification rate minimisation.

## 7.3 Density estimation for data measured with error

For all the estimation techniques exposited up till now, it has been assumed that the data sample $\boldsymbol{X}_1,\ldots,\boldsymbol{X}_n$ is free of measurement error. However, in some practical contexts we observe only a contaminated version of the data. That is, the observations from the random variable $\boldsymbol{X}$ are not directly accessible, so that any statistical analysis is made on the basis of a related, observable random variable $\boldsymbol{W}$, which is a perturbation of the variable of interest $\boldsymbol{X}$ with a measurement error. A naive application of the estimation techniques designed for error-free data to the contaminated data can produce biased results which may lead to erroneous conclusions.

Two different errors-in-variables models have been widely considered in the literature: classical errors and Berkson errors. In the classical measurement error setup, the observable variable $\boldsymbol{W}$ is related to the variable of interest $\boldsymbol{X}$ by

$$\boldsymbol{W} = \boldsymbol{X} + \boldsymbol{U}, \tag{7.4}$$

where $\boldsymbol{U}$ represents the error variable, which is assumed to be independent of $\boldsymbol{X}$ and to have a fully known distribution with density $f_{\boldsymbol{U}}$. The density $f_{\boldsymbol{W}}$ of $\boldsymbol{W}$ can be written as the convolution of the density $f_{\boldsymbol{X}}$ of $\boldsymbol{X}$ and $f_{\boldsymbol{U}}$, namely $f_{\boldsymbol{W}} = f_{\boldsymbol{X}} * f_{\boldsymbol{U}}$. As $f_{\boldsymbol{U}}$ is known and $f_{\boldsymbol{W}}$ can be estimated from the observed contaminated data $\boldsymbol{W}_1,\ldots,\boldsymbol{W}_n$, then the goal of estimating $f_{\boldsymbol{X}}$ is also known as deconvolution as we attempt to undo the convolution action of the error random variable $\boldsymbol{U}$ on $\boldsymbol{X}$.

In the alternative Berkson error setup, the model assumption is that $\boldsymbol{X} = \boldsymbol{W} + \boldsymbol{V}$, where now the role of the error variable is played by $\boldsymbol{V}$, which is assumed to be independent of $\boldsymbol{W}$. This is the key difference with Equation (7.4), where the error was assumed to be independent of the variable of interest $\boldsymbol{X}$. If in the classical error model $\boldsymbol{W}$ is understood as a version of $\boldsymbol{X}$ contaminated by measurement error, in the Berkson model $\boldsymbol{W}$ is often seen as a proxy for $\boldsymbol{X}$: it is a different variable, but linearly related to $\boldsymbol{X}$. This situation arises in experimental studies where ideally the researcher would like to measure $\boldsymbol{X}$ but, for different reasons, only the surrogate $\boldsymbol{W}$ is accessible. A common example of this context is found in exposure studies, where the researcher aims to study the exposure of individuals to a toxic agent, but the exposure is measured only at certain monitoring stations. As for classical errors, in the Berkson error model it is also possible to pose density and regression estimation problems. Moreover, the proxy variable $\boldsymbol{W}$ in the Berkson model can also be assumed to be observed with classical measurement error, resulting in a model which combines both types of error. A comprehensive review of nonparametric techniques for data measured with error can be found in Delaigle (2014).

For brevity, we consider only the classical error setup for density deconvolution. To appreciate how it may affect scale-related parameters, when Equation (7.4) holds, for example, the variance matrices of $\boldsymbol{\Sigma_W}$ and $\boldsymbol{\Sigma_X}$ of $\boldsymbol{W}$ and $\boldsymbol{X}$, are related by $\boldsymbol{\Sigma_W} = \boldsymbol{\Sigma_X} + \boldsymbol{\Sigma_U}$, where $\boldsymbol{\Sigma_U}$ is the variance matrix of $\boldsymbol{U}$. This implies that the target unobservable data $\boldsymbol{X}_1, \ldots, \boldsymbol{X}_n$ are always less spread out than the contaminated data $\boldsymbol{W}_1, \ldots, \boldsymbol{W}_n$.

### 7.3.1 *Classical density deconvolution estimation*

In the univariate case, the simplest deconvolution kernel density estimator was first proposed by Stefansky & Carroll (1987) (later published as Stefansky & Carroll, 1990). The first paper which treated the multivariate case was Masry (1991). The fact that convolutions of density functions are equivalent to products of characteristic functions implies that the construction of density estimators typically start from the Fourier domain. Let $\varphi_a(\boldsymbol{t}) = \int_{\mathbb{R}^d} \exp(i\boldsymbol{t}^\top \boldsymbol{x}) a(\boldsymbol{x}) d\boldsymbol{x}$ be the Fourier transform of any integrable function $a \colon \mathbb{R}^d \to \mathbb{R}$. If $a = f_{\boldsymbol{X}}$ is the density function of $\boldsymbol{X}$, then the Fourier transform of $f_{\boldsymbol{X}}$ is denoted as $\varphi_{\boldsymbol{X}}$ rather than $\varphi_{f_{\boldsymbol{X}}}$. Equation (7.4) implies that $\varphi_{\boldsymbol{W}} = \varphi_{\boldsymbol{X}} \varphi_{\boldsymbol{U}}$, so if $\varphi_{\boldsymbol{U}}(\boldsymbol{t}) \neq 0$ for all $\boldsymbol{t} \in \mathbb{R}^d$ and $\varphi_{\boldsymbol{X}} = \varphi_{\boldsymbol{W}} / \varphi_{\boldsymbol{U}}$ is integrable, then the inversion formula ensures that the target density $f_{\boldsymbol{X}}$ can be written as

$$f_{\boldsymbol{X}}(\boldsymbol{x}) = (2\pi)^{-d} \int_{\mathbb{R}^d} \exp(-i\boldsymbol{t}^\top \boldsymbol{x}) \varphi_{\boldsymbol{W}}(\boldsymbol{t}) / \varphi_{\boldsymbol{U}}(\boldsymbol{t})\, d\boldsymbol{t}. \tag{7.5}$$

Any estimator $\hat{\varphi}_{\boldsymbol{W}}$ of $\varphi_{\boldsymbol{W}}$ such that $\hat{\varphi}_{\boldsymbol{W}}/\varphi_{\boldsymbol{U}}$ is integrable defines an estimator of $f_{\boldsymbol{X}}$ by plugging it into Equation (7.5).

A simple estimate of $\varphi_{\boldsymbol{W}}$ from the observable data $\boldsymbol{W}_1,\ldots,\boldsymbol{W}_n$ is the empirical characteristic function $\hat{\varphi}_{\boldsymbol{W}}(\boldsymbol{t}) = n^{-1}\sum_{j=1}^n \exp(i\boldsymbol{t}^\top \boldsymbol{W}_j)$. However, this estimator cannot be directly used in Equation (7.5) because $\hat{\varphi}_{\boldsymbol{W}}/\varphi_{\boldsymbol{U}}$ is not integrable, since we have that $\varphi_{\boldsymbol{U}}(\boldsymbol{t}) \to 0$ as $\|\boldsymbol{t}\| \to \infty$ due to the Riemann-Lebesgue lemma. This can be amended by replacing the empirical characteristic function by the characteristic function of the kernel density estimator of $f_{\boldsymbol{W}}$, which can be shown to be equal to $\hat{\varphi}_{\boldsymbol{W}}(\boldsymbol{t})\varphi_K(\mathbf{H}^{1/2}\boldsymbol{t})$ for a kernel $K$ and a bandwidth matrix $\mathbf{H}$. Under the assumption that $\int_{\mathbb{R}^d} |\varphi_K(\mathbf{H}^{1/2}\boldsymbol{t})/\varphi_{\boldsymbol{U}}(\boldsymbol{t})|d\boldsymbol{t}$ is finite for all $\mathbf{H}$, the deconvolution kernel density estimator of $f_{\boldsymbol{X}}$ is defined as

$$\hat{f}_{\rm dc}(\boldsymbol{x};\mathbf{H}) = (2\pi)^{-d}\int_{\mathbb{R}^d} \exp(-i\boldsymbol{t}^\top \boldsymbol{x})\hat{\varphi}_{\boldsymbol{W}}(\boldsymbol{t})\varphi_K(\mathbf{H}^{1/2}\boldsymbol{t})/\varphi_{\boldsymbol{U}}(\boldsymbol{t})\,d\boldsymbol{t}. \quad (7.6)$$

This estimator can be alternatively expressed as

$$\hat{f}_{\rm dc}(\boldsymbol{x};\mathbf{H}) = n^{-1}\sum_{j=1}^n K^{\boldsymbol{U}}_{\mathbf{H}}(\boldsymbol{x}-\boldsymbol{W}_j;\mathbf{H}) \quad (7.7)$$

where $K^{\boldsymbol{U}}(\boldsymbol{x};\mathbf{H}) = (2\pi)^{-d}\int_{\mathbb{R}^d}\exp(-i\boldsymbol{t}^\top\boldsymbol{x})\varphi_K(\boldsymbol{t})/\varphi_{\boldsymbol{U}}(\mathbf{H}^{-1/2}\boldsymbol{t})d\boldsymbol{t}$ and the scaling is now understood to be $K^{\boldsymbol{U}}_{\mathbf{H}}(\boldsymbol{x};\mathbf{H}) = |\mathbf{H}|^{-1/2}K^{\boldsymbol{U}}(\mathbf{H}^{-1/2}\boldsymbol{x};\mathbf{H})$.

The formulation in Equation (7.7) resembles the usual form of the kernel density estimator based on $\boldsymbol{W}_1,\ldots,\boldsymbol{W}_n$, but using the deconvolution kernel $K^{\boldsymbol{U}}(\cdot;\mathbf{H})$ rather than $K$. It can be shown that $K^{\boldsymbol{U}}(\cdot;\mathbf{H})$ is a real-valued function, that it is symmetric if the error distribution is symmetric, and that $\int_{\mathbb{R}^d} K^{\boldsymbol{U}}(\boldsymbol{x};\mathbf{H})\,d\boldsymbol{x} = 1$ for all $\mathbf{H}$, so that it has many of the properties of the usual kernels. However, it is not a usual fixed, second order kernel since its shape varies with $\mathbf{H}$ and that it may take negative values.

Following analogous calculations to the error-free case, under suitable regularity conditions, the $\mathrm{MISE}\{\hat{f}_{\rm dc}(\cdot;\mathbf{H})\}$ is asymptotically equivalent to

$$\begin{aligned}\mathrm{AMISE}\{\hat{f}_{\rm dc}(\cdot;\mathbf{H})\} &= n^{-1}|\mathbf{H}|^{-1/2}R\{K^{\boldsymbol{U}}(\cdot;\mathbf{H})\}\\ &\quad + \tfrac{1}{4}m_2(K)^2\{\mathrm{vec}^\top \mathbf{R}(\mathsf{D}^{\otimes 2}f)\}(\mathrm{vec}\,\mathbf{H})^{\otimes 2}.\end{aligned}$$

The second term is the asymptotic integrated squared bias of the deconvolution density estimator $\hat{f}_{\rm dc}$, which is the same as that for the density estimator $\hat{f}$ based on the unobservable error-free data. This justifies the use and the nomenclature of the deconvolution kernel $K^{\boldsymbol{U}}$ in Equation (7.7). The first term is the asymptotic integrated variance $n^{-1}|\mathbf{H}|^{-1/2}R\{K^{\boldsymbol{U}}(\cdot;\mathbf{H})\}$; it has the same form as its counterpart $n^{-1}|\mathbf{H}|^{-1/2}R(K)$ for the classical density estimator $\hat{f}$, but they can be vastly different in value.

To better appreciate the effect of the measurement errors on the variance, assuming that the error distribution is symmetric, it is useful to use Parseval's identity to write

$$R\{K^{\boldsymbol{U}}(\cdot;\mathbf{H})\} = (2\pi)^{-d} \int_{\mathbb{R}^d} \varphi_K(\boldsymbol{t})^2 / \varphi_U(\mathbf{H}^{-1/2}\boldsymbol{t})^2 \, d\boldsymbol{t}. \tag{7.8}$$

This shows that the variance inflation due to measurement errors depends on the tail behaviour of the characteristic function of the error distribution. For example, suppose that $\boldsymbol{U}$ follows a centred Laplace distribution with variance matrix $\boldsymbol{\Sigma}$, so that $\varphi_U(\boldsymbol{t}) = (1 + \boldsymbol{t}^\top \boldsymbol{\Sigma} \boldsymbol{t}/2)^{-1}$. Then the integrated variance of $\hat{f}_{\text{dc}}(\boldsymbol{x};\mathbf{H})$ is asymptotically equivalent to $\frac{1}{4}n^{-1}|\mathbf{H}|^{-1/2}(2\pi)^{-d}(\text{vec}^\top \boldsymbol{\Sigma})^{\otimes 2}(\mathbf{H}^{-1/2})^{\otimes 4} \int_{\mathbb{R}^d} \boldsymbol{t}^{\otimes 4} \varphi_K(\boldsymbol{t})^2 \, d\boldsymbol{t}$. To balance this term with the integrated squared bias it is necessary to take $\mathbf{H}$ to be of order $n^{-2/(d+8)}$, resulting in an optimal MISE of order $n^{-4/(d+8)}$, which is slower than the error-free rate $n^{-4/(d+4)}$. This reflects the added difficulty in density estimation in the presence of data contaminated with Laplace error.

The situation is even worse under the common assumption that the error distribution is $N(\mathbf{0},\boldsymbol{\Sigma})$. Then, $\varphi_U(\boldsymbol{t})^{-2} = \exp(\boldsymbol{t}^\top \boldsymbol{\Sigma} \boldsymbol{t})$, so to ensure that the integral in Equation (7.8) is finite, it is usual to employ kernels such that $\varphi_K(\boldsymbol{t}) = 0$ for $\|\boldsymbol{t}\| \geq 1$. In this case, the optimal $\mathbf{H}$ can be shown to be of order $(\log n)^{-1}$, resulting in a slow convergence rate of $(\log n)^{-2}$ for the optimal MISE: a detailed derivation of this can be found in Stefansky (1990). This very slow rate is not due to the use of kernel estimators, since Carroll & Hall (1988) showed that this is the fastest possible rate achievable by any estimator, so it illustrates how intrinsically difficult is the problem of density deconvolution with normal measurement errors.

In the univariate setup, many of the bandwidth selection methods for the error-free kernel estimator have been generalised for the deconvolution estimator (see Delaigle & Gijbels, 2004). In contrast, the problem of bandwidth selection for multivariate kernel density deconvolution has been scarely addressed in the literature so far, see Youndjé & Wells (2008) for a cross validation proposal. So in the following section we present a different deconvolution density estimator $\hat{f}_{\text{wdc}}$, for which it makes sense to use the existing error-free bandwidth selectors.

### 7.3.2 *Weighted density deconvolution estimation*

A weighted kernel density estimator is

$$\hat{f}_{\text{wt}}(\boldsymbol{x};\boldsymbol{\alpha},\mathbf{H}) = n^{-1} \sum_{i=1}^{n} \alpha_i K_{\mathbf{H}}(\boldsymbol{x} - \boldsymbol{X}_i)$$

with a kernel $K$, a fixed bandwidth $\mathbf{H}$, the weights $\alpha_i$ are non-negative and $\alpha_1 + \cdots + \alpha_n = n$, and $\boldsymbol{\alpha} = (\alpha_1, \ldots, \alpha_n) \in \mathbb{R}^n$. This weighted density estimator was originally introduced with the aim of bias reduction for uncontaminated data $\boldsymbol{X}_1, \ldots, \boldsymbol{X}_n$ by Jones et al. (1995) and Hall & Turlach (1999), as an alternative to the methodologies examined in Chapter 4. If the weights are identically equal to 1, then $\hat{f}_{\text{wt}}$ becomes the classical density estimator $\hat{f}$ in Equation (2.2). Whilst the weighted density estimator $\hat{f}_{\text{wt}}$ can be used in any context where different importance is associated with different data points, it was Hazelton & Turlach (2009) who introduced its application to density estimation for contaminated data. These authors defined the deconvolution weighted density estimator as

$$\hat{f}_{\text{wdc}}(\boldsymbol{x}; \boldsymbol{\alpha}, \mathbf{H}) = n^{-1} \sum_{i=1}^{n} \alpha_i K_{\mathbf{H}}(\boldsymbol{x} - \boldsymbol{W}_i). \tag{7.9}$$

**Example 7.4** The Châtelet underground train station is a major hub in the Paris metro network and the air quality inside the station is continuously monitored. The local transport authority has made public the hourly mean air quality measurements from 01 January 2013 to 31 December 2016 (RATP, 2016). We focus on the concentrations of carbon dioxide $CO_2$ (g/m$^3$) and of particulate matter less than 10 $\mu$m in diameter $PM_{10}$ (parts per million). The concentrations of $CO_2$ indicate the renewal rate of fresh air, and of $PM_{10}$ the potential to affect adversely respiratory health. We analyse the concentrations at 8 p.m. There are $n = 1300$ days with fully observed measurements for these time points, as shown in the scatter plot in Figure 7.4(a). A standard density estimate $\hat{f}$ of the 8 p.m. concentrations is illustrated in Figure 7.4(b) and the deconvolution kernel density estimate $\hat{f}_{\text{wdc}}$ in Figure 7.4(c). The standard density estimate is unimodal, whereas the deconvolution density estimate is bimodal and also has narrower, smoother contours, as a consequence of taking into account the smaller (unknown) dispersion of the true error-free data. Both density estimates are computed with the plug-in bandwidth $\hat{\mathbf{H}}_{\text{PI}} = [440.0, 79.2; 79.2, 66.9]$. □

We now elaborate an algorithm for the optimal selection of the weights $\boldsymbol{\alpha}$, as utilised in Figure 7.4. The assertion by Hazelton & Turlach (2009), that if $\hat{f}_{\text{wdc}}$ is a reasonable estimator of $f_{\boldsymbol{X}}$ then $\hat{f}_{\text{wdc}}(\cdot; \boldsymbol{\alpha}, \mathbf{H}) * f_{\boldsymbol{U}}$ would be reasonably close to $f_{\boldsymbol{W}}$, led them to propose the discrepancy for selecting the weight vector $\boldsymbol{\alpha}$ as

$$\text{ISE}\{\hat{f}_{\text{wdc}}(\cdot; \mathbf{H}, \boldsymbol{\alpha})\} = \int_{\mathbb{R}^d} \{\hat{f}_{\text{wdc}}(\cdot; \boldsymbol{\alpha}, \mathbf{H}) * f_{\boldsymbol{U}} - f_{\boldsymbol{W}}(\boldsymbol{x})\}^2 \, d\boldsymbol{x}.$$

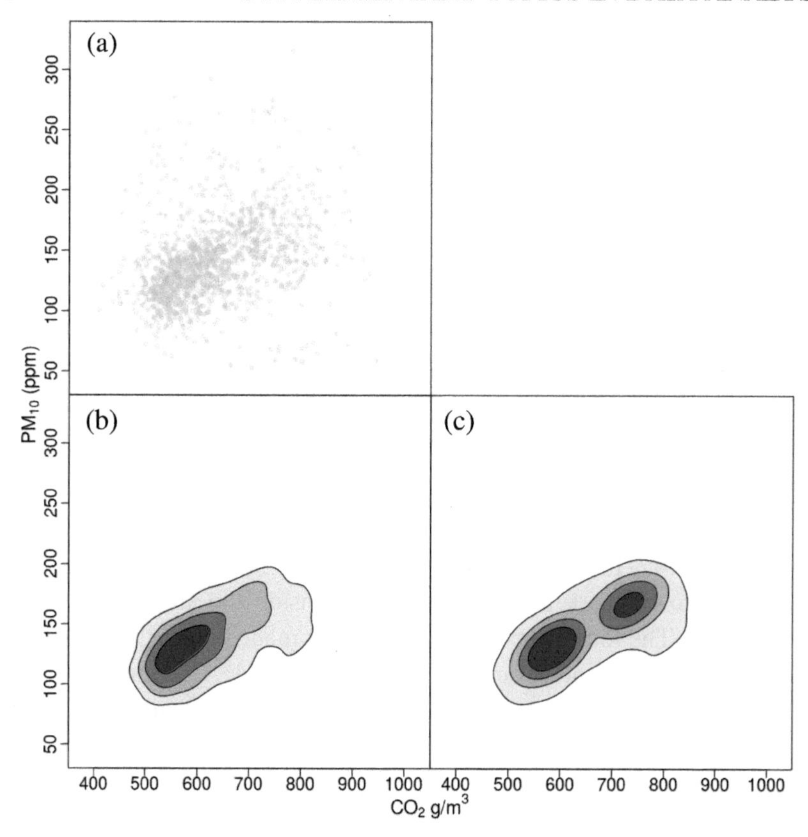


Figure 7.4 *Deconvolution density estimate for the air quality data. The horizontal axis is the $CO_2$ concentration (g/m$^3$) and the vertical axis is the $PM_{10}$ concentration (ppm). (a) Scatter plot of the $n = 1300$ data points. (b) Standard density estimate. (c) Weighted deconvolution density estimate. The contours are the quintiles (20%, 40%, 60%, 80%). Both density estimates are computed with the same bandwidth* $\hat{\mathbf{H}}_{\text{PI}} = [440.0, 79.2; 79.2, 66.9]$.

A plug-in estimator of this was obtained by replacing the unknown $f_{\boldsymbol{W}}(\boldsymbol{x})$ by its usual classical density estimator $\hat{f}_{\boldsymbol{W}}(\boldsymbol{x};\mathbf{H}) = n^{-1}\sum_{i=1}^{n} K_{\mathbf{H}}(\boldsymbol{x} - \boldsymbol{W}_i)$ as

$$Q(\boldsymbol{\alpha}) = \int_{\mathbb{R}^d} \{\hat{f}_{\text{wdc}}(\cdot;\boldsymbol{\alpha},\mathbf{H}) * f_U - \hat{f}_{\boldsymbol{W}}(\boldsymbol{x};\mathbf{H})\}^2 \, d\boldsymbol{x}.$$

The selected weights are the minimisers $\hat{\boldsymbol{\alpha}} = \operatorname{argmin}_{\boldsymbol{\alpha}>0, |\boldsymbol{\alpha}|=n} Q(\boldsymbol{\alpha})$. Hazelton & Turlach (2009) proposed to compute the optimal selected weights $\hat{\boldsymbol{\alpha}}$ via an interior point algorithm as the solution of the quadratic program

$$\text{minimise}_{\boldsymbol{\alpha}} \quad \tfrac{1}{2}\boldsymbol{\alpha}^\top(\mathbf{Q} + n^{-1}\eta\mathbf{I}_n) - \boldsymbol{b}^\top\boldsymbol{\alpha}$$

$$\text{subject to} \quad \sum_{i=1}^{n} \alpha_i = n$$
$$\alpha_i \geq 0, i = 1, \ldots, n \tag{7.10}$$

where $\mathbf{Q} \in \mathcal{M}_{n\times n}$ with $(i, j)$-th entry $n^{-2}(K_\mathbf{H} * K_\mathbf{H} * f_U * f_U)(\boldsymbol{W}_i - \boldsymbol{W}_j)$, $\boldsymbol{b} \in \mathbb{R}^n$ with $i$-th coordinate $n^{-2}\sum_{j=1}^{n}(K_\mathbf{H} * K_\mathbf{H} * f_U)(\boldsymbol{W}_i - \boldsymbol{W}_j)$ and $\eta$ is a regularisation parameter. Suppose that the observed data $\boldsymbol{W}_1, \ldots, \boldsymbol{W}_n$ are partitioned into $q$ classes $\{C_1, \ldots, C_q\}$ and that $\hat{f}_{\text{wdc},-\gamma(i)}$ is the weighted deconvolution density estimator based on the reduced sample which excludes all the data with the same class label $\gamma(i)$ as $\boldsymbol{W}_i$. These authors recommend setting the value of the regularisation parameter $\hat{\eta}$ as the minimiser of a $q$-fold cross validation criterion

$$\text{CV}(\eta) = \sum_{i=1}^{n} \log\{(\hat{f}_{\text{wdc},-\gamma(i)}(\cdot; \boldsymbol{\alpha}', \mathbf{H}) * f_U)(\boldsymbol{W}_i)\}.$$

Observe that the length of $\boldsymbol{\alpha}'$ is less than $n$ so it is not the same as $\boldsymbol{\alpha}$ in Equation (7.10).

Unlike the classical deconvolution density estimator $\hat{f}_{\text{dc}}$ (introduced in Section 7.3.1), which has poor performance in the presence of the widespread case of a normal measurement error density $\phi_{\boldsymbol{\Sigma}_U}$, for the weighted deconvolution density estimator $\hat{f}_{\text{wdc}}$ based on the normal kernel $\phi$ the assumption of normal measurement error yields a simplification of the discrepancy, leading to

$$Q(\boldsymbol{\alpha}) = n^{-2} \sum_{i,j=1}^{n} (\alpha_i \alpha_j \phi_{2\mathbf{H}+2\boldsymbol{\Sigma}_U} - 2\alpha_i \phi_{2\mathbf{H}+\boldsymbol{\Sigma}_U} + \phi_{2\mathbf{H}})(\boldsymbol{W}_i - \boldsymbol{W}_j), \tag{7.11}$$

which closely resembles the squared bias component of the smoothed cross validation for selecting $\mathbf{H}$ for $\hat{f}$ in Equation (3.18); and the coefficients in the quadratic program are composed of $[\mathbf{Q}]_{i,j} = n^{-2}\phi_{2\mathbf{H}+2\boldsymbol{\Sigma}_U}(\boldsymbol{W}_i - \boldsymbol{W}_j)$ and $b_i = n^{-2}\sum_{j=1}^{n} \phi_{2\mathbf{H}+\boldsymbol{\Sigma}_U}(\boldsymbol{W}_i - \boldsymbol{W}_j)$. Similarly the $q$-fold cross validation for the regularisation parameter simplifies to

$$\text{CV}(\eta) = \sum_{i=1}^{n} \log\left\{ \sum_{\gamma(j)\neq\gamma(i)} \alpha'_j \phi_{\mathbf{H}+\boldsymbol{\Sigma}_U}(\boldsymbol{W}_i - \boldsymbol{W}_j) \right\}.$$

To compute these components $\mathbf{Q}, \boldsymbol{b}, \hat{\eta}$ of the quadratic program in Equation (7.10) to obtain $\hat{\boldsymbol{\alpha}}$, we require an estimator the error variance $\boldsymbol{\Sigma}_U$. Suppose that for each observed, contaminated data point $\boldsymbol{W}_i$, we have the paired replicates $\boldsymbol{W}'_i, \boldsymbol{W}''_i$, $i = 1, \ldots, n$. According to Carroll et al. (2006), an estimator of $\boldsymbol{\Sigma}_U$ is $\mathbf{S}_U = \text{diag}(S^2_{U,1}, \ldots, S^2_{U,d})$ where $S^2_{U,j}$ is the $j$-th marginal sample variance of the paired differences $W'_{1,j} - W''_{1,j}, \ldots, W'_{n,j} - W''_{n,j}$, $j = 1, \ldots, d$.

The weighted deconvolution density estimate in Figure 7.4(c) is detailed in Algorithm 15. The inputs are the contaminated data $\{\boldsymbol{W}_1, \ldots, \boldsymbol{W}_n\}$ and the paired contaminated data $\{\boldsymbol{W}'_1, \ldots, \boldsymbol{W}'_n\}, \{\boldsymbol{W}''_1, \ldots, \boldsymbol{W}''_n\}$. The tuning parameters are the bandwidth matrix $\mathbf{H}$ and the $q$ for $q$-fold CV. In Line 1, the sample error variance is computed from the paired contaminated data. In Line 2, the regularisation parameter is the solution of the $q$-fold CV. In Line 3, the sample error variance and regularisation parameter are substituted into the quadratic program, resulting in the optimal weights. In Line 4, these weights are utilised in the computation of the weighted deconvolution density estimator.

---

**Algorithm 15** Weighted deconvolution density estimator

**Input:** $\{\boldsymbol{W}_1, \ldots, \boldsymbol{W}_n\}, \{\boldsymbol{W}'_1, \ldots, \boldsymbol{W}'_n\}, \{\boldsymbol{W}''_1, \ldots, \boldsymbol{W}''_n\}, \mathbf{H}, q$
**Output:** $\hat{f}_{\text{wdc}}(\boldsymbol{x}; \hat{\boldsymbol{\alpha}}, \mathbf{H})$

1: Compute sample error variance $\mathbf{S}_U$ from the paired differences $\boldsymbol{W}'_i - \boldsymbol{W}''_i$
2: Find regularisation parameter $\hat{\eta}$ as the solution of the $q$-fold CV
3: Find weights $\hat{\boldsymbol{\alpha}}$ as the solution to quadratic program with $\mathbf{S}_U$ and $\hat{\eta}$
4: Compute weighted deconvolution density estimator $\hat{f}_{\text{wdc}}(\boldsymbol{x}; \hat{\boldsymbol{\alpha}}, \mathbf{H})$

---

For the tuning parameters, Hazelton & Turlach (2009) recommended setting $q = 5$ based on their empirical evidence. What remains is the selection of the bandwidth. Whilst it is possible to develop weighted versions of the plug-in and cross validation criteria in Chapters 2–3, we follow the advice of Hazelton & Turlach (2009) that these unweighted selectors also perform well for weighted deconvolution density estimators.

Following Algorithm 15, we obtain the error sample variance $\mathbf{S}_U = \text{diag}(6705.8, 957.7)$ in utilising the air quality values at 7 p.m. and 9 p.m. as replicated measurements, and the regularisation parameter $\hat{\eta} = 0.00021$. Both density estimates are computed with the plug-in bandwidth $\hat{\mathbf{H}}_{\text{PI}} = [440.0, 79.2; 79.2, 66.9]$, so the observed differences in Figure 7.4(b)–(c), e.g., unimodal versus bimodal density estimates, are induced by the non-uniform weights.

### 7.3.3 *Manifold estimation*

Manifold estimation is, at the time of writing, an active area of research in the machine learning community, which is closely connected to various different subjects previously exposed here, such as support estimation (Section 6.1.2), ridge estimation (Section 6.3) and deconvolution problems.

The observed sample $\boldsymbol{W}_1, \ldots, \boldsymbol{W}_n$ is $D$-dimensional, but it is supposed to approximately lie on a $d$-dimensional manifold $M$ with $d$ smaller (sometimes, much smaller) than $D$. In the additive noise model it is assumed that $\boldsymbol{W}_i = \boldsymbol{X}_i + \boldsymbol{U}_i$, where the distribution $\mathbb{P}$ of $\boldsymbol{X}_1, \ldots, \boldsymbol{X}_n$ is supported on $M$ and the distribution of the noise variables $\boldsymbol{U}_1, \ldots, \boldsymbol{U}_n$ is considered to be fully known. Then, the goal is to estimate the manifold $M$, which is immediately recognized as a support estimation problem under measurement error.

For the case where the distribution $\mathbb{P}$ is absolutely continuous with respect to the $D$-dimensional Lebesgue measure (so that $d = D$), the first procedure to estimate $M$ consistently was proposed by Meister (2006). However, as noted above, in many circumstances the main focus is on the case $d \ll D$, which means that $\mathbb{P}$ is a singular distribution, since it is supported on a manifold whose dimension is smaller than that of the embedding space. This problem is studied in detail in Genovese et al. (2012), where further references and a literature review are also provided.

Genovese et al. (2012) also showed that the best possible convergence rate to estimate $M$, in terms of the Hausdorff distance, when the error distribution is $D$-variate normal, is of order $(\log n)^{-1}$. This very slow rate is the same as that of density deconvolution with normal measurement errors. A possible way to overcome this problem is suggested by the same authors in Genovese et al. (2014): instead of estimating $M$, change the goal to estimating the ridge $\mathcal{P}$ of the distribution of $\boldsymbol{W}_1, \ldots, \boldsymbol{W}_n$. Genovese et al. (2014) showed that if the noise level is not too high, then $\mathcal{P}$ is reasonably close to and retains all of the topological properties of $M$, i.e., $\mathcal{P}$ is nearly homotopic to $M$. Thus, $\mathcal{P}$ may be considered as a surrogate for $M$ even though it is a biased approximation; recall from Section 6.3 that it can be estimated more accurately than $M$.

## 7.4 Nearest neighbour estimation as variable kernel estimation

Kernel estimators with a fixed, global bandwidth can be considered as the most fundamental case for data smoothing. They are important in their own right, though their importance extends beyond their direct application to data analysis as the knowledge gained from them can be readily extended to other non-parametric data smoothers. To illustrate this, we investigate the intimate relationship between kernel and nearest neighbour estimators. The former are most useful for low dimensional analysis whereas the sparse nature of the latter ensure that they are highly useful for high dimensional analyses, so these two classes of estimators are complementary to each other.

The nearest neighbour estimator of a density function, as introduced by Loftsgaarden & Quesenberry (1965) and elaborated by Mack & Rosenblatt (1979), is

$$\hat{f}_{\rm NN}(\boldsymbol{x};k) = n^{-1}\delta_{(k)}(\boldsymbol{x})^{-d}\sum_{i=1}^{n} K((\boldsymbol{x}-\boldsymbol{X}_i)/\delta_{(k)}(\boldsymbol{x})) \tag{7.12}$$

where $\delta_{(k)}(\boldsymbol{x})$ as the $k$-th nearest neighbour distance to $\boldsymbol{x}$, i.e., $\delta_{(k)}(\boldsymbol{x})$ is the $k$-th order statistic of the (Euclidean) distances $\|\boldsymbol{x}-\boldsymbol{X}_1\|,\ldots,\|\boldsymbol{x}-\boldsymbol{X}_n\|$. Equation (7.12) is the most general form of a nearest neighbour density estimator.

The mathematical analysis of $\hat{f}_{\rm NN}$ is simplified if we recast it as a variable balloon kernel estimator. Generalising the balloon variable kernel estimator $\hat{f}_{\rm ball}$ from Section 4.1, to the density derivative case yields $\mathsf{D}^{\otimes r}\hat{f}_{\rm ball}(\boldsymbol{x};\mathbf{H}(\boldsymbol{x})) = n^{-1}|\mathbf{H}(\boldsymbol{x})|^{-1/2}(\mathbf{H}(\boldsymbol{x})^{-1/2})^{\otimes r}\sum_{i=1}^{n}\mathsf{D}^{\otimes r}K(\mathbf{H}(\boldsymbol{x})^{-1/2}(\boldsymbol{x}-\boldsymbol{X}_i))$. The connection between the nearest neighbour $\mathsf{D}^{\otimes r}\hat{f}_{\rm NN}$ and the variable balloon kernel $\mathsf{D}^{\otimes r}\hat{f}_{\rm ball}$ estimators appears when $\mathbf{H}(\boldsymbol{x}) = \delta_{(k)}(\boldsymbol{x})^2\mathbf{I}_d$ is substituted into the latter. This implies that the nearest neighbour estimator of the $r$-th derivative of $f$ follows as

$$\mathsf{D}^{\otimes r}\hat{f}_{\rm NN}(\boldsymbol{x};k) = n^{-1}\delta_{(k)}(\boldsymbol{x})^{-d-r}\sum_{i=1}^{n}\mathsf{D}^{\otimes r}K(\delta_{(k)}(\boldsymbol{x})^{-1}(\boldsymbol{x}-\boldsymbol{X}_i)). \tag{7.13}$$

It was established in Loftsgaarden & Quesenberry (1965) that the beta family kernels are computationally efficient for estimating $f$ and $\mathsf{D}f$. To see this, note that the nearest neighbour density estimator becomes

$$\hat{f}_{\rm NN}(\boldsymbol{x};k) = n^{-1}\sum_{i=1}^{n}\mathbf{1}\{\boldsymbol{X}_i \in B_d(\boldsymbol{x},\delta_{(k)}(\boldsymbol{x}))\} = k/[v_0 n\delta_{(k)}(\boldsymbol{x})^d] \tag{7.14}$$

when using the zeroth beta kernel $K(\boldsymbol{x};0) = v_0^{-1}\mathbf{1}\{\boldsymbol{x}\in B_d(\mathbf{0},1)\}$, which is the uniform kernel on the unit $d$-ball $B_d(\mathbf{0},1)$. The summation counts the number of data points which fall inside the ball $B_d(\boldsymbol{x},\delta_{(k)}(\boldsymbol{x}))$, which is equal to $k$ from the definition of $\delta_{(k)}(\boldsymbol{x})$ as the $k$-th nearest neighbour distance to $\boldsymbol{x}$. In Equation (7.14), the nearest neighbour density estimator at a point $\boldsymbol{x}$ reduces essentially to computing $\delta_{(k)}(\boldsymbol{x})$, and no explicit kernel function evaluations are required.

The nearest neighbour estimator for the density gradient is

$$\begin{aligned}\mathsf{D}\hat{f}_{\rm NN}(\boldsymbol{x};k) &= \hat{f}_{\rm NN}(\boldsymbol{x};k)\frac{d+2}{\delta_{(k)}(\boldsymbol{x})^2}\left[k^{-1}\sum_{i=1}^{n}\boldsymbol{X}_i\mathbf{1}\{\boldsymbol{X}_i\in B_d(\boldsymbol{x},\delta_{(k)}(\boldsymbol{x}))\}-\boldsymbol{x}\right]\\ &= \hat{f}_{\rm NN}(\boldsymbol{x};k)\frac{d+2}{\delta_{(k)}(\boldsymbol{x})^2}\left[k^{-1}\sum_{\boldsymbol{X}_i\in{\rm NN}_k(\boldsymbol{x})}\boldsymbol{X}_i-\boldsymbol{x}\right]\end{aligned} \tag{7.15}$$

where $\text{NN}_k(\boldsymbol{x}) = \{\boldsymbol{X}_i : \boldsymbol{X}_i \in B_d(\boldsymbol{x}, \delta_{(k)}(\boldsymbol{x}))\}$ is the set of the $k$ nearest neighbours to $\boldsymbol{x}$. This density gradient estimator uses the first beta kernel $K(\boldsymbol{x};1) = [(d+2)/(2v_0)](1-\boldsymbol{x}^T\boldsymbol{x})\mathbf{1}\{\boldsymbol{x} \in B_d(\mathbf{0},1)\}$, i.e., the Epanechnikov kernel, which has derivative $\mathsf{D}K(\boldsymbol{x};1) = -[(d+2)/v_0]\boldsymbol{x}\mathbf{1}\{\boldsymbol{x} \in B_d(\mathbf{0},1)\}$. Equation (7.15) is more complicated than Equation (7.14) as the former requires the specification of the set of nearest neighbours rather than only the nearest neighbour distance, though evaluating the kernel function at all estimation points $\boldsymbol{x}$ is still not required.

Since the nearest neighbour density estimator contains jump discontinuities due to the discreteness of the nearest neighbour distance $\delta_{(k)}$, it does not admit the pleasing smooth visualisations of the kernel counterparts. So we do not focus on nearest neighbour estimators for direct data density visualisations. Instead we focus on their suitability for clustering for higher dimensional data due to their sparse nature and reduced computational load. Recall that the target mean shift recurrence relation in Equation (6.3) is $\boldsymbol{x}_{j+1} = \boldsymbol{x}_j + \mathbf{A}\mathsf{D}f(\boldsymbol{x}_j)/f(\boldsymbol{x}_j)$. Replacing $\mathsf{D}f(\boldsymbol{x})/f(\boldsymbol{x})$ by its nearest neighbour estimator $\mathsf{D}\hat{f}(\boldsymbol{x};k)/\hat{f}(\boldsymbol{x};k)$, and setting $\mathbf{A} = (d+2)^{-1}\delta_{(k)}(\boldsymbol{x})^2\mathbf{I}_d$, we have that nearest neighbour mean shift recurrence is

$$\boldsymbol{x}_{j+1} = \boldsymbol{x}_j + \boldsymbol{\eta}_{\text{NN}}(\boldsymbol{x}_j;k) = k^{-1} \sum_{\boldsymbol{X}_i \in \text{NN}_k(\boldsymbol{x}_j)} \boldsymbol{X}_i \tag{7.16}$$

where $\boldsymbol{\eta}_{\text{NN}}(\boldsymbol{x};k) = k^{-1}\sum_{\boldsymbol{X}_i \in \text{NN}_k(\boldsymbol{x}_j)} \boldsymbol{X}_i - \boldsymbol{x}$ is the nearest neighbour mean shift. In comparison to the kernel mean shift in Equation (6.3), we only have to consider sample means of the nearest neighbours of the current iterate $\boldsymbol{x}_j$ rather than the computationally intensive evaluations of the kernel function centred at $\boldsymbol{x}_j$. The recurrence relation in Equation (7.16) was introduced by Fukunaga & Hostetler (1975), beginning from a different starting point. Furthermore, the convergence of the sequence $\{\boldsymbol{x}_0, \boldsymbol{x}_1, \dots\}$ to a local mode has been asserted for the kernel version of Equation (7.16) was established by Comaniciu & Meer (2002, Theorem 1) for a fixed bandwidth $\mathbf{H}$. Their proofs remain valid when $\mathbf{H}$ is replaced with the nearest neighbour distance $\delta_{(k)}$, which decreases as the iteration number increases.

Image segmentation is a procedure of assigning a label to each pixel in an image so that similar pixels are grouped together. Within each group/segment, all pixels are assigned the same colour so this reduces the complexity of the colour information stored in the image. Comaniciu & Meer (2002) recognised its mathematical equivalence to statistical clustering, and they popularised mean shift clustering for image segmentation.

The colour information in an 8-bit RGB digital colour image is commonly represented as a triplet of red, green and blue colour levels $(R, G, B)$ of integral values in the interval $[0, 255]$. Since an image is a 2-dimensional array of pixels, let $(x, y)$ be the row and column index of a pixel. The spatial and colour information of a pixel can thus be represented as a 5-dimensional vector $(x, y, R, G, B)$. As the proximity of values in the RGB colour space do not correspond well to human perceptions of colour closeness, a common choice to replace the RGB colours is the $L^*u^*v^*$ colour space (Pratt, 2001, Equations 3.5-1a, 3.5-8a–f) for these image segmentation tasks. We omit these equations as they are difficult to express concisely due to the many but slightly different colour space parametrisations: in any case, it is recommended to employ implementations provided by expert colour analysts.

**Example 7.5** In Figure 7.5(a) is a $481 \times 321$ RGB image of two elephants in a field: it is Test Image #296059 from the Berkeley Segmentation Dataset (Martin et al., 2001). In addition to providing freely available images for image analysis, the Berkeley Segmentation Dataset also supplies manual image segmentations against which the computational algorithms can be compared. The manually segmented image by User #1130 of the elephants is shown in Figure 7.5(b) which distinguishes the sky, the vegetation and the elephants: for the elephants, this user has further segmented the two different animals as well their eyes and tusks. For an automatic segmentation, mutually separating the sky, the vegetation and the elephants is fairly straightforward as they are widely differing colours. What is more difficult is to segment (a) the two elephants from each other as they form adjacent regions with subtly different colours and their (b) eyes and tusks from the skin as they are smaller regions. The nearest neighbour mean shift segmentation with $k = 780$ nearest neighbours is displayed in Figure 7.5(c). Whilst there are 28491 unique colours in the original image, the mean shift segmented image consists of only 109 colours, indicating an important reduction in the image complexity. From this segmented image, the edges of the segmented regions can be found. These agree mostly with the manual edges in the separation of the sky, vegetation and elephants, even though the former are not as smooth or as connected as the latter. The manual segmentation more cleanly separates the elephant eyes and tusks, as well as the two different animals. We do not claim that this is an optimal edge detection for the segmented image: we merely wish to illustrate the potential gains offered by nearest neighbour mean shift clustering even with our sub-optimal, ad hoc edge detection. □

Comaniciu & Meer (2002) initially used the kernel mean shift for image segmentation. More recently Duong et al. (2016) adapted this to the nearest neighbour mean shift as it is suitable for clustering higher dimensional data.

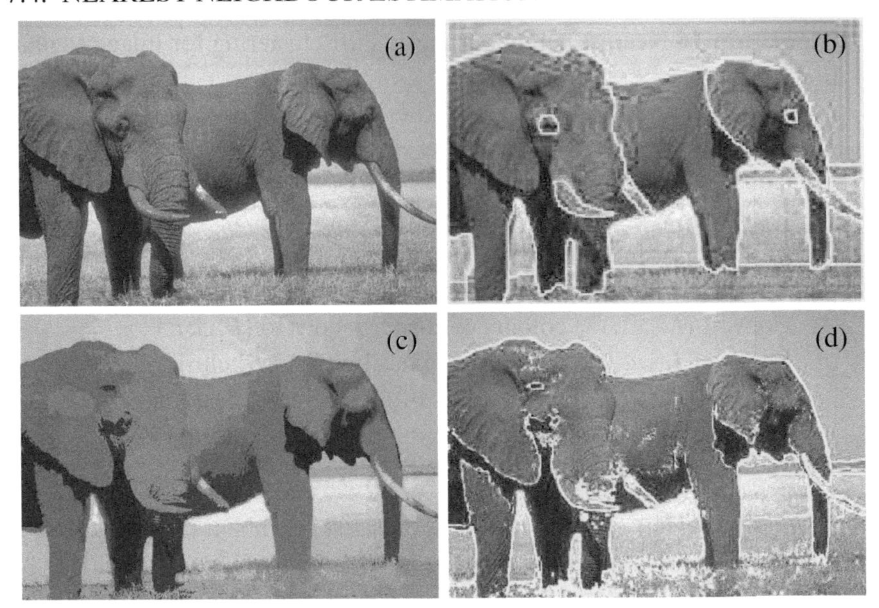


Figure 7.5 *Segmentations of the elephants image. 1108 (a) Image is* $481 \times 321$ *pixels with 8-bit RGB colours. (b) Manual segmentation with edge detection by User #1130. (c) Nearest neighbour mean shift segmentation with* $k = 780$ *nearest neighbours. (d) Nearest neighbour mean shift edge detection.*

This is detailed in Algorithm 16. The inputs are an RGB image and the tuning parameters for the nearest neighbour mean shift: the number of nearest neighbours $k$, the tolerance $\varepsilon_1$ under which subsequent iterations in the mean shift update are considered convergent, the maximum number of iterations $j_{\max}$, and the tolerance $\varepsilon_2$ under which two cluster centres are considered to form a single cluster.

Lines 1-2 convert the RGB colours in the $n \times m$ RGB image to the $L^*u^*v^*$ colour space and, with the spatial indices $(x, y)$, initialise the concatenated $(x, y, L^*, u^*, v^*)$ data matrix of dimensions $nm \times 5$. The nearest neighbour mean shift clustering itself is carried out in Line 3. This is carried out by suitably modifying the kernel mean shift clustering in Algorithm 10 by replacing the kernel mean shift $\boldsymbol{\eta}$ with its nearest neighbour counterpart $\boldsymbol{\eta}_{\mathrm{NN}}$, and the bandwidth with the number of nearest neighbours $k$. The other tuning parameters play the same role. The outputs are clusters with cluster labels. In Line 4, the $L^*u^*v^*$ colours of the cluster modes are converted to RGB colours. In Lines 5-6, the $L^*u^*v^*$ colour for each pixel is replaced by the RGB colours of its cluster mode and then converted to an $n \times m$ RGB segmented image. Line 7 is an optional step of edge detection in the segmented image.

**Algorithm 16** Nearest neighbour mean shift clustering for image segmentation

**Input:** RGB image, $k, \varepsilon_1, \varepsilon_2, j_{\max}$
**Output:** Segmented RGB image
1: Convert $(R,G,B)$ to $(L^*,u^*,v^*)$ colours
2: Initialise $(x,y,L^*,u^*,v^*)$ data matrix
3: Carry out nearest neighbour mean shift with $k, \varepsilon_1, \varepsilon_2, j_{\max}$ with modified Algorithm 10 (kernel mean shift clustering)
4: Convert $(L^*,u^*,v^*)$ colours of cluster modes to $(R,G,B)$
5: Replace $(L^*,u^*,v^*)$ colours of pixels by $(R,G,B)$ cluster mode
6: Convert $(x,y,R,G,B)$ data matrix to RGB segmented image
7: (Optional) Carry out edge detection of RGB segmented image

For the selection of the number of nearest neighbours, most authors have focused on cross validation approaches (Li, 1984; Biau et al., 2011; Kung et al., 2012) in the direct form where each of the $n$ leave-one-out estimators is explicitly calculated for each $k$ taken from a set of candidate values, and then the optimal $k$ is taken to be the value that yields the minimal cross validation error. A procedure which avoids this explicit enumeration, like for the cross validation selectors for kernel estimation, remains unknown. One of the main reasons for this is the lack of an explicit closed form expression of the AMISE of $\mathsf{D}^{\otimes r}\hat{f}_{\rm NN}$ or $\mathsf{D}^{\otimes r}\hat{f}_{\rm ball}$ since the latter are not guaranteed to be (square) integrable.

An alternative error measure is to start with the AMISE of the fixed bandwidth kernel estimator $\mathsf{D}^{\otimes r}\hat{f}(\cdot;\mathbf{H})$ from Equation (5.16) and to replace $\mathbf{H}$ with $\mathbf{H}(\boldsymbol{x}) = \delta_{(k)}(\boldsymbol{x})^2\mathbf{I}_d$, see Duong et al. (2016). This results in a random quantity, so we compute its expectation to give an AMISE-like quantity for the nearest neighbour estimator

$$\mathrm{A}\{\mathsf{D}^{\otimes r}\hat{f}_{\rm NN}(\boldsymbol{x};k)\} = \mathbb{E}\{\mathrm{AMISE}\{\mathsf{D}^{\otimes r}\hat{f}(\cdot;\delta_{(k)}(\boldsymbol{x})^2\mathbf{I}_d)\}\}.$$

For this expectation to be convergent, suppose that (B1)–(B2) in Conditions B hold, and further suppose

**(B3')** $k = k_n$ is a sequence of the number of nearest neighbours such that $k \to \infty$, $k/n \to$ const as $n \to \infty$.

Thus we have

$$\begin{aligned}\mathrm{A}\{&\mathsf{D}^{\otimes r}\hat{f}_{\rm NN}(\boldsymbol{x};k)\}\\ &= \mathrm{tr}(\mathbf{R}(\mathsf{D}^{\otimes r}K))[v_0 f(\boldsymbol{x})]^{(d+2r)/d} n^{2r/d} k^{-(d+2r)/d}\\ &\quad + (-1)^r \tfrac{1}{4} m_2(K)^2 \boldsymbol{\psi}_{2r+4}^\top (\mathrm{vec}\,\mathbf{I}_d)^{\otimes(r+2)} [v_0 f(\boldsymbol{x})]^{-4/d} n^{-4/d} k^{4/d} \qquad (7.17)\end{aligned}$$

where $v_0 = \pi^{d/2}\Gamma((d+2)/d)$ is the hyper-volume of the unit $d$-ball. The first term is the integrated variance and the second term is the integrated squared bias of $\mathsf{D}^{\otimes r}\hat{f}_{\rm NN}$; the role of $k$ in a bias-variance trade-off is established in Equation (7.17), analogous to that for kernel estimators in Equation (5.16).

The error measure $\mathrm{A}\{\mathsf{D}^{\otimes r}\hat{f}_{\rm NN}(\boldsymbol{x};k)\}$ still depends on $\boldsymbol{x}$ so it remains a local measure. However its integral is not necessarily finite and so direct integration does not lead to a global measure. On the other hand, integrating after optimising does lead to a feasible selector, i.e., $k_{\mathrm{A},r} = \int_{\mathbb{R}^d}\{\mathrm{argmin}_{k>0}\,\mathrm{A}\{\mathsf{D}^{\otimes r}\hat{f}(\boldsymbol{x};k)\}\}\,d\boldsymbol{x}$ is finite. Moreover, its explicit formula is

$$k_{\mathrm{A},r} = v_0\left[\frac{(d+2r)\,\mathrm{tr}(\mathbf{R}(\mathsf{D}^{\otimes r}K))}{(-1)^r m_2(K)^2\boldsymbol{\psi}_{2r+4}^\top(\mathrm{vec}\,\mathbf{I}_d)^{\otimes(r+2)}}\right]^{d/(d+2r+4)} n^{(2r+4)/(d+2r+4)}$$

which serves as an optimal number of nearest neighbours.

For data-based selection of $k_{\mathrm{A},r}$, the estimation of $\boldsymbol{\psi}_{2r+4}$ and higher order functionals is required. As these nearest neighbours estimators are as yet unknown, we employ instead the normal scale estimator $\hat{\boldsymbol{\psi}}_{\mathrm{NS},2r+4}$ to give a normal scale selector for $k_{\mathrm{A},r}$ as

$$\hat{k}_{\mathrm{NS},r} = v_0\left[\frac{4|\mathbf{S}|^{1/2}\prod_{j=1}^r(d+2j)}{v_{r+2}(\mathbf{S}^{-1})}\right]^{d/(d+2r+4)} n^{(2r+4)/(d+2r+4)}. \tag{7.18}$$

When $\mathbf{S} = \mathbf{I}_d$ this reduces to $v_0[4/(d+2r+2)]^{d/(d+2r+4)}n^{(2r+4)/(d+2r+4)}$ as $v_{r+2}(\mathbf{I}_d) = \prod_{j=1}^{r+1}(d+2j)$, which closely resembles the normal scale bandwidth selector $\hat{\mathbf{H}}_{\mathrm{NS},r} = [4/(d+2r+2)]^{2/(d+2r+4)}n^{-2/(d+2r+4)}\mathbf{I}_d$ for the kernel estimator $\mathsf{D}^{\otimes r}\hat{f}$ in Section 5.6. The number of nearest neighbours grows at rate $n^{(2r+4)/(d+2r+4)}$ as $n\to\infty$, in comparison to the bandwidth which decreases to 0 at rate $n^{-2/(d+2r+4)}$.

In terms of optimal selection of the number of nearest neighbours, we are at an embryonic stage since we have only exhibited a normal scale selector. For more sophisticated data-based selectors of $k_{\mathrm{A},r}$, the required estimators of $\boldsymbol{\psi}_{2r+4}$ and higher order functionals by nearest neighbour methods remain an open problem.

So we are restricted to selecting the number of nearest neighbours for the mean shift as $\hat{k}_{\mathrm{NS},1} = v_0\left[4|\mathbf{S}|^{1/2}d(d+2)/v_3(\mathbf{S}^{-1})\right]^{d/(d+6)}n^{6/(d+6)}$. Recall that bandwidth selection based on the density gradient was demonstrated to be more suitable than that based on the density itself for kernel mean shift clustering in Section 6.2. For the elephants image in Figure 7.5, this normal scale selector is $\hat{k}_{\mathrm{NS},1} = 780$ (for a pre-scaled $(x,y,L^*,u^*,v^*)$ matrix) for the image segmentation.

To conclude, for a standard normal density, the AMISE$\{\hat{f}_{\rm NN}(\cdot;k)\}$ is convergent and was computed by Terrell & Scott (1992). These authors also computed the efficiency of a fixed bandwidth density estimator and the nearest neighbour density estimator, as a function of the data dimension $d$, to be $\mathrm{Eff}(\hat{f},\hat{f}_{\rm NN}) = \min_{h>0}\mathrm{AMISE}\{\hat{f}(\cdot;h^2\mathbf{I}_d)\}/\min_{k>0}\mathrm{AMISE}\{\hat{f}_{\rm NN}(\cdot;k)\} = 2^{2d/(d+4)}[(d-2)/d]^{d(d+2)/(2d+8)}[(d^2-4)/(d^2-6d+16)]^{d/(d+4)}\mathbf{1}\{d>2\}$. To achieve the same AMISE of the kernel estimator $\hat{f}$ with a sample size $n$, the nearest neighbour estimator $\hat{f}_{\rm NN}$ requires a sample size of $n/\mathrm{Eff}(\hat{f},\hat{f}_{\rm NN})$. The efficiency curve monotonically increases from $d = 2$ to the maximum $\mathrm{Eff}(\hat{f},\hat{f}_{\rm NN})$ at $d = 15.27$, and afterwards decreases monotonically to its asymptote value of 1.47 as $d \to \infty$. An augmented version of Terrell & Scott (1992, Table 3) is reproduced in Table 7.2. This provides some mathematical justification for the heuristic observation that nearest neighbour estimators are more efficient than kernel density estimators for moderate to high dimensions ($d > 4$).

| $d$ | 1 | 2 | 3 | 4 | 5 | 6 | 10 | 15 | 20 | 100 |
|---|---|---|---|---|---|---|---|---|---|---|
| $\mathrm{Eff}(\hat{f},\hat{f}_{\rm NN})$ | 0 | 0 | 0.48 | 0.87 | 1.15 | 1.32 | 1.52 | 1.54 | 1.54 | 1.49 |

Table 7.2 *Efficiency of a nearest neighbour ($\hat{f}_{\rm NN}$) and a kernel density ($\hat{f}$) estimator as a function of dimension d.*

## 7.5 Further mathematical analysis

### 7.5.1 *Squared error analysis for deconvolution kernel density estimators*

We begin by showing that the expected value of the deconvolution kernel density estimator based on the contaminated data $\boldsymbol{W}_1,\ldots,\boldsymbol{W}_n$ is the same as for the usual kernel density estimator based on the unobservable, error-free data $\boldsymbol{X}_1,\ldots,\boldsymbol{X}_n$:

$$\mathbb{E}\{\hat{f}_{\rm dc}(\boldsymbol{x};\mathbf{H})\} = \mathbb{E}\{K_{\mathbf{H}}^{\boldsymbol{U}}(\boldsymbol{x}-\boldsymbol{W};\mathbf{H})\} = \mathbb{E}\{K_{\mathbf{H}}(\boldsymbol{x}-\boldsymbol{X})\} = K_{\mathbf{H}} * f_{\boldsymbol{X}}(\boldsymbol{x}).$$

First, taking into account that $\varphi_{K_{\mathbf{H}}*f_{\boldsymbol{X}}}(\boldsymbol{t}) = \varphi_K(\mathbf{H}^{1/2}\boldsymbol{t})\varphi_{\boldsymbol{X}}(\boldsymbol{t})$, it follows by the inversion formula that

$$K_{\mathbf{H}} * f_{\boldsymbol{X}}(\boldsymbol{x}) = (2\pi)^{-d}\int_{\mathbb{R}^d} e^{-i\boldsymbol{t}^\top\boldsymbol{x}}\varphi_K(\mathbf{H}^{1/2}\boldsymbol{t})\varphi_{\boldsymbol{X}}(\boldsymbol{t})d\boldsymbol{t}. \qquad (7.19)$$

Analogously, note that $\mathbb{E}\{K_{\mathbf{H}}^{\boldsymbol{U}}(\boldsymbol{x}-\boldsymbol{W};\mathbf{H})\} = \{K_{\mathbf{H}}^{\boldsymbol{U}}(\cdot;\mathbf{H}) * f_{\boldsymbol{W}}\}(\boldsymbol{x})$ and that, by definition, $\varphi_{K^{\boldsymbol{U}}(\cdot;\mathbf{H})}(\boldsymbol{t}) = \varphi_K(\boldsymbol{t})/\varphi_{\boldsymbol{U}}(\mathbf{H}^{-1/2}\boldsymbol{t})$, so that

$$\varphi_{\{K_{\mathbf{H}}^{\boldsymbol{U}}(\cdot;\mathbf{H})*f_{\boldsymbol{W}}\}}(\boldsymbol{t}) = \varphi_{K^{\boldsymbol{U}}(\cdot;\mathbf{H})}(\mathbf{H}^{1/2}\boldsymbol{t})\varphi_{\boldsymbol{W}}(\boldsymbol{t}) = \varphi_K(\mathbf{H}^{1/2}\boldsymbol{t})\varphi_{\boldsymbol{W}}(\boldsymbol{t})/\varphi_{\boldsymbol{U}}(\boldsymbol{t}).$$

Recalling that $\varphi_W = \varphi_X \varphi_U$, then in conjunction with Equation (7.19), the result follows immediately.

The integrated variance of $\hat{f}_X(\boldsymbol{X};\mathbf{H})$ is $n^{-1} \int_{\mathbb{R}^d} \mathbb{E}\{K_{\mathbf{H}}^{\boldsymbol{U}}(\boldsymbol{x}-\boldsymbol{W};\mathbf{H})^2\}\, d\boldsymbol{x} - n^{-1}\{\int_{\mathbb{R}^d} \mathbb{E}\{K_{\mathbf{H}}^{\boldsymbol{U}}(\boldsymbol{x}-\boldsymbol{W};\mathbf{H})\}\, d\boldsymbol{x}\}^2$. From the calculations in the previous paragraph and the bias expansions in the error-free case, we can establish that the first term, provided that $\int_{\mathbb{R}^d} K^{\boldsymbol{U}}(\boldsymbol{z};\mathbf{H})^2 d\boldsymbol{z}$ is finite, is dominant over the second term which is of order $n^{-1}$. Also, the usual change of variables $\boldsymbol{z} = \mathbf{H}^{-1/2}(\boldsymbol{x}-\boldsymbol{w})$ allows us to express the first term as

$$n^{-1} \int_{\mathbb{R}^d} \mathbb{E}\{K_{\mathbf{H}}^{\boldsymbol{U}}(\boldsymbol{x}-\boldsymbol{W};\mathbf{H})^2\}\, d\boldsymbol{x} = n^{-1}|\mathbf{H}|^{-1/2} \int_{\mathbb{R}^d} K^{\boldsymbol{U}}(\boldsymbol{z};\mathbf{H})^2\, d\boldsymbol{z}$$

thus completing the proof of the asymptotic expansion of the MISE of $\hat{f}_{\text{dc}}(\boldsymbol{x};\mathbf{H})$.

### 7.5.2 *Optimal selection of the number of nearest neighbours*

The properties of the $k$-th nearest neighbour distance $\delta_{(k)}(\boldsymbol{x}), k = 1,\dots,n$ as the $k$-th order statistic are difficult to establish explicitly. The usual approach is to analyse the order statistic $U_{(k)}$ for the simpler case where $U_1,\dots,U_n$ is a random sample drawn from a common standard univariate uniform distribution Unif$[0,1]$. The connection to the general $k$-th order distance statistic $\delta_{(k)}(\boldsymbol{x})$ is provided by the random variable $R \equiv R(\boldsymbol{x}) = \mathbf{1}\{\boldsymbol{X} \in \partial B_d(\boldsymbol{x},r)\}$ where $\boldsymbol{X} \sim f$ and $B_d(\boldsymbol{x},r) = \{\boldsymbol{y} : \|\boldsymbol{x}-\boldsymbol{y}\| < r\}$ is the $d$-dimensional ball centred at $\boldsymbol{x}$ with radius $r \geq 0$. Since the distribution of $R$ is $F_R(r;\boldsymbol{x}) = \mathbb{P}(\boldsymbol{X} \in B_d(\boldsymbol{x},r))$ is a well-defined distribution function, following Mack & Rosenblatt (1979); Hall (1983), we have $\delta_{(k)}(\boldsymbol{x}) = F_R^{-1}(U_{(k)};\boldsymbol{x})$. This equation allows us to derive immediately the relationship between the distributions of $\delta(k)(\boldsymbol{x})$ and $U_{(k)}$,

$$F_{\delta(k)}(r;\boldsymbol{x}) = \mathbb{P}(F_R^{-1}(U_{(k)};\boldsymbol{x}) \leq r) = \mathbb{P}(U_{(k)} \leq F_R(r;\boldsymbol{x})) = F_{U_{(k)}}(F_R(r;\boldsymbol{x})).$$

From standard results, e.g., Johnson et al. (1994, p. 280), the $k$-th order uniform statistic $U_{(k)}$ is approximately Beta$(k, n-k+1)$ distributed. The density of $\delta_{(k)}$ is thus a suitably transformed Beta density

$$f_{\delta(k)}(r;\boldsymbol{x}) = \frac{n!}{(j-1)!(n-j)!} F_R(r;\boldsymbol{x})^{k-1}\{1-F_R(r;\boldsymbol{x})\}^{n-k} f_R(r;\boldsymbol{x})\{1+o(1)\}$$

where $f_R = F_R'$ is the derivative of $F_R$. Whilst this closed form for $f_{\delta(k)}$ is aesthetically pleasing, its main role in selecting $k$ is via the asymptotic approximations of the moments of $\delta_{(k)}(\boldsymbol{x})^\alpha$, that is, $\mathbb{E}\{\delta_{(k)}(\boldsymbol{x})^\alpha\} =$

$[k/(nv_0 f(\boldsymbol{x}))]^{\alpha/d}\{1+o(1)\}$ for any $\alpha$ and $f(\boldsymbol{x})>0$, as established by Hall (1983, Equation (2.2))

Substituting $\mathbf{H}(\boldsymbol{x}) = \delta_{(k)}(\boldsymbol{x})^2\mathbf{I}_d$ for $\mathbf{H}$ in AMISE$\{\mathsf{D}^{\otimes r}\hat{f}(\cdot;\mathbf{H})\}$, we obtain

$$\begin{aligned}\text{AMISE}\{\mathsf{D}^{\otimes r}\hat{f}_{\text{NN}}(\boldsymbol{x};k)\} &= n^{-1}\operatorname{tr}(\mathbf{R}(\mathsf{D}^{\otimes r}K))\delta_{(k)}(\boldsymbol{x})^{-d-2r}\\ &\quad+\tfrac{1}{4}(-1)^r m_2(K)^2\delta_{(k)}(\boldsymbol{x})^4\boldsymbol{\psi}_{2r+4}^\top(\operatorname{vec}\mathbf{I}_d)^{\otimes(r+2)}\end{aligned}$$

following similar reasoning to Chacón et al. (2011, Lemma 3(ii)) and Schott (2003, Theorem 1(iv)). Taking its expected value yields our proposed optimality criterion

$$\begin{aligned}\text{A}\{\mathsf{D}^{\otimes r}\hat{f}_{\text{NN}}(\boldsymbol{x};k)\} &= \operatorname{tr}(\mathbf{R}(\mathsf{D}^{\otimes r}K))[v_0 f(\boldsymbol{x})]^{(d+2r)/d}n^{2r/d}k^{-(d+2r)/d}\\ &\quad+(-1)^r\tfrac{1}{4}m_2(K)^2\boldsymbol{\psi}_{2r+4}^\top(\operatorname{vec}\mathbf{I}_d^{\otimes(r+2)})[v_0 f(\boldsymbol{x})]^{-4/d}n^{-4/d}k^{4/d}.\end{aligned}$$

Solving the differential equation $(\partial/\partial k)\text{A}\{\mathsf{D}^{\otimes r}\hat{f}_{\text{ball}}(\boldsymbol{x};k)\}=0$, and noting that the resulting exponent of $v_0 f(\boldsymbol{x})$ is $(d+2r)/d+4/d = 1+(2r+4)/d$ which is exactly the same as that of $k$, the solution is $k^*_{\text{A},r}(\boldsymbol{x}) = C_r v_0 f(\boldsymbol{x})n^{(2r+4)/(d+2r+4)}$ where

$$C_r = \left[\frac{(d+2r)\operatorname{tr}(\mathbf{R}(\mathsf{D}^{\otimes r}K))}{(-1)^r m_2(K)^2\boldsymbol{\psi}_{2r+4}^T(\operatorname{vec}\mathbf{I}_d)^{\otimes(r+2)}}\right]^{d/(d+2r+4)}$$

and thus $k_{\text{A},r} = \int_{\mathbb{R}^d} k^*_{\text{A},r}(\boldsymbol{x})\,d\boldsymbol{x}$ follows immediately.

For the normal scale selector, let $f=\phi_{\boldsymbol{\Sigma}}(\cdot-\boldsymbol{\mu})$ and $K=\phi$, then $m_2(\phi)=1$ and $\operatorname{tr}(\mathbf{R}(\mathsf{D}^{\otimes r}\phi)) = 2^{-d-r}\pi^{-/d2}\nu_r(\mathbf{I}_d) = 2^{-r}(4\pi)^{-d/2}\prod_{j=0}^{r-1}(d+2j)$ using Chacón et al. (2011, Lemma 3, Corollary 7), and $\boldsymbol{\psi}_{\text{NS},2r+4}^\top(\operatorname{vec}\mathbf{I}_d)^{\otimes(r+2)} = (-1)^{r+2}2^{-r-2}(4\pi)^{-d/2}|\boldsymbol{\Sigma}|^{-1/2}\nu_{r+2}(\boldsymbol{\Sigma}^{-1})$ in conjunction with Chacón & Duong (2010, Equation (7)). See also the calculations in Section 5.9.

Chapter 8

# Computational algorithms

The data analysis procedures in the preceding chapters have been presented as concise mathematical equations or pseudo-code algorithms, which facilitate the comprehension of the underlying statistical reasoning. A direct translation of these equations or pseudo-codes into a computer programming language normally yields efficient software, but this is not always the case. In this chapter, we detail the computational algorithms which are markedly different and less statistically intuitive than their concise descriptions, but which lead to important gains in computational efficiency, in terms of execution time and memory management.

Section 8.1 outlines the R package and associated R scripts which implement the algorithms and generate the figures presented in this monograph. Section 8.2 elaborates binned estimation as a method of computing kernel estimators based on Fast Fourier Transform methods. Sections 8.3–8.4 explore recursive algorithms for the exact computation of the derivatives and functionals of the multivariate normal density. Section 8.5 considers numerical optimisation for matrix-valued inputs.

## 8.1 R implementation

All the data analysis for the experimental and simulated data in this monograph has been carried out in the R statistical programming environment (R Core Team, 2017), and in particular with the `ks` package (Duong, 2007). R has established itself as one of the leading platforms for data analysis due to the breadth and depth of its coverage of the statistical algorithms from its decentralised open source community of contributors. It is not our intent here to provide the complete R script that was utilised to create the figures in order that the reader can reproduce them. For this, see the web page `http://mvstat.net/mvksa`. Rather, we present some of the high level commands (from `ks` version $\geq$ 1.11.0) to indicate a typical user interface.

For kernel density estimation, kde computes $\hat{f}$ in Equation (2.2). The command

```
> fhat.tempb <- kde(tempb)
```

computes a density estimate with a default plug-in bandwidth $\hat{\mathbf{H}}_{\text{PI}}$, for tempb the daily temperature data ($21908 \times 2$ matrix). Since tempb is a large data set, binned approximations from Section 8.2 are invoked in the density estimate kde and the bandwidth Hpi. The contour plot in Figure 2.4(a) and the perspective plot in (b) are produced by

```
> plot(fhat.tempb, display="filled.contour")
> plot(fhat.tempb, display="persp")
```

All the plotting functions in the ks package are overloaded by the same function name plot, but since R is an object-oriented programming language, the same plot command automatically calls the appropriate plotting method.

In comparison to tempb, the grevillea data set is a smaller $222 \times 2$ matrix so the exact calculations can be used for the density estimation and bandwidth selection. To use a bandwidth other than the default Hpi in kde, it needs to be called explicitly, e.g., for the SCV bandwidth

```
> fhat.scv <- kde(x=grevillea, H=Hscv(grevillea))
```

Figure 2.11(a) is created using Hns, (b) Hnm, (c) Hucv, (d) Hbcv, (e) Hpi, (f) Hscv, as listed in Table 8.1.

| **Method** | **Notation** | **ks function** | **Text reference** |
|---|---|---|---|
| Normal scale | $\hat{\mathbf{H}}_{\text{NS}}$ | Hns | Equation (3.2) |
| Normal mixture | $\hat{\mathbf{H}}_{\text{NM}}$ | Hnm | Algorithm 1 |
| Unbiased cross validation | $\hat{\mathbf{H}}_{\text{UCV}}$ | Hucv | Algorithm 2 |
| Biased cross validation | $\hat{\mathbf{H}}_{\text{BCV}}$ | Hbcv | Algorithm 3 |
| Plug-in | $\hat{\mathbf{H}}_{\text{PI}}$ | Hpi | Algorithm 4 |
| Smoothed cross validation | $\hat{\mathbf{H}}_{\text{SCV}}$ | Hscv | Algorithm 5 |

Table 8.1 *R functions for bandwidth selection for density estimation in* ks *package, along with their references in this text.*

The selectors employed in Figures 2.11(a)–(f) are class $\mathcal{F}$ unconstrained matrices. They are also available as class $\mathcal{D}$ diagonal matrices, as denoted by the .diag suffix, e.g., Hpi.diag. Figure 2.8(b) shows Hpi and (c) shows Hpi.diag.

For variable density estimation in Section 4.1, the balloon estimator $\hat{f}_{\text{ball}}$ is kde.balloon and the sample point one $\hat{f}_{\text{SP}}$ is kde.SP. The logarithm transformation density estimator $\hat{f}_{\text{trans}}$ in Equation (4.3) is implemented as an extra parameter in kde, namely, kde(,positive=TRUE).

For the boundary kernel estimators in Section 4.3, the basic command is kde.boundary, and the beta boundary estimator $\hat{f}_{\text{beta}}$ is invoked by setting boundary.kernel="beta" and the linear boundary one $\hat{f}_{\text{LB}}$ by boundary.kernel="linear".

The full worldbank data is a $218 \times 7$ matrix, with wb$i$ being $218 \times 2$ matrices comprising the appropriate columns from worldbank. The balloon and sample point variable density estimators in Figures 4.1(d) and (f) are called by

```
> fhat.wb1.ball <- kde.balloon(x=wb1)
> fhat.wb1.sp <- kde.sp(x=wb1)
```

and the transformation density estimator in Figures 4.2(b) and (c) by

```
> fhat.wb2.trans <- kde(x=wb2, adj.positive=c(0,0),
+     positive=TRUE)
```

and the beta and linear boundary density estimators in Figures 4.3(d) and (f) by

```
> xmin <- c(0,0); xmax <- c(100,100)
> fhat.wb3.beta <- kde.boundary(x=wb3, xmin=xmin,
+     xmax=xmax, boundary.kernel="beta")
> fhat.wb3.LB <- kde.boundary(x=wb3, xmin=xmin, xmax=xmax,
+     boundary.kernel="linear")
```

For kernel density derivative estimation, kdde computes $\mathsf{D}^{\otimes r}\hat{f}$ in Equation (5.1). The command

```
> fhat1.tempb <- kdde(tempb, deriv.order=1)
```

computes a density gradient estimate $\mathsf{D}\hat{f}$ with a plug-in bandwidth $\hat{\mathbf{H}}_{\text{PI},1}$. Figures 5.1(a)–(b) are the contour plots for each partial derivative. The quiver plot in (c) is produced by

```
> plot(fhat1.tempb, display="quiver")
```

The commands

```
> fhat2.tempb <- kdde(tempb, deriv.order=2)
> fhat2.tempb.curv <- kcurv(fhat2.tempb)
```

compute a density second derivative estimate $\mathsf{D}^{\otimes 2}\hat{f}$ with a plug-in bandwidth $\hat{\mathbf{H}}_{\text{PI},2}$ and its summary curvature $\hat{s}$. Figures 5.2(a)–(d) show the contour plots for each partial derivative and the summary curvature.

The available bandwidth selectors for density derivative estimation are: normal scale $\hat{\mathbf{H}}_{\text{NS},r}$ (Equation (5.18)), unbiased cross validation $\hat{\mathbf{H}}_{\text{UCV},r}$ (Algorithm 7), plug-in $\hat{\mathbf{H}}_{\text{PI},r}$ (Algorithm 8), and smoothed cross validation $\hat{\mathbf{H}}_{\text{SCV},r}$ (Algorithm 9), and the functions to compute them have the same name, with a `deriv.order` option to set the derivative order $r$, e.g., $\hat{\mathbf{H}}_{\text{PI},1}$ is `Hpi(,deriv.order=1)`. For the *Grevillea* data, Figure 5.3(a) is generated by `Hns`, (b) by `Hucv`, (c) by `Hpi`, (d) by `Hscv` for the gradient, and similarly for Figures 5.6(a)–(d) for the curvature. To use a bandwidth other than `Hpi` in `kdde`, it should be called explicitly,

```
> H <- Hscv(grevillea, deriv.order=1)
> fhat1.tempb.scv <- kdde(x=grevillea, deriv.order=1, H=H)
```

For level set estimation, the `plot` method displays them from a density estimate. The modal region $\hat{\mathcal{L}}(\hat{f}_{0.5})$ of the daily temperature data in Figure 6.1(a) is

```
> plot(fhat.tempb, cont=50)
```

The density support estimate $\hat{\mathcal{L}}(\hat{f}_{0.0005})$ is

```
> plot(fhat.tempb, cont=99.95)
```

and its convex hull can be obtained by applying the native R command `chull`.

For mean shift clustering, `kms` computes the estimated cluster labels $\hat{\gamma}$ in Algorithm 10, where the default tuning parameter values are as described in Section 6.2.2 and the default plug-in bandwidth $\hat{\mathbf{H}}_{\text{PI},1}$ as given by `Hpi(,deriv.order=1)`. The scatter plot with the cluster labels for the hematopoietic stem cell data `hsct6` (the data matrix for the mouse subject #12 of size $6236 \times 3$) in Figure 6.8 is produced from

```
> ms.hsct <- kms(hsct12)
```

For density ridge estimation, `kdr` computes $\hat{\mathcal{P}}$ in Algorithm 11, with the default values of the tuning parameters, and plug-in bandwidth $\hat{\mathbf{H}}_{\text{PI},2}$ obtained through `Hpi(,deriv.order=2)`. The 1-dimensional density ridge estimate for the quake data ($2646 \times 2$ matrix) in Figure 6.10 is computed by

```
> dr.quake <- kdr(quake)
```

For feature significance, `kfs` computes the significant modal regions $\hat{\mathcal{M}}$ in Algorithm 12, where the default tuning parameter values are as described in Section 6.3 and the default plug-in bandwidth $\hat{\mathbf{H}}_{\text{PI},2}$ is computed with `Hpi(,deriv.order=2)`. The significant modal region estimates for the `hsct12` data in Figure 6.12 are produced from

```
> fs.hsct <- kfs(hsct12)
```

For significant density difference region estimation, `kde.loc.test` computes $\hat{\mathcal{U}}^+, \hat{\mathcal{U}}^-$ in Algorithm 13, with default plug-in bandwidths $\hat{\mathbf{H}}_{1,\text{PI}}, \hat{\mathbf{H}}_{2,\text{PI}}$ as computed by `Hpi`. The significant density difference regions between the control subject #6 (`hsct6`) and treatment subject #12 (`hsct12`) from the stem cell data in Figure 7.1 is calculated by the command

```
> loc.test.hsct <- kde.local.test(x1=hsct6, x2=hsct12)
```

For kernel classification, `kda` computes the estimated group labels $\hat{\gamma}$ in Algorithm 14, with plug-in bandwidths `Hpi` for each group. The classifier for the training foetal cardiotocographic data (`cardio.train`, $532 \times 2$ matrix) with labels (`cardio.train.lab`, vector of length 532) in Figure 7.3(a) is

```
> kda.cardio <- kda(x=cardio.train, x.gr=cardio.train.lab)
```

and the scatter plot in Figure 7.3(b) of the estimated group labels for the test cardiotocographic data (`cardio.test`, $1594 \times 2$ matrix) is based on

```
> cardio.test.lab.est <- predict(kda.cardio, x=cardio.test)
```

For weighted density estimation, `kde` allows for non-uniform weights `w` via `kde(,w)` to compute $\hat{f}_{\text{wt}}$. For weighted deconvolution density estimation, `dckde` computes $\hat{f}_{\text{wdc}}$ in Algorithm 15. Once the air quality data is suitably reshaped into a $1300 \times 2$ matrix `air`, Figure 7.4(c) is produced by

```
> fhat.air.dec <- dckde(x=air,Sigma=Sigma.air,reg=0.00021)
```

where the values of the sample error variance $\mathbf{S}_U$ (`Sigma.air`) and the regularisation parameter $\hat{\eta}$ (`reg`) are computed according to Section 7.3.

For nearest neighbour mean shift clustering, `nnms` computes the estimated group labels $\hat{\gamma}$ in Algorithm 16. For image segmentation `image.nnms` further converts these group labels for each pixel to a colour segmented image. From the original $481 \times 321$ JPEG image `elephant`, then `elephant.luv` is the $154401 \times 5$ matrix in the transformed $(x, y, L^*, u^*, v^*)$ coordinates. The segmented image in Figure 7.5(c) is generated by

```
> nnms.elephant <- image.nnms(x=elephant.luv,
+   x.orig=elephant, min.clust.size=16*24)
```

These nearest neighbour commands are not included in the `ks` package.

## 8.2 Approximate binned estimation

### 8.2.1 *Approximate density estimation*

From Equation (2.2), a kernel density estimator at an single estimation point $\boldsymbol{x}$ is $\hat{f}(\boldsymbol{x};\mathbf{H}) = n^{-1}\sum_{i=1}^{n} K_{\mathbf{H}}(\boldsymbol{x} - \boldsymbol{X}_i)$. For visualisation purposes, we are required to evaluate $\hat{f}$ over a grid of estimation points $\boldsymbol{x}_1, \dots, \boldsymbol{x}_M$. For a given grid, a direct implementation of this, by explicitly looping over the grid points and data points leads to computationally intensive calculations for large $n$ and/or $M$. A common approach to reduce this computational burden for large $n$ is to seek an approximate estimator. One of the most effective approximations is known as the binned kernel density estimator (Silverman, 1986, Chapter 3.5), as it is computed directly on a hyper-rectangular grid, where each of the hyper-rectangles are known as bins. Whilst they share the hyper-rectangular grid structure of histograms, unlike the latter, the binwidths do not play a role in data smoothing, but only in controlling the approximation error (Hall & Wand, 1996).

There are three main steps in constructing a binned estimator. First, given a grid of size $M$, the data sample $\boldsymbol{X}_1, \ldots, \boldsymbol{X}_n$ are converted from $n$ points to the $M$ counts at each of the grid points. These counts are then embedded in a larger matrix $\mathbf{C}$. Second, the kernel function is evaluated at the same grid points and also embedded into a larger matrix $\mathbf{K}$ which has the same dimensions as $\mathbf{C}$. Third, the binned density estimator $\tilde{f}$ is obtained from a sequence of discrete convolutions of $\mathbf{C}$ and $\mathbf{K}$. The key steps thus concern how (a) to embed the binning counts and kernel function evaluations to obtain $\mathbf{C}$ and $\mathbf{K}$ and (b) to extract the elements from the convolution of $\mathbf{C}$ and $\mathbf{K}$.

The first step is to convert the data points into counts on the binning grid. The procedure for simple binning assigns a weight of 1 to the nearest grid point to the data point $\boldsymbol{X}_i$, though this loses too much information in the discretisation. More accurate is the linear binning which assigns the weights proportional to the subtended hyper-rectangles to the $2^d$ nearest grid points (Hall & Wand, 1996). To illustrate this in more detail, we focus on the bivariate case.

Write $\boldsymbol{x}_{j_1,j_2}$ as an abbreviation of $(x_{j_1}, x_{j_2})$ and suppose that the $i$-th data point $\boldsymbol{X}_i$ falls in the bin whose vertices are $\boldsymbol{x}_{j_1,j_2}, \boldsymbol{x}_{j_1+1,j_2}, \boldsymbol{x}_{j_1,j_2+1}$ and $\boldsymbol{x}_{j_1+1,j_2+1}$. Further suppose that the area of the bin is $A$ and the areas of the rectangles subtended from $\boldsymbol{X}_i$ are $A_1, A_2, A_3, A_4$. The count assigned to each vertex grid point is then equal to the area of the diagonally opposite rectangle divided by the total area of the bin, i.e., $c_{j_1,j_2} = A_4/A, c_{j_1+1,j_2} = A_3/A, c_{j_1,j_2+1} = A_2/A, c_{j_1+1,j_2+1} = A_1/A$, and $A = A_1 + A_2 + A_3 + A_4$, as illustrated in Figure 8.1. This is repeated for all data points and the individual counts at the grid points are thus accumulated.

Suppose that the binning grid size is $M = M_1 M_2$. Let the binning counts be $c_{\ell_1,\ell_2}$ for $\ell_1 = 1, \ldots, M_1, \ell_2 = 1, \ldots, M_2$. The binned density estimator at a grid point $\boldsymbol{x}_{j_1,j_2}$ is the discrete convolution of the binning counts and the kernel evaluations, as introduced by Wand (1994):

$$\hat{f}_{\text{bin}}(\boldsymbol{x}_{j_1,j_2}) = n^{-1} \sum_{\ell_1=1}^{M_1} \sum_{\ell_2=1}^{M_2} c_{\ell_1,\ell_2} K_{\mathbf{H}}(\boldsymbol{x}_{j_1,j_2} - \boldsymbol{x}_{\ell_1,\ell_2}). \tag{8.1}$$

The second step for the evaluation of the kernel function at the same binning grid is characterised in the following manner. Let the binning grid be

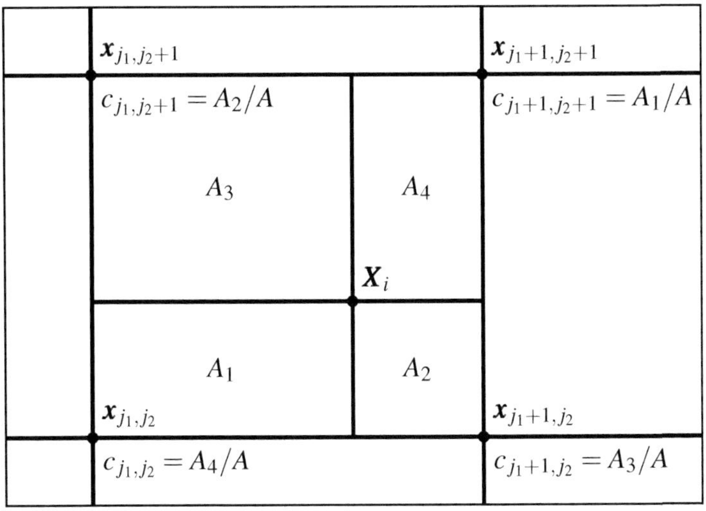

Figure 8.1 *Linear binning counts. A bivariate point* $\boldsymbol{X}_i$ *is converted to the counts assigned to its 4 nearest grid points* $\boldsymbol{x}_{j_1,j_2}, \boldsymbol{x}_{j_1,j_2+1}, \boldsymbol{x}_{j_1+1,j_2}, \boldsymbol{x}_{j_1+1,j_2+1}$. *Their respective counts are equal to the area of the diagonally opposite rectangle divided by the total area of bin A.*

covered exactly by $[a_1, b_1] \times [a_2, b_2]$, where $a_1, a_2$ are less than the marginal sample minima, and $b_1, b_2$ are greater than the marginal sample maxima. Hence, $\boldsymbol{x}_{j_1,j_2} = (a_1 + j_1\delta_1, a_2 + j_2\delta_2)$, where the binwidths are $\delta_1 = (b_1 - a_1)/(M_1 - 1), \delta_2 = (b_2 - a_2)/(M_2 - 1)$. Then the required kernel function evaluations are $k_{\ell_1,\ell_2} = n^{-1}K_{\mathbf{H}}(\delta_1\ell_1, \delta_2\ell_2)$. This notation allows us to re-write Equation (8.1) as a bidimensional discrete convolution

$$\hat{f}_{\text{bin}}(\boldsymbol{x}_{j_1,j_2}) = \sum_{\ell_1=-(M_1-1)}^{M_1-1} \sum_{\ell_2=-(M_2-1)}^{M_2-1} c_{j_1-\ell_1,j_2-\ell_2} k_{\ell_1,\ell_2} \tag{8.2}$$

where we understand that $c_{\ell_1,\ell_2} = 0$ for $(\ell_1, \ell_2) \notin \{1, \dots, M_1\} \times \{1, \dots, M_2\}$.

The third step is more complicated as it is computationally intensive to compute the double sum in Equation (8.2) as a double loop, so the usual approach is to embed the $M_1M_2$ binning counts $c_{\ell_1,\ell_2}$ and the kernel function evaluations $k_{\ell_1,\ell_2}$ into larger matrices and to take advantage of faster matrix-based calculations. The embeddings posited by Wand (1994) were suitable only for the constrained classes of scalar matrices $\mathcal{A}$ and diagonal matrices $\mathcal{D}$, and were extended to the unconstrained class $\mathcal{F}$ by Gramacki & Gramacki (2017*a*,*b*).

Since discrete convolutions are most effective when operating on matrices whose dimensions are highly composite numbers, we set $P_1, P_2$ to be a power of 2 greater than the grid sizes $M_1, M_2$. Writing $\mathbf{0}_{m,n}$ for the $(m \times n)$ zero matrix, the zero-padded version of the binning counts is

$$\mathbf{C} = \left[\begin{array}{c|ccc|c}
\mathbf{0}_{M_1-1,M_2-1} & & \mathbf{0}_{M_1,M_2-1} & & \mathbf{0}_{P_1-2M_1+1,M_2-1} \\
\hline
 & c_{1,1} & \cdots & c_{1,M_2} & \\
\mathbf{0}_{M_1-1,M_2} & \vdots & & \vdots & \mathbf{0}_{M_1 P_1-2M_1+1,M_2} \\
 & c_{M_1,1} & \cdots & c_{M_1,M_2} & \\
\hline
\mathbf{0}_{M_1-1,P_2-2M_2+1} & & \mathbf{0}_{M_1,P_2-2M_2+1} & & \mathbf{0}_{P_1-2M_1+1,P_2-2M_2+1}
\end{array}\right]$$

and of the kernel evaluations is

$$\mathbf{K} = \left[\begin{array}{ccccc|c}
k_{-M_1,-M_2} & \cdots & k_{-M_1,0} & \cdots & k_{-M_1,M_2} & \\
\vdots & & \vdots & & \vdots & \\
k_{0,-M_2} & \cdots & k_{0,0} & \cdots & k_{0,M_2} & \mathbf{0}_{2M_1+1,P_2-2M_2-1} \\
\vdots & & \vdots & & \vdots & \\
k_{M_1,-M_2} & \cdots & k_{M_1,0} & \cdots & k_{M_1,M_2} & \\
\hline
 & & \mathbf{0}_{P_1-2M_1-1,2M_2+1} & & & \mathbf{0}_{P_1-2M_1-1,P_2-2M_2-1}
\end{array}\right]$$

where $\mathbf{C}, \mathbf{K} \in \mathcal{M}_{P_1 \times P_2}$. We then compute $\mathbf{F} = \varphi^{-1}(\varphi(\mathbf{C})\varphi(\mathbf{K}))$ where $\varphi$ is a discrete Fourier transform and $\varphi^{-1}$ is its inverse transform. These are efficiently computed using FFT (Fast Fourier Transform) methods. As $\mathbf{F}$ is a $P_1 \times P_2$ matrix, the appropriate normalised submatrix of $\mathbf{F}$ which gives the approximate density estimator evaluated on the $M_1 \times M_2$ binning grid is

$$\hat{f}_{\text{bin}}(\cdot; \mathbf{H}) = (P_1 P_2)^{-1} \mathbf{F}[(2M_1 - 1) : (3M_1 - 2), (2M_2 - 1) : (3M_2 - 2)]$$

where $\mathbf{F}[r_1 : r_2, c_1 : c_2]$ denotes the submatrix of $\mathbf{F}$ formed by selecting the elements in rows $r_1$ to $r_2$ and columns $c_1$ to $c_2$.

The implementation steps for this binned density estimator $\hat{f}_{\text{bin}}$ are presented in Algorithm 17. The inputs are the data $\boldsymbol{X}_1, \ldots, \boldsymbol{X}_n$, and the tuning parameters are the bandwidth matrix $\mathbf{H}$, the minima $\boldsymbol{a}$, the maxima $\boldsymbol{b}$ and the vector $\mathbf{M}$ of coordinate sizes of the binning grid. The output is the binned density estimator $\hat{f}_{\text{bin}}$ evaluated on the grid. Lines 1–2 initialise the binning grid and matrix sizes. Lines 3–5 perform the binning counts, and Lines 6–9, the kernel function evaluations. Lines 9–10 carry out the zero-padding and FFT operations to create $\mathbf{F}$. Line 11 normalises and takes the appropriate subset of $\mathbf{F}$ to be the binned density estimator.

The algorithm described above can be further expedited if we realise that it is not required to evaluate the binning counts and kernel functions at all

**Algorithm 17** Binned density estimator

**Input:** $\{\boldsymbol{X}_1,\ldots,\boldsymbol{X}_n\}, \mathbf{H}, \boldsymbol{a}, \boldsymbol{b}, \mathbf{M}$
**Output:** $\hat{f}_{\text{bin}}(\cdot;\mathbf{H})$
1: Initialise $M_1 \times \cdots \times M_d$ binning grid on $[a_1,b_1] \times \cdots \times [a_d,b_d]$
2: Initialise $P_1,\ldots,P_d := 2^{\lceil \log_2(2M_1)\rceil},\ldots,2^{\lceil \log_2(2M_d)\rceil}$
3: **for** $\ell_1,\ldots,\ell_d := 1$ **to** $M_1,\ldots M_d$ **do**
4: Compute binning counts $c_{\ell_1,\ldots,\ell_d}$
5: **end for**
6: **for** $\ell_1,\ldots,\ell_d := -M_1,\ldots,-M_d$ **to** $M_1,\ldots M_d$ **do**
7: Evaluate kernel functions $k_{\ell_1,\ldots,\ell_d}$
8: **end for**
9: Create zero-padded counts and kernel matrices $\mathbf{C},\mathbf{K}$
10: Perform FFT operations $\mathbf{F} := \varphi^{-1}(\varphi(\mathbf{C})\varphi(\mathbf{K}))$
11: $\hat{f}_{\text{bin}}(\cdot;\mathbf{H}) :=$ normalised subset of $\mathbf{F}$

the grid points, due to the finite effective support of the kernel. Gramacki & Gramacki (2017*a*) suggest that $M_j$ in the loops in Lines 3–8 of Algorithm 17 can be replaced by, e.g., $L_j = \min(M_j - 1, \lceil \tau\lambda_1^{1/2}/\delta_j \rceil)$ for $j = 1,\ldots,d$. Here, $\tau = 3.7$ is a commonly used value, and $\lambda_1$ is the largest eigenvalue of $\mathbf{H}$. These adjustments usually lead to secondary gains in efficiency compared to primary gains due to the FFT operations in Line 10.

Here it is convenient to note that, from experimental evidence, for $d = 1$, a grid size $M = 401$ is sufficiently dense to ensure an approximation that is visually indistinguishable from the exact estimator $\hat{f}$ for the vast majority of cases. Likewise, for $d = 2$, the value $M = 151^2$ is widely used. For $d = 3$, there is less consensus for a generally suitable grid size but $M = 51^3$ gives a reasonably close approximation in a reasonable time frame. This is illustrated empirically for the temperature data in Figure 8.2. In (a), the dotted black lines are the density estimate with an $11 \times 11$ grid, which is overlaid with the density estimate in the solid purple lines with a $151 \times 151$ grid. The former grid is too sparse and leads to insufficiently smooth contours. In (b), the dotted black lines are the density estimate with a $301 \times 301$ grid, which is visually indistinguishable from the default density estimate in the dotted purple lines, despite that the higher resolution grid contains $301^2 = 90601$ points which is a 4-fold increase on the $151^2 = 22801$ points.

The squared error analysis of $\hat{f}_{\text{bin}}$ was comprehensively studied by Hall & Wand (1996) and Holmström (2000). Theorem 2.5 in Holmström (2000) asserted that $\text{MISE}\{\hat{f}_{\text{bin}}(\cdot;\mathbf{H})\}$ converges at the rate $n^{-1}|\mathbf{H}|^{-1/2} + \|\text{vec}\,\mathbf{H}\|^2 + \delta^4$ as $n \to \infty$, where $\delta = \max(\delta_1,\ldots,\delta_d)$. The third term accounts for the bin-

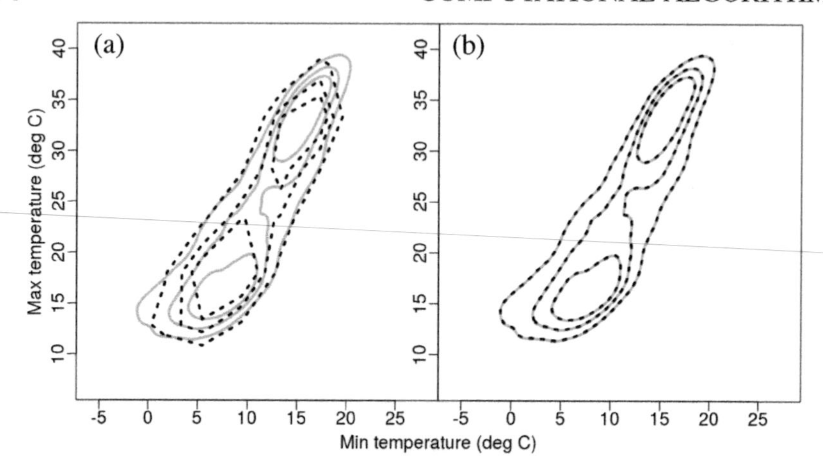


Figure 8.2 *Different grid sizes for the kernel density estimates of the temperature data. The density estimate on the default* $151 \times 151$ *grid is in solid purple. (a) The density estimate on an* $11 \times 11$ *grid is in dotted black. (b) The density estimate on a* $301 \times 301$ *grid is in dotted black.*

ning grid binwidths in comparison to the exact density estimator which has $\text{MISE}\{\hat{f}(\cdot;\mathbf{H})\} = O(n^{-1}|\mathbf{H}|^{-1/2} + \|\text{vec}\,\mathbf{H}\|^2)$. For the usual case where $\mathbf{H} = O(n^{-2/(d+4)})$, if the maximum binwidth $\delta$ is of order $n^{-1/(d+4)}$, then $\text{MISE}\{\hat{f}_{\text{bin}}(\cdot;\mathbf{H})\}$ converges to 0 at the same rate as $\text{MISE}\{\hat{f}(\cdot;\mathbf{H})\}$.

### *8.2.2 Approximate density derivative and functional estimation*

For each $r$-th order partial derivative of the density $f^{(\boldsymbol{r})}$ indexed by $\boldsymbol{r} = (r_1, \ldots, r_d)$, the binned estimator is analogously defined as

$$\hat{f}^{(\boldsymbol{r})}_{\text{bin}}(\boldsymbol{x}_{j_1,\ldots,j_d};\mathbf{H}) = n^{-1} \sum_{\ell_1=1}^{P_1} \cdots \sum_{\ell_d=1}^{P_d} c_{\ell_1,\ldots,\ell_d} K^{(\boldsymbol{r})}_{\mathbf{H}}(\boldsymbol{x}_{j_1,\ldots,j_d} - \boldsymbol{x}_{\ell_1,\ldots,\ell_d}). \tag{8.3}$$

The binning counts previously computed for the density estimator can be re-utilised without modification, so this discretisation needs only to be performed once per data set. Moreover, the FFT operations remain the same, and so the only difference arises from the kernel function evaluations. We compute the partial derivative $k^{(\boldsymbol{r})}_{\ell_1,\ldots,\ell_d} = n^{-1} K^{(\boldsymbol{r})}_{\mathbf{H}}(\delta_1\ell_1, \ldots, \delta_d\ell_d)$ for binwidths $\delta_j = (b_j - a_j)/(M_j - 1), j = 1, \ldots, d$. Zero-padding the binning counts and the kernel derivatives to produce $\mathbf{C}, \mathbf{K}$, the binned estimator is the normalised submatrix of $\mathbf{F} = \varphi^{-1}(\varphi(\mathbf{C})\varphi(\mathbf{K}))$

$$\hat{f}^{(\boldsymbol{r})}_{\text{bin}}(\cdot;\mathbf{H}) = (P_1 \cdots P_d)^{-1} \mathbf{F}[(2M_1 - 1) : (3M_1 - 2), \ldots, (2M_d - 1) : (3M_d - 2)].$$

If $\mathbf{A}$ is a $P_1 \times \cdots \times P_d$ array, then the $(\ell_1, \ldots, \ell_d)$-th element of its discrete Fourier transform $\mathbf{A}' = \varphi(\mathbf{A})$ is

$$a'_{\ell_1,\ldots,\ell_d} = \sum_{j_1=0}^{P_1} \cdots \sum_{j_d=0}^{P_d} a_{\ell_1,\ldots,\ell_d} \exp[2\pi i(\ell_1 j_1/P_1 + \cdots + \ell_d j_d/P_d)]$$

where $a_{\ell_1,\ldots,\ell_d}$ is the corresponding element of $\mathbf{A}$ and $i$ is the pure imaginary number. The inverse transform $\varphi^{-1}(\mathbf{A}')$ which recovers the original elements from $\mathbf{A}$ is $a_{\ell_1,\ldots,\ell_d} = \sum_{j_1=0}^{P_1} \cdots \sum_{j_d=0}^{P_d} a'_{\ell_1,\ldots,\ell_d} \exp[-2\pi i(\ell_1 j_1/P_1 + \cdots + \ell_d j_d/P_d)]$.

For the individual density derivative functionals $\psi_{\boldsymbol{r}} = \int_{\mathbb{R}^d} f^{(\boldsymbol{r})}(\boldsymbol{x}) f(\boldsymbol{x})\, d\boldsymbol{x}$, the binned estimator is

$$\hat{\psi}_{\mathrm{bin},\boldsymbol{r}}(\mathbf{G}) = n^{-1} \sum_{j_1=1}^{P_1} \cdots \sum_{j_d=1}^{P_d} c_{j_1,\ldots,j_d} \hat{f}^{(\boldsymbol{r})}_{\mathrm{bin}}(\boldsymbol{x}_{j_1,\ldots,j_d}; \mathbf{H}) \tag{8.4}$$

where we re-utilise the partial density derivatives estimators above.

The goal is to produce estimators of the vectorised derivative forms $\mathsf{D}^{\otimes r} f$ and $\boldsymbol{\psi}_r$, whilst the binned estimators only act on individual partial derivatives. It is computationally efficient to calculate the entire normal density derivative $\mathsf{D}^{\otimes r} \phi_{\boldsymbol{\Sigma}}(\boldsymbol{x})$ initially, enumerate over the individual partial derivatives to compute all the required $\hat{f}^{(\boldsymbol{r})}_{\mathrm{bin}}(\cdot; \mathbf{H})$, and then collate them into $\mathsf{D}^{\otimes r} \hat{f}_{\mathrm{bin}}(\cdot; \mathbf{H})$ or $\hat{\boldsymbol{\psi}}_{\mathrm{bin},r}(\mathbf{G})$.

## 8.3 Recursive computation of the normal density derivatives

Binned estimators can vastly improve the computational efficiency of kernel estimators for large sample sizes via FFT operations. To maintain this computational gain, we require an efficient method for computing higher order derivatives of the normal density as they are one of the inputs for the FFTs.

For the normal density $\phi_{\boldsymbol{\Sigma}}(\boldsymbol{x}) = (2\pi)^{-d/2} |\Sigma|^{-1/2} \exp(-\frac{1}{2} \boldsymbol{x}^\top \boldsymbol{\Sigma}^{-1} \boldsymbol{x})$ we have $\mathsf{D}^{\otimes r} \phi_{\boldsymbol{\Sigma}}(\boldsymbol{x}) = (-1)^r (\boldsymbol{\Sigma}^{-1})^{\otimes r} \mathcal{H}_r(\boldsymbol{x}; \boldsymbol{\Sigma}) \phi_{\boldsymbol{\Sigma}}(\boldsymbol{x})$, where we recall from Equation (5.5) that $\mathcal{H}_r$ is the $r$-th order Hermite polynomial in $\boldsymbol{x}$, given by

$$\mathcal{H}_r(\boldsymbol{x}; \boldsymbol{\Sigma}) = r! \mathcal{S}_{d,r} \sum_{j=0}^{\lceil r/2 \rceil} \frac{(-1)^j}{j!(r-2j)!2^j} \{\boldsymbol{x}^{\otimes(r-2j)} \otimes (\operatorname{vec} \boldsymbol{\Sigma})^{\otimes j}\},$$

with $\mathcal{S}_{d,r}$ standing for the symmetriser matrix. An important special case is $\boldsymbol{x} = \mathbf{0}$, since the normal scale functionals which are the starting point for the plug-in and smoothed cross validation bandwidth selectors can be expressed as $\boldsymbol{\psi}_{\mathrm{NS},r} = \mathsf{D}^{\otimes r}\phi_{2\boldsymbol{\Sigma}}(\mathbf{0})$. The general case is required for the other stages in these bandwidth selection algorithms.

Whilst the previous explicit formula for the $r$-th order Hermite polynomial allows for the study of the theoretical properties for arbitrary dimension $d$ and derivative order $r$, some authors have noticed the difficulties that the computation of the symmetriser matrix $\mathcal{S}_{d,r}$ entails (Triantafyllopoulos, 2003; Kan, 2008). Here we present an alternative recursive formula that avoids the explicit computation of the symmetriser matrix $\mathcal{S}_{d,r}$, and also re-utilises the results from lower order polynomials.

This recursion is outlined in Algorithm 18, adapted from Chacón & Duong (2015). It is based on the vector $\mathfrak{H}_r(\boldsymbol{x};\boldsymbol{\Sigma})$ containing the unique elements of $(\boldsymbol{\Sigma}^{-1})^{\otimes r}\boldsymbol{\mathcal{H}}_r(\boldsymbol{x};\boldsymbol{\Sigma})$ in a coherent order that facilitates recursive computation. This vector $\mathfrak{H}_r$ can be shown to have length $N_{d,r} = \binom{r+d-1}{r}$, which for most values of $d$ and $r$ is much smaller than $d^r$, which is the length of $\boldsymbol{\mathcal{H}}_r$. For a multi-index $\boldsymbol{m} = (m_1,\dots,m_d)$ with $|\boldsymbol{m}| = m_1 + \cdots + m_d = r$ we introduce the scalar Hermite polynomial $\mathcal{H}^{(\boldsymbol{m})}(\boldsymbol{x};\boldsymbol{\Sigma})$ such that the partial derivative of $\phi_{\boldsymbol{\Sigma}}(\boldsymbol{x})$ indexed by $\boldsymbol{m}$ is $\phi_{\boldsymbol{\Sigma}}^{(\boldsymbol{m})}(\boldsymbol{x}) = (-1)^r\phi_{\boldsymbol{\Sigma}}(\boldsymbol{x})\mathcal{H}^{(\boldsymbol{m})}(\boldsymbol{x};\boldsymbol{\Sigma})$. The inputs of Algorithm 18 are the evaluation point $\boldsymbol{x}$, the variance matrix $\boldsymbol{\Sigma}$ and the order $r$ of the derivative. The output is the full derivative $\mathsf{D}^{\otimes r}\phi_{\boldsymbol{\Sigma}}(\boldsymbol{x})$. Line 1 initialises the recursion with $\mathfrak{H}_0(\boldsymbol{x};\boldsymbol{\Sigma}), \mathfrak{H}_1(\boldsymbol{x};\boldsymbol{\Sigma})$. In Lines 2–6, the recursive sub-routine in Algorithm 19 is employed to compute $\mathfrak{H}_\ell(\boldsymbol{x};\boldsymbol{\Sigma})$, for $\ell = 2,\dots,r$. Lines 7–8 rearrange the elements of $\mathfrak{H}_r(\boldsymbol{x};\boldsymbol{\Sigma})$ to form the required $\mathsf{D}^{\otimes r}\phi_{\boldsymbol{\Sigma}}(\boldsymbol{x})$.

**Algorithm 18** Recursive computation of the normal density derivative

**Input:** $\boldsymbol{x}, \boldsymbol{\Sigma}, r$
**Output:** $\mathsf{D}^{\otimes r}\phi_{\boldsymbol{\Sigma}}(\boldsymbol{x})$
1: Initialise $\mathfrak{H}_0(\boldsymbol{x};\boldsymbol{\Sigma}) := 1$; $\mathfrak{H}_1(\boldsymbol{x};\boldsymbol{\Sigma}) := \boldsymbol{\Sigma}^{-1}\boldsymbol{x}$
2: **if** $r \geq 2$ **then**
3: **for** $\ell := 2$ **to** $r$ **do** /* Algorithm 19 */
4: Obtain $\mathfrak{H}_\ell(\boldsymbol{x};\boldsymbol{\Sigma})$ from $\mathfrak{H}_{\ell-1}(\boldsymbol{x};\boldsymbol{\Sigma})$ and $\mathfrak{H}_{\ell-2}(\boldsymbol{x};\boldsymbol{\Sigma})$
5: **end for**
6: **end if**
7: Distribute elements of $\mathfrak{H}_r(\boldsymbol{x};\boldsymbol{\Sigma})$ to form $(\boldsymbol{\Sigma}^{-1})^{\otimes r}\boldsymbol{\mathcal{H}}_r(\boldsymbol{x};\boldsymbol{\Sigma})$
8: $\mathsf{D}^{\otimes r}\phi_{\boldsymbol{\Sigma}}(\boldsymbol{x}) := (-1)^r(\boldsymbol{\Sigma}^{-1})^{\otimes r}\boldsymbol{\mathcal{H}}_r(\boldsymbol{x};\boldsymbol{\Sigma})\phi_{\boldsymbol{\Sigma}}(\boldsymbol{x})$

To describe the mathematical justifications for Algorithms 18–19, we require some notation that characterises precisely the systematic computation of $\mathsf{D}^{\otimes r}\phi_{\boldsymbol{\Sigma}}(\boldsymbol{x})$. Let $\mathcal{PR}_{d,r} = \{1,\dots,d\}^r$ be the set of all $r$-length multi-indices of $\{1,\dots,d\}$. Each element of $\mathsf{D}^{\otimes r}$ can be written as $\mathsf{D}_{(\boldsymbol{i})} = \partial^r/(\partial x_{i_1}\cdots\partial x_{i_r})$ for some $\boldsymbol{i} = (i_1,\dots,i_r) \in \mathcal{PR}_{d,r}$, so that for each $j \in \{1,\dots,r\}$ the in-

**Algorithm 19** Recursive computation of the minimal Hermite polynomial

**Input:** $\mathfrak{H}_r(\boldsymbol{x};\boldsymbol{\Sigma}), \mathfrak{H}_{r-1}(\boldsymbol{x};\boldsymbol{\Sigma})$
**Output:** $\mathfrak{H}_{r+1}(\boldsymbol{x};\boldsymbol{\Sigma})$
1: $j := 1; \ell := 1$
2: **for** $\boldsymbol{m} \in \mathcal{I}_{d,r}$ **do**
3: $\quad \mathfrak{H}_{r+1}(\boldsymbol{x};\boldsymbol{\Sigma})[\ell] := \mathcal{H}^{(\boldsymbol{m}+\boldsymbol{e}_1)}(\boldsymbol{x};\boldsymbol{\Sigma}); \ell := \ell + 1$
4: **end for**
5: **for** $j := 2$ **to** $d-1$ **do**
6: $\quad$ **for** $\boldsymbol{m} \in$ last $N_{d-j+1,r}$ elements of $\mathfrak{H}_r(\boldsymbol{x};\boldsymbol{\Sigma})$ **do**
7: $\quad\quad \mathfrak{H}_{r+1}(\boldsymbol{x};\boldsymbol{\Sigma})[\ell] := \mathcal{H}^{(\boldsymbol{m}+\boldsymbol{e}_\ell)}(\boldsymbol{x};\boldsymbol{\Sigma}); \ell := \ell + 1$
8: $\quad$ **end for**
9: **end for**
10: $j := d; \mathfrak{H}_{r+1}(\boldsymbol{x};\boldsymbol{\Sigma})[\ell] := \mathcal{H}^{(0,\ldots,0,r+1)}(\boldsymbol{x};\boldsymbol{\Sigma})$

dex coordinate $i_j$ refers to the coordinate of $\boldsymbol{x}$ with respect to which the $j$-th partial derivative is performed. Noting that the cardinality of $\mathcal{PR}_{d,r}$ is $d^r$, Chacón & Duong (2015, Appendix 2) constructed a bijective map $p\colon \mathcal{PR}_{d,r} \to \{1,2,\ldots,d^r\}$ that allows us to order all the elements of $\mathcal{PR}_{d,r}$ as $\boldsymbol{i}_1 = p^{-1}(1),\ldots,\boldsymbol{i}_{d^r} = p^{-1}(d^r)$ in a way such that the $k$-th element of $\mathsf{D}^{\otimes r} \in \mathbb{R}^{d^r}$ is precisely $\mathsf{D}_{\boldsymbol{i}_k}$; that is, $\mathsf{D}^{\otimes r} = (\mathsf{D}_{(\boldsymbol{i}_1)},\ldots,\mathsf{D}_{(\boldsymbol{i}_{d^r})})$. This provides a well-organised procedure to compute all the partial derivatives to form $\mathsf{D}^{\otimes r}$. Nevertheless, the fact that the cardinality $d^r$ rapidly increases with $d$ and (especially) $r$ implies that computational bottlenecks are reached if the brute enumeration of all of its elements is carried out.

Since $\phi$ is infinitely differentiable, most of the entries of $\mathsf{D}^{\otimes r}\phi_{\boldsymbol{\Sigma}}(\boldsymbol{x})$ are duplicated at different positions, due to Schwarz's theorem for the interchangeability of the partial derivative order, and in this sense the number of unique elements of $\mathsf{D}^{\otimes r}\phi_{\boldsymbol{\Sigma}}(\boldsymbol{x})$ is lower than its length $d^r$. To take advantage of this fact, let $\mathcal{I}_{d,r} = \{(m_1,\ldots,m_d) \in \{0,1,\ldots,r\}^d\colon |\boldsymbol{m}| = r\}$ be the set of all non-negative integer $d$-vectors whose sum is $r$. Then, any coordinate of $\mathsf{D}^{\otimes r}$ can be written as $\mathsf{D}^{(\boldsymbol{m})}$ for some $\boldsymbol{m} \in \mathcal{I}_{d,r}$, where $\mathsf{D}^{(\boldsymbol{m})} = \partial^{|\boldsymbol{m}|}/(\partial x_1^{m_1}\cdots\partial x_d^{m_d})$. Here, for each $\ell \in \{1,\ldots,d\}$, the index $m_\ell$ refers to the number of times that the operator partially differentiates with respect to $x_\ell$. This implies that $\{\mathsf{D}^{(\boldsymbol{m})}\colon \boldsymbol{m} \in \mathcal{I}_{d,r}\}$ contains the unique elements of $\mathsf{D}^{\otimes r}$. An efficient way to obtain $\mathsf{D}^{\otimes r}$ is to compute $\{\mathsf{D}^{(\boldsymbol{m})}\colon \boldsymbol{m} \in \mathcal{I}_{d,r}\}$ and then rearrange them to form $\mathsf{D}^{\otimes r}$. Thus we require $\mathfrak{D}^{(r)}$, an ordering of $\{\mathsf{D}^{(\boldsymbol{m})}\colon \boldsymbol{m} \in \mathcal{I}_{d,r}\}$, which facilitates this rearrangement.

For a given $\boldsymbol{m} \in \mathcal{I}_{d,r}$, there are possibly many $\boldsymbol{i} \in \mathcal{PR}_{d,r}$ such that $\mathsf{D}^{(\boldsymbol{m})} = \mathsf{D}_{(\boldsymbol{i})}$, so it suffices to have $m_\ell = \sum_{j=1}^r \mathbf{1}\{i_j = \ell\}$, for $\ell = 1,\ldots,d$. Since the elements in $\mathcal{PR}_{d,r}$ have been given a natural order via the map $p^{-1}$, we associate each $\boldsymbol{m} \in \mathcal{I}_{d,r}$ with the first $\boldsymbol{i} \in \mathcal{PR}_{d,r}$ for which $\mathsf{D}^{(\boldsymbol{m})} = \mathsf{D}_{(\boldsymbol{i})}$ holds, i.e., $\min\{k : \mathsf{D}^{(\boldsymbol{m})} = \mathsf{D}_{(\boldsymbol{i}_k)}, k \in \{1,\dots,d^r\}\}$. By doing this for all the elements $\boldsymbol{m} \in \mathcal{I}_{d,r}$ we obtain an unambiguous ordering of the unique elements of $\mathsf{D}^{\otimes r}$, which are thus collected in a vector $\mathfrak{D}^{(r)}$ of length $|\mathcal{I}_{d,r}| = N_{d,r}$. This correspondence between $\mathfrak{D}^{(r)}$ and $\mathsf{D}^{\otimes r}$ facilitates the rearrangement of the former to the latter in Line 7 of Algorithm 18.

We follow the ordering of $\mathfrak{D}^{(r)}$ to construct the minimal Hermite polynomial $\boldsymbol{\mathfrak{H}}_r(\boldsymbol{x};\boldsymbol{\Sigma})$ such that $\mathfrak{D}^{(r)}\phi_{\boldsymbol{\Sigma}}(\boldsymbol{x}) = (-1)^r \phi_{\boldsymbol{\Sigma}}(\boldsymbol{x})\boldsymbol{\mathfrak{H}}_r(\boldsymbol{x};\boldsymbol{\Sigma})$. The definition of $\boldsymbol{\mathfrak{H}}_r(\boldsymbol{x};\boldsymbol{\Sigma})$ is not suitable for computation, so it would be desirable to have a more convenient, recursive form. Given the $N_{d,r-1}$-vector $\boldsymbol{\mathfrak{H}}_{r-1}(\boldsymbol{x};\boldsymbol{\Sigma})$ and the $N_{d,r}$-vector $\boldsymbol{\mathfrak{H}}_r(\boldsymbol{x};\boldsymbol{\Sigma})$, the goal is to compute the $N_{d,r+1}$-vector $\boldsymbol{\mathfrak{H}}_{r+1}(\boldsymbol{x};\boldsymbol{\Sigma})$, with the first two recursion evaluations $\boldsymbol{\mathfrak{H}}_0(\boldsymbol{x};\boldsymbol{\Sigma}) = 1$ and $\mathfrak{H}_1(\boldsymbol{x};\boldsymbol{\Sigma}) = \boldsymbol{\Sigma}^{-1}\boldsymbol{x}$. To proceed, we draw on the recursive form of the scalar-valued Hermite polynomials (Savits, 2006, Theorem 4.1):

$$\mathcal{H}^{(\boldsymbol{m}+\boldsymbol{e}_j)}(\boldsymbol{x};\boldsymbol{\Sigma}) = z_j \mathcal{H}^{(\boldsymbol{m})}(\boldsymbol{x};\boldsymbol{\Sigma}) - \sum_{\ell=1}^{d} v_{j\ell} m_\ell \mathcal{H}^{(\boldsymbol{m}-\boldsymbol{e}_\ell)}(\boldsymbol{x};\boldsymbol{\Sigma}) \tag{8.5}$$

where $\mathbf{V} = \boldsymbol{\Sigma}^{-1} = [v_{j\ell}]_{j,\ell=1}^{d}$ and $\boldsymbol{z} = \mathbf{V}\boldsymbol{x} = (z_1,\dots,z_d)$, for $j = 1,\dots,d$, and we follow the convention that $\mathcal{H}^{(\boldsymbol{m}-\boldsymbol{e}_\ell)}(\boldsymbol{x};\boldsymbol{\Sigma}) = 1$ if $m_\ell = 0$.

For $j = 1$, with Equation (8.5), we obtain all $N_{d,r}$ scalar-valued Hermite polynomials corresponding to the derivative indices $\{\boldsymbol{m}+\boldsymbol{e}_1 : \boldsymbol{m} \in \mathcal{I}_{d,r}\} = \{\boldsymbol{m} \in \mathcal{I}_{d,r+1} : m_1 \geq 1\}$. These are placed in the first $N_{d,r}$ positions of $\boldsymbol{\mathfrak{H}}_{r+1}(\boldsymbol{x};\boldsymbol{\Sigma})$. This is carried out in Lines 1–4 in Algorithm 19. The remaining $N_{d-1,r+1}$ positions will be occupied by polynomials of the form $\{\boldsymbol{m} \in \mathcal{I}_{d,r+1} : m_1 = 0\} = \{(0,m_2,\dots,m_d) : m_2 + \dots + m_d = r+1\}$.

For $j = 2$, with Equation (8.5), we obtain all $N_{d-1,r}$ scalar-valued Hermite polynomials of the form $\{(0,m_2,\dots,m_d) : m_2 + \dots + m_d = r+1, m_2 \geq 1\} = \{\boldsymbol{m}+\boldsymbol{e}_2 : \boldsymbol{m} = (0,m_2,\dots,m_d), \boldsymbol{m} \in \mathcal{I}_{d,r}\}$. As $m_1 = 0$, then Equation (8.5) simplifies to $\mathcal{H}^{(\boldsymbol{m}+\boldsymbol{e}_2)}(\boldsymbol{x};\boldsymbol{\Sigma}) = z_2 \mathcal{H}^{(\boldsymbol{m})}(\boldsymbol{x};\boldsymbol{\Sigma}) - \sum_{\ell=2}^{d} v_{2\ell} m_\ell \mathcal{H}^{(\boldsymbol{m}-\boldsymbol{e}_\ell)}(\boldsymbol{x};\boldsymbol{\Sigma})$. Since the first $N_{d,r-1}$ elements of the $\boldsymbol{\mathfrak{H}}_r(\boldsymbol{x};\boldsymbol{\Sigma})$ correspond to the multi-indices $\boldsymbol{m} \in \mathcal{I}_{d,r}$ with $m_1 \geq 1$, then Equation (8.5) is applied to the last $N_{d-1,r}$ elements of $\boldsymbol{\mathfrak{H}}_r(\boldsymbol{x};\boldsymbol{\Sigma})$. The results are then placed in $\boldsymbol{\mathfrak{H}}_{r+1}(\boldsymbol{x};\boldsymbol{\Sigma})$ immediately after those already in the first $N_{d,r}$ positions from $j = 1$, which maintains a coherent ordering of the elements of $\boldsymbol{\mathfrak{H}}_{r+1}(\boldsymbol{x};\boldsymbol{\Sigma})$. At this stage, we have placed $N_{d,r} + N_{d-1,r}$ elements of $\boldsymbol{\mathfrak{H}}_{r+1}(\boldsymbol{x};\boldsymbol{\Sigma})$. The remaining $N_{d-2,r+1}$ elements will be occupied by polynomials of the form $\{\boldsymbol{m} \in \mathcal{I}_{d,r+1} : m_1 = m_2 = 0\} = \{(0,0,m_3,\dots,m_d) : m_3 + \dots + m_d = r+1\}$.

For $j = 3,\dots,d-1$, we repeat the analogous calculations which lead to the placement of $N_{d,r} + N_{d-1,r} + \cdots + N_{1,r}$ elements of $\mathfrak{H}_{r+1}(\boldsymbol{x};\boldsymbol{\Sigma})$. Lines 5–9 in Algorithm 19 cover the cases $j = 2, \ldots, d-1$.

For $j = d$, as $N_{d,r+1} = N_{d,r} + N_{d-1,r} + \cdots + N_{1,r} + 1$ from Pascal's rule for binomial coefficients, there remains one element to compute. Equation (8.5) simplifies to $\mathcal{H}^{(0,\ldots,0,r+1)}(\boldsymbol{x};\boldsymbol{\Sigma}) = z_d \mathcal{H}^{(0,\ldots,0,r)}(\boldsymbol{x};\boldsymbol{\Sigma}) - v_{dd} r \mathcal{H}^{(0,\ldots,0,r-1)}(\boldsymbol{x};\boldsymbol{\Sigma})$. This is carried out in Line 10 in Algorithm 19.

These computations, developed in Chacón & Duong (2015), improve on the algorithm provided by Savits (2006, Theorem 4.1) as the former maintains a coherent ordering in $\mathfrak{H}^{(r+1)}(\boldsymbol{x};\boldsymbol{\Sigma})$ which is straightforward to rearrange to the complete derivative $\mathsf{D}^{\otimes(r+1)}\phi_{\boldsymbol{\Sigma}}(\boldsymbol{x})$.

## 8.4 Recursive computation of the normal functionals

For the bandwidth selection algorithms, the $\nu$ functionals from Section 5.1.3 are a crucial component of the normal scale, normal mixture, plug-in and smoothed cross validation selectors. Whilst Holmquist (1996*b*, Theorem 8) exhibited their concise expression as $\nu_r(\mathbf{A}) = \mathbb{E}\{(\boldsymbol{Y}\mathbf{A}^\top\boldsymbol{Y})^r\} = (2r)!(\operatorname{vec}^\top\mathbf{A})^{\otimes r}\mathcal{S}_{d,2r}\sum_{i=0}^{r}[\boldsymbol{\mu}^{\otimes(2r-2i)} \otimes (\operatorname{vec}\boldsymbol{\Sigma})^{\otimes i}]/[i!(2r-2i)!2^i]$ where $\boldsymbol{Y} \sim N(\boldsymbol{\mu},\boldsymbol{\Sigma})$, this does not lend itself to efficient calculation. An alternative approach is based on the recursive relation between cumulants and lower order $\nu$ functionals. For a real random variable $Y$, its cumulant generating function is given by $\psi(t) = \log\mathbb{E}\{\exp(tY)\}$ and, if this function is $r$-times differentiable, the $r$-th cumulant of $Y$ is defined as $\psi^{(r)}(0)$ for $r \geq 1$. Mathai & Provost (1992, Theorem 3.2b.2) asserted that, for $r \geq 1$,

$$\nu_r(\mathbf{A}) = \sum_{i=0}^{r-1}\binom{r-1}{i}\kappa_{r-i}(\mathbf{A})\nu_i(\mathbf{A}),$$

with the recursion starting with $\nu_0(\mathbf{A}) = 1$, where $\kappa_r(\mathbf{A})$ is the $r$-th cumulant of the random quadratic form $\boldsymbol{Y}^\top\mathbf{A}\boldsymbol{Y}$, which is explicitly given by the formula

$$\kappa_r(\mathbf{A}) \equiv \kappa_r(\mathbf{A};\boldsymbol{\mu},\boldsymbol{\Sigma}) = 2^{r-1}(r-1)!\{\operatorname{tr}[(\mathbf{A}\boldsymbol{\Sigma})^r] + r\boldsymbol{\mu}^\top(\mathbf{A}\boldsymbol{\Sigma})^{r-1}\mathbf{A}\boldsymbol{\mu}\}.$$

For the mixed moment $\nu_{r,s}(\mathbf{A},\mathbf{B})$, Smith (1995, Equation (10)) showed that

$$\nu_{r,s}(\mathbf{A},\mathbf{B}) = \sum_{i=0}^{r}\sum_{j=0}^{s-1}\binom{r}{i}\binom{s-1}{j}\kappa_{r-i,s-j}(\mathbf{A},\mathbf{B})\nu_{i,j}(\mathbf{A},\mathbf{B}) \tag{8.6}$$

where $\kappa_{r,s}(\mathbf{A},\mathbf{B})$ is the joint $(r,s)$-th cumulant of $\boldsymbol{Y}^\top\mathbf{A}\boldsymbol{Y}$ and $\boldsymbol{Y}^\top\mathbf{B}\boldsymbol{Y}$. For $r+s \geq 1$ this is defined as $\partial^{r+s}/(\partial t_1^r \partial t_2^s)\psi(0,0)$, where $\psi(t_1,t_2) =$

$\log \mathbb{E}\{\exp(t_1 \boldsymbol{Y}^\top \mathbf{A} \boldsymbol{Y} + t_2 \boldsymbol{Y}^\top \mathbf{B} \boldsymbol{Y})\}$ is the joint cumulant generating function, and is explicitly given by

$$\kappa_{r,s}(\mathbf{A}, \mathbf{B}) = 2^{r+s-1} r! s! \sum_{\boldsymbol{i} \in \mathcal{MP}_{r,s}} \operatorname{tr}\left\{\mathbf{F}_{i_1} \cdots \mathbf{F}_{i_{r+s}} [\mathbf{I}_d/(r+s) + \mathbf{\Sigma}^{-1} \boldsymbol{\mu}\boldsymbol{\mu}^\top]\right\} \quad (8.7)$$

where $\mathbf{F}_1 = \mathbf{A}\mathbf{\Sigma}, \mathbf{F}_2 = \mathbf{B}\mathbf{\Sigma}$, and $\mathcal{MP}_{r,s} = \left\{\boldsymbol{i} = (i_1, \ldots, i_{r+s}) \in \{1,2\}^{r+s} \colon n_1(\boldsymbol{i}) = r,\ n_2(\boldsymbol{i}) = s\right\}$ is the set of permutations of the multiset having $r$ copies of 1 and $s$ copies of 2. Here $n_\ell(\boldsymbol{i})$ denotes the number of times that $\ell$ appears in $\boldsymbol{i}$, for $\ell = 1, 2$; i.e., $n_\ell(\boldsymbol{i}) = \sum_{k=1}^{r+s} \mathbf{1}\{i_k = \ell\}$. Chacón & Duong (2015, Theorem 3) established Equation (8.7), which corrects some errors in the formula derived from Mathai & Provost (1992, Theorem 3.3.4 and Corollary 3.3.1).

The other important class of functionals in the algorithms for the plug-in, and unbiased and smoothed cross validation selectors are $\eta_{2r}(\boldsymbol{x}; \mathbf{\Sigma}) = (\operatorname{vec}^\top \mathbf{I}_d)^{\otimes r} \mathsf{D}^{\otimes 2r} \phi_{\mathbf{\Sigma}}(\boldsymbol{x})$. In the previous section, we developed recursive formulas for $\mathsf{D}^{\otimes 2r} \phi_{\mathbf{\Sigma}}(\boldsymbol{x})$ at a single point $\boldsymbol{x}$, but here we require double sums of the type $Q_r(\mathbf{\Sigma}) = n^{-2} \sum_{i,j}^n \eta_{2r}(\boldsymbol{X}_i - \boldsymbol{X}_j; \mathbf{\Sigma})$. For large $n$, these can pose two different, in some sense dual, problems. If we enumerate singly the data difference $\boldsymbol{X}_i - \boldsymbol{X}_j$, then this increases the computation time in $n^2$. If we wish to take advantage of vectorised computations offered in many software packages, then this requires storing an $n^2 \times d$ matrix in memory which is not always feasible. Thus we search for a suitable compromise between execution speed and memory usage on commonly available desktop computers.

Chacón & Duong (2015, Theorem 4) asserted that $Q_r(\mathbf{\Sigma}) = \sum_{i,j=1}^n \phi_{\mathbf{\Sigma}}(\boldsymbol{X}_i - \boldsymbol{X}_j) \nu_r(\mathbf{I}_d; \mathbf{\Sigma}^{-1}(\boldsymbol{X}_i - \boldsymbol{X}_j), -\mathbf{\Sigma}^{-1})$, which requires the evaluations of the product of a $\nu_r$ functional for a non-constant mean with the normal density in a double sum. The presence of the $\nu_r$ functional allows for a recursive computation. The corresponding cumulant is $\kappa_r(\mathbf{I}_d; \mathbf{\Sigma}^{-1}(\boldsymbol{X}_i - \boldsymbol{X}_j), -\mathbf{\Sigma}^{-1}) = (-2)^{r-1}(r-1)!\{-\operatorname{tr}(\mathbf{\Sigma}^{-1}) + r(\boldsymbol{X}_i - \boldsymbol{X}_j)^\top \mathbf{\Sigma}^{-r-1}(\boldsymbol{X}_i - \boldsymbol{X}_j)$. This cumulant, along with the normal density $\phi_{\mathbf{\Sigma}}(\boldsymbol{X}_i - \boldsymbol{X}_j) = (2\pi)^{-d/2}|\mathbf{\Sigma}|^{-1/2} \exp\{-\frac{1}{2}(\boldsymbol{X}_i - \boldsymbol{X}_j)^\top \mathbf{\Sigma}^{-1}(\boldsymbol{X}_i - \boldsymbol{X}_j)\}$ are the two most computationally intensive operations in $Q_r(\mathbf{\Sigma})$, due to the calculation of the quadratic forms $(\boldsymbol{X}_i - \boldsymbol{X}_j)^\top \mathbf{\Sigma}^{-\ell}(\boldsymbol{X}_i - \boldsymbol{X}_j)$ for $1 \le \ell \le r+1$. If these are decomposed as $(\boldsymbol{X}_i - \boldsymbol{X}_j)^\top \mathbf{\Sigma}^{-\ell}(\boldsymbol{X}_i - \boldsymbol{X}_j) = \boldsymbol{X}_i^\top \mathbf{\Sigma}^{-\ell} \boldsymbol{X}_i + \boldsymbol{X}_j^\top \mathbf{\Sigma}^{-\ell} \boldsymbol{X}_j - 2\boldsymbol{X}_i^\top \mathbf{\Sigma}^{-\ell} \boldsymbol{X}_j$, then each term is efficiently handled by the vectorised computations in many software packages, in terms of execution speed but with memory requirements only slightly larger than storing the original sample $\boldsymbol{X}_1, \ldots, \boldsymbol{X}_n$, since the differences $\boldsymbol{X}_i - \boldsymbol{X}_j$, $j = 1, \ldots, n$, are kept in memory for each $i$ singly rather than for all $i$ as we loop over $i$.

This is outlined in Algorithm 20. Lines 1–5 perform this decomposition. Line 6 computes the normal density $\phi_{\mathbf{\Sigma}}$ from the decomposed quadratic from with $\boldsymbol{\Sigma}^{-1}$. Lines 7–9 recursively compute $\nu_\ell(\mathbf{I}_d;\boldsymbol{\Sigma}^{-1}(\boldsymbol{X}_i-\boldsymbol{X}_j),-\boldsymbol{\Sigma}^{-1})$, $\ell = 0,\ldots,r$, from the lower order functionals $\nu_k$ and cumulants $\kappa_k$, $k=0,\ldots,\ell$. Line 10 forms $Q_r$ from the inner product of $\phi_{\boldsymbol{\Sigma}}$ and $\nu_r$.

---

**Algorithm 20** Recursive computation of $Q_r$

---

**Input:** $\{\boldsymbol{X}_{,}\ldots,\boldsymbol{X}_n\},\boldsymbol{\Sigma},r$
**Output:** $Q_r(\boldsymbol{\Sigma})$
1: **for** $\ell := 1$ **to** $r+1$ **do**
2:     **for** $i,j := 1$ **to** $n$ **do**
3:         Decompose quadratic form $(\boldsymbol{X}_i-\boldsymbol{X}_j)^\top\boldsymbol{\Sigma}^{-\ell}(\boldsymbol{X}_i-\boldsymbol{X}_j)$
4:     **end for**
5: **end for**
6: Compute $\phi_{\boldsymbol{\Sigma}}(\boldsymbol{X}_i-\boldsymbol{X}_j)$ from decomposed quadratic form
7: **for** $\ell := 1$ **to** $r$ **do**
8:     Obtain $\nu_\ell$ from $\nu_0,\ldots,\nu_{\ell-1}$ and $\kappa_0,\ldots,\kappa_{\ell-1}$ from decomposed quadratic forms
9: **end for**
10: $Q_r(\boldsymbol{\Sigma}) := n^{-2}\sum_{i,j=1}^n \phi_{\boldsymbol{\Sigma}}(\boldsymbol{X}_i-\boldsymbol{X}_j)\nu_r(\mathbf{I}_d;\boldsymbol{\Sigma}^{-1}(\boldsymbol{X}_i-\boldsymbol{X}_j),-\boldsymbol{\Sigma}^{-1})$

---

Recall that from Equation (5.29) the SCV criterion for the $r$-th density derivative is

$$\begin{aligned}\mathrm{SCV}_r(\mathbf{H}) &= 2^{-(d+r)}\pi^{-d/2}n^{-1}|\mathbf{H}|^{-1/2}\nu_r(\mathbf{H}^{-1})\\ &\quad + (-1)^r n^{-2}\{Q_r(2\mathbf{H}+2\mathbf{G}) - 2Q_r(\mathbf{H}+2\mathbf{G}) + Q_r(2\mathbf{G})\}.\end{aligned}$$

The $\nu_r(\mathbf{H}^{-1})$ is efficiently computed using Equations (8.6)–(8.7) since $\nu_r \equiv \nu_{r,0}$. Using the binned approximations $\hat{\boldsymbol{\psi}}_{\mathrm{bin},2r}(\boldsymbol{\Sigma})$ of $\hat{\boldsymbol{\psi}}_{2r}(\boldsymbol{\Sigma})$ from Section 8.2 leads to $\hat{Q}_{\mathrm{bin},r}(\boldsymbol{\Sigma}) = (\mathrm{vec}^\top\mathbf{I}_d)^{\otimes r}\hat{\boldsymbol{\psi}}_{2r}(\boldsymbol{\Sigma})$. So the binned estimator of $\mathrm{SCV}_r(\mathbf{H})$ is $\mathrm{SCV}_{\mathrm{bin},r}(\mathbf{H}) = 2^{-(d+r)}\pi^{-d/2}n^{-1}|\mathbf{H}|^{-1/2}\nu_r(\mathbf{H}^{-1}) + (-1)^r n^{-2}\{\hat{Q}_{\mathrm{bin},r}(2\mathbf{H}+2\mathbf{G}) - 2\hat{Q}_{\mathrm{bin},r}(\mathbf{H}+2\mathbf{G}) + \hat{Q}_{\mathrm{bin},r}(2\mathbf{G})\}$. Otherwise the exact estimator can be computed using the procedures described in this section.

## 8.5 Numerical optimisation over matrix spaces

To compute data-based bandwidth selectors, most methods require a numerical optimisation of the appropriate objective function. As most numerical optimisation algorithms focus on scalar or vector-valued inputs, computing the constrained bandwidth selectors from the class $\mathcal{A}$ or $\mathcal{D}$ is straightforward. However, we have focused on unconstrained bandwidths from class $\mathcal{F}$, e.g., the plug-in bandwidth from Equation (5.24) is $\hat{\mathbf{H}}_{\mathrm{PI},r} = \mathrm{argmin}_{\mathbf{H}\in\mathcal{F}}\mathrm{PI}_r(\mathbf{H};\mathbf{G})$.

There are many fewer optimisation algorithms dedicated to matrix-valued inputs, especially for those with a special structure like those which comprise the class $\mathcal{F}$.

Fortunately we are not required to develop special matrix optimisation algorithms for our purposes, as we can adapt the existing vector optimisation ones. Since we require a symmetric, positive definite $d \times d$ bandwidth matrix, we carry out the optimisation over a $d^*$-vector $\boldsymbol{\eta}$, to obtain

$$\hat{\boldsymbol{\eta}} = \underset{\boldsymbol{\eta} \in \mathbb{R}^{d^*}}{\text{argmin}} \, \text{PI}_r\big((\text{vech}^{-1} \boldsymbol{\eta})(\text{vech}^{-1} \boldsymbol{\eta})^\top; \mathbf{G}\big)$$

where $d^* = d(d+1)/2$ and $\text{vech}^{-1}$ is the inverse vector half operator, i.e., it forms a symmetric $d \times d$ matrix from a vector of length $d^*$, such that $\text{vech}^{-1}(\text{vech}\,\mathbf{A}) = \mathbf{A}$ for any symmetric $d \times d$ matrix $\mathbf{A}$. The resulting plug-in bandwidth matrix $\hat{\mathbf{H}}_{\text{PI},r} = \hat{\boldsymbol{\eta}}\hat{\boldsymbol{\eta}}^\top$ is by construction symmetric and positive definite. We can thus take advantage of the many, efficiently coded optimisation algorithms for vector inputs, e.g., a Newton-type algorithm in Schnabel et al. (1985).

A convenient initial value $\boldsymbol{\eta}_0$ for this optimisation is based on the normal scale bandwidth, by setting $\boldsymbol{\eta}_0 = \text{vech}(\hat{\mathbf{H}}_{\text{NS},r}^{1/2})$ for a matrix square root of the normal scale selector $\hat{\mathbf{H}}_{\text{NS},r}$ in Equation (5.18).

Appendix A

# Notation

$\mathbf{1}_d$ is the $d$-vector whose elements are all ones

Ab is the Abramson selector

$[\boldsymbol{a}]_i$ is the $i$-th element of a vector $\boldsymbol{a}$

$\mathbf{1}\{\mathbf{A}\}$ is the indicator function for $\mathbf{A}$

$[\mathbf{A}]_{i,j}$ is the $(i,j)$-th element of a matrix $\mathbf{A}$

$\mathbf{A}^{\otimes r}$ is the $r$-fold Kronecker product of a matrix $\mathbf{A}$

$\mathbf{A}^{1/2}$ is the matrix square root of a matrix $\mathbf{A}$

$\mathbf{A}^\top$ is the transpose of a matrix $\mathbf{A}$

$\bar{A}$ is the closure of set $A$

$A \triangle B$ is the symmetric difference between two sets $A$ and $B$

$\mathcal{A} = \{h^2\mathbf{I}_d : h > 0\}$ is the class of scalar matrices

AMISE is the asymptotic mean integrated squared error

AMSE is the asymptotic mean squared error

ARE is the asymptotic relative efficiency

Bias is the bias

$B(\alpha_1, \alpha_2) = \Gamma(\alpha_1)\Gamma(\alpha_2)/\Gamma(\alpha_1 + \alpha_2)$ is the complete beta function with parameters $\alpha_1, \alpha_2$

$B_d(\boldsymbol{x}, r)$ is the $d$-dimensional ball centred at $\boldsymbol{x}$ with radius $r$

$\hat{\boldsymbol{b}}$ is a data-based binwidth

BCV is biased cross validation

$\mathscr{C} = \{C_1, \dots, C_q\}$ is a clustering partition with $q$ classes

CV is cross validation

$d$ is the data dimension

$\mathsf{d}$ is the differential operator

$\mathcal{D}$ is the class of diagonal matrices

$\mathsf{D}$ is the vector differential operator

$\mathsf{D}^{\otimes r}$ is the $r$-th order differential operator

$\mathsf{D}_{\mathbf{H}}$ is the differential operator with respect to $\mathbf{H}$

$\mathsf{D}^{\otimes r}\hat{f}$ is an $r$-th order kernel density derivative estimator

$\mathsf{D}^{\otimes r}\hat{f}_{\mathrm{NN}}$ is an $r$-th order nearest neighbour density derivative estimator

$\partial/\partial x_i$ is the partial derivative operator with respect to the $i$-th coordinate

$\delta_{(k)}$ is the $k$-th nearest neighbour distance

$\mathbb{E}$ is the expected value operator

$f, f_{\boldsymbol{X}}$ is a probability density function, of the random variable $\boldsymbol{X}$

$f^{(\boldsymbol{r})}$ is a single $\boldsymbol{r}$-th order partial derivative of $f$

$f_+^{(\boldsymbol{r})}, f_-^{(\boldsymbol{r})}$ are the positive and negative parts of $f^{(\boldsymbol{r})}$

$f_\tau$ is the height of the $100\tau\%$ probability contour of a density $f$

$\hat{f}, \hat{f}(\cdot;\mathbf{H})$ is a kernel density estimator, with bandwidth matrix $\mathbf{H}$

$\tilde{f}, \tilde{f}(\cdot;\mathbf{G})$ is a pilot kernel density estimator, with pilot bandwidth matrix $\mathbf{G}$

$\hat{f}_{-i}$ is a leave out $i$ kernel density estimator

$\hat{f}_{\mathrm{ball}}$ is a balloon variable kernel density estimator

$\hat{f}_{\mathrm{beta}}$ is a beta boundary kernel density estimator

$\hat{f}_{\mathrm{dc}}$ is a deconvolution density estimator

$\hat{f}_{\mathrm{hist}}, \hat{f}_{\mathrm{hist}}(\cdot;\boldsymbol{b})$ is a histogram density estimator, with binwidth $\boldsymbol{b}$

$\hat{f}_{\mathrm{LB}}$ is a linear boundary kernel density estimator

$\hat{f}_{\mathrm{NN}}, \hat{f}_{\mathrm{NN}}(\cdot;k)$ is a nearest neighbour density estimator, with $k$ nearest neighbours

$\hat{f}_{\mathrm{SP}}$ is a sample point variable kernel density estimator

$\hat{f}_{\mathrm{trans}}$ is a transformation kernel density estimator

$\hat{f}_{\mathrm{wdc}}$ is a weighted deconvolution density estimator

$F, f_{\boldsymbol{X}}$ is a cumulative distribution function, of the random variable $\boldsymbol{X}$

$F(\mathbf{A})$ is the probability that $\boldsymbol{X}$ belongs to $\mathbf{A}$ for $\boldsymbol{X} \sim F$

$\mathcal{F}$ is the class of positive definite, symmetric matrices

$g$ is a scalar pilot bandwidth

$\mathbf{G}$ is a pilot bandwidth matrix

$\gamma$ is a cluster labelling function

$\gamma_{\mathrm{Bayes}}$ is the cluster labelling function of the Bayes classifier

$h$ is a scalar bandwidth

$\hat{h}$ is a data-based bandwidth selector

$\mathbf{H}$ is a bandwidth (or smoothing parameter) matrix

$\hat{\mathbf{H}}$ is a data-based bandwidth matrix selector

$\mathsf{H}$ is the Hessian operator

$\mathsf{H}_{\mathbf{H}}$ is the Hessian operator with respect to $\mathbf{H}$

$\mathcal{H}_r$ is the $r$-th order multivariate Hermite polynomial

$\mathbf{I}_d$ is the $d \times d$ identity matrix

ISB is the integrated squared bias

ISE is the integrated squared error

IV is the integrated variance

$\mathbf{J}_d$ is the $d \times d$ matrix of ones

$K_{(1)}$ is a univariate kernel function

$K_{\text{beta}(1)}, K_{(1)}(\cdot; 1)$ is the univariate beta kernel function

$K$ is a multivariate kernel function

$K_h(x)$ is a scaled univariate kernel function

$K_{\mathbf{H}}(\boldsymbol{x})$ is a scaled multivariate kernel function

$K_{\text{LB}}$ is a linear boundary kernel function

$K^P$ is a multivariate product kernel

$K^S$ is a multivariate spherically symmetric kernel

$K^S(\cdot; r)$ is the $r$-th beta family spherically symmetric kernel

$K^{\boldsymbol{U}}$ is a deconvolution kernel

$\mathcal{L}(c), \mathcal{L}(f; c)$ is the level set of $f$ at height $c$

$\hat{\mathcal{L}}(c), \hat{\mathcal{L}}(\hat{f}; c)$ is the level set of the kernel density estimator $\hat{f}$ at height $c$

$\mathcal{L}^*$ is a level set with strict inequality

$\boldsymbol{\Lambda}$ is a diagonal matrix of eigenvalues

$\|\cdot\|$ is the Euclidean norm

$\mathcal{M}, \mathcal{M}(f)$ is the modal region of the density function $f$

$\mathcal{M}_{r \times s}$ is the class of $r \times s$ matrices

$m_r(K)$ is the $r$-th order moment of a kernel $K$

$\boldsymbol{m}_r(K)$ is the $r$-th order vector moment of a kernel $K$

$\boldsymbol{m}_r(\mathbf{A}; K)$ is the $r$-th order vector moment of a kernel $K$, restricted to $\mathbf{A}$

$\boldsymbol{\mu}_r$ is the $r$-th vector moment of a standard normal distribution

MISE is the mean integrated squared error

MR is the misclassification rate

MS is maximal smoothing

MSE is the mean squared error

$N(\boldsymbol{\mu}, \boldsymbol{\Sigma})$ is the normal distribution with mean $\boldsymbol{\mu}$ and variance $\boldsymbol{\Sigma}$

NM is normal mixture

NS is normal scale

$\nu_r, \nu_{r,s}$ are the $r$-th and $(r,s)$-th order functionals of quadratic forms of normal random variables

$o, O$ are the small and big O orders

$o_P, O_P$ are the small and big O orders in probability

$\Omega$ is a finite data support

$\otimes$ is the Kronecker product operator between two matrices

OF is the odd factorial

$\phi$ is the standard $d$-variate normal density

$\phi_{\boldsymbol{\Sigma}}(\cdot - \boldsymbol{\mu})$ is the $d$-variate normal density with mean $\boldsymbol{\mu}$ and variance $\boldsymbol{\Sigma}$

$\mathbb{P}$ is the probability operator

$\mathcal{P}, \mathcal{P}(f)$ is the ridge of the density $f$

PI is plug-in

$\varphi_a$ is the characteristic function of function $a$

$\varphi_{\mathbf{X}}$ is the characteristic function of density $f_{\mathbf{X}}$

$\boldsymbol{\psi}_r$ is the $r$-th order integrated density functional

$\hat{\boldsymbol{\psi}}_r$ is the kernel integrated density functional estimator

$r$ is the derivative order

$R(a) = \int_{\mathbb{R}^d} a(\boldsymbol{x})^2 \, d\boldsymbol{x}$

$\mathbf{R}(\boldsymbol{a}) = \int_{\mathbb{R}^d} \boldsymbol{a}(\boldsymbol{x})\boldsymbol{a}(\boldsymbol{x})^\top \, d\boldsymbol{x}$

$\mathbf{R}(b, \boldsymbol{a}) = \int_{\mathbb{R}^d} b * \boldsymbol{a}(\boldsymbol{x})\boldsymbol{a}(\boldsymbol{x})^\top \, d\boldsymbol{x}$

$\mathbb{R}^d$ is the $d$-dimensional Euclidean space

$s$ is the summary curvature function

$S, S(f)$ is the support of density function $f$

$\mathbf{S}$ is a sample variance matrix

$\mathbb{S}_{d,r}$ is the $d^r \times d^r$ symmetriser matrix

SCV is smoothed cross validation

$\operatorname{tr}\mathbf{A}$ is the trace of a square matrix $\mathbf{A}$

$\triangle$ is the Laplacian operator

$\mathbf{U}$ is a matrix of eigenvectors

$\mathbf{U}^{(p)}, \mathbf{U}_{(p)}$ are matrices of the $p$ eigenvectors with the largest/smallest $p$ eigenvalues

$\mathcal{U}^+, \mathcal{U}^-$ are significant positive/negative regions of a density difference

UCV is unbiased cross validation

Var is the variance

vec is the vector operator

vech is the vector half operator

$W_+^s, W_-^u$ are stable/unstable manifolds

$\boldsymbol{\xi}$ is a critical point of a function

$\boldsymbol{x}, \boldsymbol{y}, \boldsymbol{z}$ etc. are $d$-dimensional free variables

$\boldsymbol{X}, \boldsymbol{Y}, \boldsymbol{Z}$ etc. are $d$-dimensional random vectors

$\sim$ is the operator which indicates asymptotic equivalence

$\sim$ is the operator which indicates distribution

$*$ is the convolution operator of two functions

Appendix B

# Matrix algebra

For completeness, this appendix collects some well-known matrix algebra results that are used throughout the monograph. More details of them can be found in Magnus & Neudecker (1999) and Schott (2005).

## B.1 The Kronecker product

For any pair of matrices $\mathbf{A} = (a_{ij}) \in \mathcal{M}_{m\times n}$ and $\mathbf{B} \in \mathcal{M}_{p\times q}$, their Kronecker product is defined as the matrix $\mathbf{A}\otimes\mathbf{B} \in \mathcal{M}_{(mp)\times(nq)}$ formed by $(m\times n)$ blocks, with the $(i,j)$th block given by $a_{ij}\mathbf{B} \in \mathcal{M}_{p\times q}$; that is,

$$\mathbf{A}\otimes\mathbf{B} = \left[\begin{array}{c|c|c} a_{11}\mathbf{B} & \cdots & a_{1n}\mathbf{B} \\ \hline \vdots & \ddots & \vdots \\ \hline a_{m1}\mathbf{B} & \cdots & a_{mn}\mathbf{B} \end{array}\right].$$

Similarly, the $r$th Kronecker power of a matrix $\mathbf{A} \in \mathcal{M}_{m\times n}$ is defined as $\mathbf{A}^{\otimes r} = \bigotimes_{i=1}^{r}\mathbf{A} = \mathbf{A}\otimes\mathbf{A}\otimes\cdots\otimes\mathbf{A} \in \mathcal{M}_{m^r\times n^r}$.

For conformable matrices $\mathbf{A}$, $\mathbf{B}$, $\mathbf{C}$, $\mathbf{D}$, some of the basic properties of the Kronecker product are:

$$\begin{gathered}
\mathbf{A}\otimes\mathbf{B} \neq \mathbf{B}\otimes\mathbf{A} \\
\mathbf{A}\otimes\mathbf{B}\otimes\mathbf{C} = (\mathbf{A}\otimes\mathbf{B})\otimes\mathbf{C} = \mathbf{A}\otimes(\mathbf{B}\otimes\mathbf{C}) \\
(\mathbf{A}+\mathbf{B})\otimes\mathbf{C} = \mathbf{A}\otimes\mathbf{C}+\mathbf{B}\otimes\mathbf{C} \\
(\mathbf{A}\otimes\mathbf{B})(\mathbf{C}\otimes\mathbf{D}) = (\mathbf{AC})\otimes(\mathbf{BD}) \\
(\mathbf{A}\otimes\mathbf{B})^\top = \mathbf{A}^\top\otimes\mathbf{B}^\top \\
(\mathbf{A}\otimes\mathbf{B})^{-1} = \mathbf{A}^{-1}\otimes\mathbf{B}^{-1} \\
\operatorname{tr}(\mathbf{A}\otimes\mathbf{B}) = (\operatorname{tr}\mathbf{A})(\operatorname{tr}\mathbf{B}).
\end{gathered}$$

Also, $\alpha\otimes\mathbf{A} = \alpha\mathbf{A} = \mathbf{A}\alpha = \mathbf{A}\otimes\alpha$ for $\alpha \in \mathbb{R}$, and $|\mathbf{A}\otimes\mathbf{B}| = |\mathbf{A}|^p|\mathbf{B}|^m$ for $\mathbf{A} \in \mathcal{M}_{m\times m}$, $\mathbf{B} \in \mathcal{M}_{p\times p}$.

The Kronecker product also makes sense for vectors, provided they are understood as single-column matrices. Hence, for $\boldsymbol{a} = (a_1, \dots, a_p) \in \mathbb{R}^p$ and $\boldsymbol{b} \in \mathbb{R}^q$ we have $\boldsymbol{a} \otimes \boldsymbol{b} = (a_1\boldsymbol{b}; \dots; a_p\boldsymbol{b}) \in \mathbb{R}^{pq}$ and it is straightforward to verify that $\boldsymbol{a} \otimes \boldsymbol{b}^\top = \boldsymbol{a}\boldsymbol{b}^\top = \boldsymbol{b}^\top \otimes \boldsymbol{a} \in \mathcal{M}_{p\times q}$.

## B.2 The vec operator

If $\mathbf{A} = (a_{ij}) \in \mathcal{M}_{m\times n}$ then $\operatorname{vec}\mathbf{A} \in \mathbb{R}^{mn}$ is the vector constructed by stacking the columns of $\mathbf{A}$ one underneath the other; that is,

$$\operatorname{vec}\mathbf{A} = (a_{11}, \dots, a_{m1}; \dots; a_{1n}, \dots, a_{mn}).$$

For conformable matrices **A**, **B**, **C** and vectors $\boldsymbol{a}$, $\boldsymbol{b}$, some of the properties of the vec operator which are extensively used in this monograph are:

$$\begin{aligned}
\operatorname{vec}\boldsymbol{a} &= \operatorname{vec}\boldsymbol{a}^\top = \boldsymbol{a} \\
\operatorname{vec}(\mathbf{ABC}) &= (\mathbf{C}^\top \otimes \mathbf{A}) \operatorname{vec}\mathbf{B} \\
\operatorname{vec}(\boldsymbol{a}\boldsymbol{b}^\top) = \operatorname{vec}(\boldsymbol{a}\otimes\boldsymbol{b}^\top) &= \operatorname{vec}(\boldsymbol{b}^\top \otimes \boldsymbol{a}) = \boldsymbol{b}\otimes\boldsymbol{a} \\
\operatorname{tr}(\mathbf{A}^\top\mathbf{B}) &= (\operatorname{vec}\mathbf{A})^\top \operatorname{vec}\mathbf{B}.
\end{aligned}$$

## B.3 The commutation matrix

Let $\mathbf{A} \in \mathcal{M}_{m\times n}$ and $\mathbf{B} \in \mathcal{M}_{p\times q}$. The commutation matrix $\mathbf{K}_{m,n} \in \mathcal{M}_{(mn)\times(mn)}$ is the only matrix that transforms $\operatorname{vec}\mathbf{A}$ into $\operatorname{vec}(\mathbf{A}^\top)$; that is, $\mathbf{K}_{m,n}\operatorname{vec}\mathbf{A} = \operatorname{vec}(\mathbf{A}^\top)$.

The basic properties of the commutation matrix that are used in this book include:

$$\begin{aligned}
\mathbf{K}_{m,n}^\top = \mathbf{K}_{m,n}^{-1} &= \mathbf{K}_{n,m} \\
\mathbf{K}_{p,m}(\mathbf{A}\otimes\mathbf{B})\mathbf{K}_{n,q} &= \mathbf{B}\otimes\mathbf{A} \\
\operatorname{vec}(\mathbf{A}\otimes\mathbf{B}) &= (\mathbf{I}_n \otimes \mathbf{K}_{q,m} \otimes \mathbf{I}_p)(\operatorname{vec}\mathbf{A}\otimes\operatorname{vec}\mathbf{B}).
\end{aligned}$$

# Bibliography

Abramson, I. S. (1982), 'On bandwidth variation in kernel estimates: A square root law', *Annals of Statistics* **10**, 1217–1223.

Aghaeepour, N., Finak, G., The FlowCAP Consortium, The DREAM Consortium, Hoos, H., Mosmann, T. R., Brinkman, R., Gottardo, R. & Scheuermann, R. H. (2013), 'Critical assessment of automated flow cytometry data analysis techniques', *Nature Methods* **10**, 228–238.

Akaike, H. (1954), 'An approximation to the density function', *Annals of the Institute of Statistical Mathematics* **6**, 127–132.

Aldershof, B. K. (1991), Estimation of Integrated Squared Density Derivatives, PhD thesis, University of North Carolina, Chapel Hill, USA.

Aliyari Ghassabeh, Y. (2013), 'On the convergence of the mean shift algorithm in the one-dimensional space', *Pattern Recognition Letters* **34**, 1423–1427.

Amezziane, M. & McMurry, T. (2012), 'Bandwidth selection for adaptive density estimators', *Journal of Applied Probability and Statistics* **6**, 29–46.

Arias-Castro, E., Mason, D. & Pelletier, B. (2016*a*), 'Errata: On the estimation of the gradient lines of a density and the consistency of the mean-shift algorithm', *Journal of Machine Learning Research* **17(206)**, 1–4.

Arias-Castro, E., Mason, D. & Pelletier, B. (2016*b*), 'On the estimation of the gradient lines of a density and the consistency of the mean-shift algorithm', *Journal of Machine Learning Research* **17(43)**, 1–28.

Ayres-de Campos, D., Bernardes, J., Garrido, A., Marques-de Sá, J. & Pereira-Leite, L. (2000), 'SisPorto 2.0: A program for automated analysis of cardiotocograms', *Journal of Maternal-Fetal Medicine* **9**, 311–318.

Azzalini, A. & Torelli, N. (2007), 'Clustering via nonparametric density estimation', *Statistics and Computing* **17**, 71–80.

Baíllo, A. (2003), 'Total error in a plug-in estimator of level sets', *Statistics and Probability Letters* **65**, 411–417.

Baíllo, A. & Chacón, J. E. (2018), 'A survey and a new selection criterion for statistical home range estimation'. Unpublished manuscript.

Baíllo, A., Cuesta-Albertos, J. A. & Cuevas, A. (2001), 'Convergence rates in nonparametric estimation of level sets', *Statistics and Probability Letters* **53**, 27–35.

Baíllo, A., Cuevas, A. & Justel, A. (2000), 'Set estimation and nonparametric detection', *Canadian Journal of Statistics* **28**, 765–782.

Bartlett, M. S. (1951), 'An inverse matrix adjustment arising in discriminant analysis', *Annals of Mathematical Statistics* **22**, 107–111.

Beirlant, J., Györfi, L. & Lugosi, G. (1994), 'On the asymptotic normality of the $L_1$- and $L_2$-errors in histogram density estimation', *Canadian Journal of Statistics* **22**, 309–318.

Bellman, R. (1961), *Adaptive Control Processes: A Guided Tour*, Princeton University Press, Princeton.

Berlinet, A. & Devroye, L. (1994), 'A comparison of kernel density estimates', *Publications de l'Institute de Statistique de L'Université de Paris* **38**, 3–59.

Bhattachatya, P. (1967), 'Estimation of a probability density function and its derivatives', *Sankhyā, Series A* **29**, 373–382.

Biau, G., Chazal, F., Cohen-Steiner, D., Devroye, L. & Rodríguez, C. (2011), 'A weighted $k$-nearest neighbor density estimate for geometric inference', *Electronic Journal of Statistics* **5**, 204–237.

Bird, P. (2003), 'An updated digital model of plate boundaries', *Geochemistry, Geophysics, Geosystems* **4(3)**, 1–52. 1027. Data set accessed 2016-03-24 from `http://peterbird.name/publications/2003_PB2002/2003_PB2002.htm`.

Bobrowski, O., Mukherjee, S. & Taylor, J. E. (2017), 'Topological consistency via kernel estimation', *Bernoulli* **23**, 288–328.

Bosq, D. & Lecoutre, J.-P. (1987), *Théorie de l'Estimation Fonctionnelle*, Économica, Paris.

Bouezmarni, T. & Rombouts, J. V. K. (2010), 'Nonparametric density estimation for multivariate bounded data', *Journal of Statistical Planning and Inference* **140**, 139–152.

Bowman, A. W. (1984), 'An alternative method of cross-validation for the smoothing of density estimates', *Biometrika* **71**, 353–360.

Bowman, A. W. & Azzalini, A. (1997), *Applied Smoothing Techniques for Data Analysis*, Oxford University Press, New York.

Bowman, A. W. & Foster, P. (1993*a*), 'Adaptive smoothing and density-based tests of multivariate normality', *Journal of the American Statistical Association* **88**, 529–537.

Bowman, A. W. & Foster, P. (1993*b*), 'Density based exploration of bivariate data', *Statistics and Computing* **3**, 171–177.

Breiman, L., Meisel, W. & Purcell, E. (1977), 'Variable kernel estimates of probability density estimates', *Technometrics* **19**, 135–144.

Burt, W. H. (1943), 'Territoriality and home range concepts as applied to mammals', *Journal of Mammalogy* **24**, 346–352.

Cacoullos, T. (1966), 'Estimation of a multivariate density', *Annals of the Institute of Statistical Mathematics* **18**, 179–189.

Cadre, B. (2006), 'Kernel estimation of density level sets', *Journal of Multivariate Analysis* **97**, 999–1023.

Cadre, B., Pelletier, B. & Pudlo, P. (2013), 'Estimation of density level sets with a given probability content', *Journal of Nonparametric Statistics* **25**, 261–272.

Cao, R. (1993), 'Bootstrapping the mean integrated squared error', *Journal of Multivariate Analysis* **45**, 137–160.

Cao, R. (2001), 'Relative efficiency of local bandwidths in kernel density estimation', *Statistics* **35**, 113–137.

Cao, R., Cuevas, A. & González-Manteiga, W. (1994), 'A comparative study of several smoothing methods in density estimation', *Computational Statistics and Data Analysis* **17**, 153–176.

Carr, D. B., Littlefield, R. J., Nicholson, W. L. & Littlefield, J. S. (1987), 'Scatterplot matrix techniques for large n', *Journal of the American Statistical Association* **82**, 424–436.

Carroll, R. J. & Hall, P. (1988), 'Optimal rates of convergence for deconvolving a density', *Journal of the American Statistical Association* **83**, 1184–1186.

Carroll, R. J., Rupper, D., Stefanski, L. A. & Crainceanu, C. (2006), *Measurement Error in Nonlinear Models*, 2nd edn, Chapman and Hall/CRC, Boca Raton.

Castells, M. (2010), *The Rise of the Network Society: The Information Age: Economy, Society, and Culture*, Vol. 1, 2nd edn, John Wiley and Sons, Chichester.

Celeux, G. & Govaert, G. (1995), 'Gaussian parsimonious clustering models', *Pattern Recognition* **28**, 781–793.

Chacón, J. E. (2009), 'Data-driven choice of the smoothing parametrization for kernel density estimators', *Canadian Journal of Statistics* **37**, 249–255.

Chacón, J. E. (2015), 'A population background for nonparametric density-based clustering', *Statistical Science* **30**, 518–532.

Chacón, J. E. & Duong, T. (2010), 'Multivariate plug-in bandwidth selection with unconstrained bandwidth matrices', *Test* **19**, 375–398.

Chacón, J. E. & Duong, T. (2011), 'Unconstrained pilot selectors for smoothed cross-validation', *Australian and New Zealand Journal of Statistics* **53**, 331–351.

Chacón, J. E. & Duong, T. (2013), 'Data-driven density estimation, with applications to nonparametric clustering and bump hunting', *Electronic Journal of Statistics* **7**, 499–532.

Chacón, J. E. & Duong, T. (2015), 'Efficient recursive algorithms for functionals based on higher order derivatives of the multivariate Gaussian density', *Statistics and Computing* **25**, 959–974.

Chacón, J. E., Duong, T. & Wand, M. P. (2011), 'Asymptotics for general multivariate kernel density derivative estimators', *Statistica Sinica* **21**, 807–840.

Chacón, J. E. & Monfort, P. (2014), 'A comparison of bandwidth selectors for mean shift clustering', in C. H. Skiadas, ed., *Theoretical and Applied Issues in Statistics and Demography*, International Society for the Advancement of Science and Technology (ISAST), Athens, pp. 47–59.

Chacón, J. E., Montanero, J. & Nogales, A. G. (2007), 'A note on kernel density estimation at a parametric rate', *Journal of Nonparametric Statistics* **19**, 13–21.

Chaudhuri, P. & Marron, J. S. (1999), 'SiZer for exploration of structures in curves', *Journal of the American Statistical Association* **94**, 807–823.

Chen, S. X. (1999), 'Beta kernel estimators for density functions', *Computational Statistics and Data Analysis* **31**, 131–145.

Chen, Y.-C., Genovese, C. R. & Wasserman, L. (2016), 'A comprehensive approach to mode clustering', *Electronic Journal of Statistics* **10**, 210–241.

Ciollaro, M., Genovese, C. R. & Wang, D. (2016), 'Nonparametric clustering of functional data using pseudo-densities', *Electronic Journal of*

*Statistics* **10**, 2922–2972.

Comaniciu, D. & Meer, P. (2002), 'Mean shift: A robust approach toward feature space analysis', *IEEE Transactions on Pattern Analysis and Machine Intelligence* **24**, 603–619.

CSIRO (2016), 'Atlas of Living Australia: *Grevillea uncinulata* Diels', `http://bie.ala.org.au/species/urn:lsid:biodiversity.org.au:apni.taxon:249319`. Commonwealth Scientific and Industrial Research Organisation. Accessed 2016-03-11.

Cuevas, A. & Fraiman, R. (1997), 'A plug-in approach to support estimation', *Annals of Statistics* **25**, 2300–2312.

Cuevas, A., González-Manteiga, W. & Rodríguez-Casal, A. (2006), 'Plug-in estimation of general level sets', *Australian and New Zealand Journal of Statistics* **48**, 7–19.

Cwik, J. & Koronacki, J. (1997), 'A combined adaptive-mixtures/plug-in estimator of multivariate probability densities', *Computational Statistics and Data Analysis* **26**, 199–218.

Deheuvels, P. (1977), 'Estimation non paramétrique de la densité par histogrammes généralisés. II', *Publications de l'Institut de Statistique de l'Université de Paris* **22**, 1–23.

Delaigle, A. (2014), 'Nonparametric kernel methods with errors-in-variables: constructing estimators, computing them, and avoiding common mistakes', *Australian & New Zealand Journal of Statistics* **56**, 105–124.

Delaigle, A. & Gijbels, I. (2004), 'Practical bandwidth selection in deconvolution kernel density estimation', *Computational Statistics & Data Analysis* **45**, 249–267.

Delicado, P. & Vieu, P. (2015), 'Optimal level sets for bivariate density representation', *Journal of Multivariate Analysis* **140**, 1–18.

Devroye, L. (1983), 'The equivalence of weak, strong and complete convergence in $l_1$ for kernel density estimates', *Annals of Statistics* **11**, 896–904.

Devroye, L. (1992), 'On the usefulness of superkernels in density estimation', *Annals of Statistics* **20**, 2037–2056.

Devroye, L. & Györfi, L. (1985), *Nonparametric Density Estimation: The $L_1$ View*, John Wiley and Sons, New York.

Devroye, L., Györfi, L. & Lugosi, G. (1996), *A Probabilistic Theory of Pattern Recognition*, Springer-Verlag, New York.

Devroye, L. & Lugosi, G. (2001*a*), *Combinatorial Methods in Density Estimation*, Springer Science and Business Media, New York.

Devroye, L. & Lugosi, G. (2001*b*), 'Variable kernel estimates: on the impossibility of tuning the parameters', in E. Giné, D. M. Mason & J. A. Wellner, eds, *High-Dimensional Probability II*, Birkhäuser, Boston, pp. 405–424.

Devroye, L. & Wise, G. L. (1980), 'Detection of abnormal behavior via nonparametric estimation of the support', *SIAM Journal on Applied Mathematics* **38**, 480–488.

Dobrow, R. (1992), 'Estimating level sets of densities'. Unpublished manuscript.

Duda, R. O., Hart, P. E. & Stork, D. G. (2000), *Pattern Classification*, 2nd edn, Wiley-Interscience.

Dümbgen, L. & Walther, G. (1996), 'Rates of convergence for random approximations of convex sets', *Advances in Applied Probability* **28**, 384–393.

Duong, T. (2004), Bandwidth Selectors for Multivariate Kernel Density Estimation, PhD thesis, University of Western Australia, Australia.

Duong, T. (2007), 'ks: Kernel density estimation and kernel discriminant analysis for multivariate data in R', *Journal of Statistical Software* **21(7)**, 1–16.

Duong, T. (2013), 'Local significant differences from nonparametric two-sample tests', *Journal of Nonparametric Statistics* **25**, 635–645.

Duong, T. (2015), 'Spherically symmetric multivariate beta family kernels', *Statistics and Probability Letters* **104**, 141–145.

Duong, T., Beck, G., Azzag, H. & Lebbah, M. (2016), 'Nearest neighbour estimators of density derivatives, with application to mean shift clustering', *Pattern Recognition Letters* **80**, 224–230.

Duong, T., Cowling, A., Koch, I. & Wand, M. P. (2008), 'Feature significance for multivariate kernel density estimation', *Computational Statistics and Data Analysis* **52**, 4225–4242.

Duong, T. & Hazelton, M. L. (2003), 'Plug-in bandwidth matrices for bivariate kernel density estimation', *Journal of Nonparametric Statistics* **15**, 17–30.

Duong, T. & Hazelton, M. L. (2005*a*), 'Convergence rates for unconstrained bandwidth matrix selectors in multivariate kernel density estimation', *Journal of Multivariate Analysis* **93**, 417–433.

Duong, T. & Hazelton, M. L. (2005*b*), 'Cross-validation bandwidth matrices for multivariate kernel density estimation', *Scandinavian Journal of Statistics* **32**, 485–506.

Epanechnikov, V. A. (1969), 'Non-parametric estimation of a multivariate probability density', *Theory of Probability and Its Applications* **14**, 153–158.

Everitt, B. S., Landau, S., Leese, M. & Stahl, D. (2011), *Cluster Analysis*, 5th edn, John Wiley & Sons, Chichester.

Faraway, J. J. & Jhun, M. (1990), 'Bootstrap choice of bandwidth for density estimation', *Journal of the American Statistical Association* **85**, 1119–1122.

Fix, E. & Hodges, J. L. (1951), Discriminatory analysis–nonparametric discrimination: Consistency properties, Technical Report Project 21-49-004, Report 4, USAF School of Aviation Medicine, Randolph Field, Texas.

Fix, E. & Hodges, J. L. (1989), 'Discriminatory analysis–nonparametric discrimination: Consistency properties', *International Statistical Review* **57**, 238–247.

Fukunaga, K. & Hostetler, L. (1975), 'The estimation of the gradient of a density function, with applications in pattern recognition', *IEEE Transactions on Information Theory* **21**, 32–40.

Gasser, T. & Müller, H.-G. (1979), 'Kernel estimation of regression functions', in T. Gasser & M. Rosenblatt, eds, *Smoothing Techniques for Curve Estimation*, Springer-Verlag, Berlin, pp. 23–68.

Geenens, G. (2014), 'Probit transformation for kernel density estimation on the unit interval', *Journal of the American Statistical Association* **109**, 346–358.

Genovese, C. R., Perone-Pacifico, M., Verdinelli, I. & Wasserman, L. (2012), 'Manifold estimation and singular deconvolution under Hausdorff loss', *Annals of Statistics* **40**, 941–963.

Genovese, C. R., Perone-Pacifico, M., Verdinelli, I. & Wasserman, L. (2014), 'Nonparametric ridge estimation', *Annals of Statistics* **42**, 1511–1545.

Genovese, C. R., Perone-Pacifico, M., Verdinelli, I. & Wasserman, L. (2016), 'Non-parametric inference for density modes', *Journal of the Royal Statistical Society: Series B* **78**, 99–126.

Ghosh, A. K. & Chaudhuri, P. (2004), 'Optimal smoothing in kernel discriminant analysis', *Statistica Sinica* **14**, 457–483.

Ghosh, A. K. & Hall, P. (2008), 'On error-rate estimation in nonparametric classification', *Statistica Sinica* **18**, 1081–1100.

Givan, A. L. (2001), *Flow Cytometry: First Principles*, 2nd edn, Wiley and Sons, New York.

Godtliebsen, F., Marron, J. S. & Chaudhuri, P. (2002), 'Significance in scale space for bivariate density estimation', *Journal of Computational and Graphical Statistics* **11**, 1–21.

Good, I. J. & Gaskins, R. A. (1980), 'Density estimation and bump-hunting by the penalized likelihood method exemplified by scattering and meteorite data', *Journal of the American Statistical Association* **75**, 42–56.

Gramacki, A. & Gramacki, J. (2017*a*), 'FFT-based fast bandwidth selector for multivariate kernel density estimation', *Computational Statistics and Data Analysis* **106**, 27–45.

Gramacki, A. & Gramacki, J. (2017*b*), 'FFT-based fast computation of multivariate kernel density estimators with unconstrained bandwidth matrices', *Journal of Computational and Graphical Statistics* **26**, 459–462.

Hall, P. (1983), 'On near neighbour estimates of a multivariate density', *Journal of Multivariate Analysis* **13**, 24–39.

Hall, P. (1984), 'Central limit theorem for integrated square error of multivariate nonparametric density estimators', *Journal of Multivariate Analysis* **14**, 1–16.

Hall, P. (1993), 'On plug-in rules for local smoothing of density estimators', *Annals of Statistics* **21**, 694–710.

Hall, P. & Kang, K. H. (2005), 'Bandwidth choice for nonparametric classification', *Annals of Statistics* **33**, 284–306.

Hall, P. & Marron, J. S. (1988), 'Variable window width kernel estimates of probability densities', *Probability Theory and Related Fields* **80**, 37–49.

Hall, P. & Marron, J. S. (1991), 'Local minima in cross-validation functions', *Journal of the Royal Statistical Society Series B* **53**, 245–252.

Hall, P., Marron, J. S. & Park, B. U. (1992), 'Smoothed cross-validation', *Probability Theory and Related Fields* **92**, 1–20.

Hall, P. & Schucany, W. R. (1989), 'A local cross validation algorithm', *Statistics and Probability Letters* **8**, 109–117.

Hall, P. & Turlach, B. A. (1999), 'Reducing bias in curve estimation by use of weights', *Computational Statistics and Data Analysis* **30**, 67–86.

Hall, P. & Wand, M. P. (1988), 'On nonparametric discrimination using density differences', *Biometrika* **75**, 541–547.

Hall, P. & Wand, M. P. (1996), 'On the accuracy of binned kernel density estimators', *Journal of Multivariate Analysis* **56**, 165 – 184.

Hand, D. J. (1982), *Kernel Discriminant Analysis*, Research Studies Press [John Wiley and Sons], Chichester.

Hartigan, J. A. (1975), *Clustering Algorithms*, John Wiley and Sons, New York.

Hastie, T. & Stuetzle, W. (1989), 'Principal curves', *Journal of the American Statistical Association* **84**, 502–516.

Hazelton, M. L. (1996), 'Bandwidth selection for local density estimators', *Scandinavian Journal of Statistics. Theory and Applications* **23**, 221–232.

Hazelton, M. L. & Marshall, J. C. (2009), 'Linear boundary kernels for bivariate density estimation', *Statistics and Probability Letters* **79**, 999–1003.

Hazelton, M. L. & Turlach, B. A. (2009), 'Nonparametric density deconvolution by weighted kernel density estimators', *Statistics and Computing* **19**, 217–228.

Heidenreich, N. B., Schindler, A. & Sperlich, S. (2013), 'Bandwidth selection methods for kernel density estimation: a review of fully automatic selectors', *AStA Advances in Statistical Analysis* **97**, 403–433.

Hochberg, Y. (1988), 'A sharper Bonferroni procedure for multiple tests of significance', *Biometrika* **75**, 800–802.

Holmquist, B. (1988), 'Moments and cumulants of the multivariate normal distribution', *Stochastic Analysis and Applications* **6**, 273–278.

Holmquist, B. (1996*a*), 'The $d$-variate vector Hermite polynomial of order $k$', *Linear Algebra and Its Applications* **237/238**, 155–190.

Holmquist, B. (1996*b*), 'Expectations of products of quadratic forms in normal variables', *Stochastic Analysis and Applications* **14**, 149–164.

Holmström, L. (2000), 'The accuracy and the computational complexity of a multivariate binned kernel density estimator', *Journal of Multivariate Analysis* **72**, 264–309.

Horová, I., Koláček, J. & Zelinka, J. (2012), *Kernel Smoothing in MATLAB: Theory and Practice of Kernel Smoothing*, World Scientific, Singapore.

Horová, I., Koláček, J. & Vopatová, K. (2013), 'Full bandwidth matrix selectors for gradient kernel density estimate', *Computational Statistics and Data Analysis* **57**, 364–376.

Hyndman, R. J. (1996), 'Computing and graphing highest density regions', *American Statistician* **50**, 120–126.

Jennrich, R. I. & Turner, F. B. (1969), 'Measurement of non-circular home range', *Journal of Theoretical Biology* **22**, 227–237.

Johnson, N. L., Kotz, S. & Balakrishnan, N. (1994), *Continuous Univariate Distributions*, 2nd edn, John Wiley and Sons, New York.

Jones, M. C. (1990), 'Variable kernel density estimates and variable kernel density estimates', *Australian & New Zealand Journal of Statistics* **32**, 361–371.

Jones, M. C. (1991), 'The roles of ISE and MISE in density estimation', *Statistics and Probability Letters* **12**, 51–56.

Jones, M. C. (1992), 'Potential for automatic bandwidth choice in variations on kernel density estimation', *Statistics and Probability Letters* **13**, 351–356.

Jones, M. C. (1994), 'On kernel density derivative estimation', *Communications in Statistics: Theory and Methods* **23**, 2133–2139.

Jones, M. C. & Foster, P. J. (1993), 'Generalized jackknifing and higher order kernels', *Journal of Nonparametric Statistics* **3**, 81–94.

Jones, M. C. & Henderson, D. A. (2007), 'Kernel-type density estimation on the unit interval', *Biometrika* **94**, 977–984.

Jones, M. C. & Kappenman, R. F. (1992), 'On a class of kernel density estimate bandwidth selectors', *Scandinavian Journal of Statistics* **19**, 337–349.

Jones, M. C., Marron, J. S. & Park, B. U. (1991), 'A simple root *n* bandwidth selector', *Annals of Statistics* **19**, 1919–1932.

Jones, M. C., Marron, J. S. & Sheather, S. J. (1996*a*), 'A brief survey of bandwidth selection for density estimation', *Journal of the American Statistical Association* **91**, 401–407.

Jones, M. C., Marron, J. S. & Sheather, S. J. (1996*b*), 'Progress in data-based bandwidth selection for kernel density estimation', *Computational Statistics* **11**, 337–381.

Jones, M. C., McKay, I. J. & Hu, T. C. (1994), 'Variable location and scale kernel density estimation', *Annals of the Institute of Statistical Mathematics* **46**, 521–535.

Jones, M. C. & Sheather, S. J. (1991), 'Using nonstochastic terms to advantage in kernel-based estimation of integrated squared density derivatives', *Statistics and Probability Letters* **11**, 511–514.

Jones, M. C. & Signorini, D. F. (1997), 'A comparison of higher-order bias kernel density estimators', *Journal of the American Statistical Association* **92**, 1063–1073.

Jones, M., Linton, O. & Nielsen, J. (1995), 'A simple bias reduction method for density estimation', *Biometrika* **82**, 327–338.

Kan, R. (2008), 'From moments of sum to moments of product', *Journal of Multivariate Analysis* **99**, 542–554.

Klemelä, J. (2004), 'Visualization of multivariate density estimates with level set trees', *Journal of Computational and Graphical Statistics* **13**, 599–620.

Klemelä, J. (2009), *Smoothing of Multivariate Data: Density Estimation and Visualization*, Wiley, New York.

Kollo, T. & von Rosen, D. (2005), *Advanced Multivariate Statistics with Matrices*, Springer, Dordrecht.

Korostelev, A. P. & Tsybakov, A. B. (1993), *Minimax Theory of Image Reconstruction*, Springer-Verlag, New York.

Kung, Y.-H., Lin, P.-S. & Kao, C.-H. (2012), 'An optimal $k$-nearest neighbor for density estimation', *Statistics and Probability Letters* **82**, 1786–1791.

Lee, M.-L. T. (1996), 'Properties and applications of the Sarmanov family of bivariate distributions', *Communications in Statistics: Theory and Methods* **25**, 1207–1222.

Li, J., Ray, S. & Lindsay, B. G. (2007), 'A nonparametric statistical approach to clustering via mode identification', *Journal of Machine Learning Research* **8**, 1687–1723.

Li, K. C. (1984), 'Consistency for cross-validated nearest neighbour estimates in nonparametric regression', *Annals of Statistics* **12**, 230–240.

Li, X., Hu, Z. & Wu, F. (2007), 'A note on the convergence of the mean shift', *Pattern Recognition* **40**, 1756–1762.

Lichman, M. (2013), 'UCI machine learning repository', `http://archive.ics.uci.edu/ml/datasets/Cardiotocography`. University of California, Irvine, School of Information and Computer Sciences. Accessed 2017-05-18.

Loftsgaarden, D. O. & Quesenberry, C. P. (1965), 'A nonparametric estimate of a multivariate density function', *Annals of Mathematical Statistics* **36**, 1049–1051.

Lugosi, G. & Nobel, A. (1996), 'Consistency of data-driven histogram methods for density estimation and classification', *Annals of Statistics* **24**, 687–706.

Mack, Y. P. & Rosenblatt, M. (1979), 'Multivariate $k$-nearest neighbor density estimates', *Journal of Multivariate Analysis* **9**, 1–15.

Magnus, J. R. & Neudecker, H. (1999), *Matrix Differential Calculus with Applications in Statistics and Econometrics: Revised edition*, John Wiley and Sons, Chichester.

Mardia, K. V., Kent, J. T. & Bibby, J. M. (1979), *Multivariate Analysis*, Academic Press, London.

Marron, J. (1994), 'Visual understanding of higher-order kernels', *Journal of Computational and Graphical Statistics* **3**, 447–458.

Marron, J. S. (1992), 'Bootstrap bandwidth selection', in R. LePage & L. Billard, eds, *Exploring the Limits of Bootstrap*, Wiley, New York, pp. 249–262.

Marron, J. S. & Wand, M. P. (1992), 'Exact mean integrated squared error', *Annals of Statistics* **20**, 712–736.

Marshall, J. C. & Hazelton, M. L. (2010), 'Boundary kernels for adaptive density estimators on regions with irregular boundaries', *Journal of Multivariate Analysis* **101**, 949–963.

Martin, D., Fowlkes, C., Tal, D. & Malik, J. (2001), 'A database of human segmented natural images and its application to evaluating segmentation algorithms and measuring ecological statistics', in *Proceedings of the 8th International Conference on Computer Vision*, Vol. 2, pp. 416–423. `https://www2.eecs.berkeley.edu/Research/Projects/CS/vision/bsds`. Accessed 2017-05-29.

Masry, E. (1991), 'Multivariate probability density deconvolution for stationary random processes', *IEEE Transactions On Information Theory* **37**, 1105–1115.

Mathai, A. M. & Provost, S. B. (1992), *Quadratic Forms in Random Variables: Theory and Applications*, Marcel Dekker, New York.

Meister, A. (2006), 'Estimating the support of multivariate densities under measurement error', *Journal of Multivariate Analysis* **97**, 1702–1717.

Menne, M. J., Durre, I., Vose, R. S., Gleason, B. E. & Houston, T. (2012), 'An overview of the global historical climatology network-daily database', *Journal of Atmospheric and Oceanic Technology* **429**, 897–910. `https://climexp.knmi.nl/selectdailyseries.cgi`. Accessed 2016-10-20.

Molchanov, I. S. (1991), 'Empirical estimation of distribution quantiles of random closed sets', *Theory of Probability and its Applications* **35**, 594–560.

Müller, H. G. & Stadtmüller, U. (1999), 'Multivariate boundary kernels and a continuous least squares principle', *Journal of the Royal Statistical Society: Series B* **61**, 439–458.

Myers, N., Mittermeier, R. A., Mittermeier, C. G., Da Fonseca, G. A. B. & Kent, J. (2000), 'Biodiversity hotspots for conservation priorities', *Nature* **403**, 853–858.

NGDC/WDS (2017), 'Global significant earthquake database. National Geophysical Data Center, NOAA', `doi:10.7289/V5TD9V7K`. National Geophysical Data Center/World Data Service. Accessed 2017-03-30.

OKI (2016), 'Open knowledge: What is open?', `https://okfn.org/opendata`. Open Knowledge International. Accessed 2016-03-02.

Olkin, I. & Liu, R. (2003), 'A bivariate beta distribution', *Statistics and Probability Letters* **62**, 407–412.

Ozertem, U. & Erdogmus, D. (2011), 'Locally defined principal curves and surfaces', *Journal of Machine Learning Research* **12**, 1249–1286.

Park, B. U. & Marron, J. S. (1992), 'On the use of pilot estimators in bandwidth selection', *Journal of Nonparametric Statistics* **1**, 231–240.

Parzen, E. (1962), 'On estimation of a probability density function and mode', *Annals of Mathematical Statistics* **33**, 1065–1076.

Politis, D. N. (2003), 'Adaptive bandwidth choice', *Journal of Nonparametric Statistics* **15**, 517–533.

Politis, D. N. & Romano, J. P. (1999), 'Multivariate density estimation with general flat-top kernels of infinite order', *Journal of Multivariate Analysis* **68**, 1–25.

Polonik, W. & Wang, Z. (2005), 'Estimation of regression contour clusters: An application of the excess mass approach to regression', *Journal of Multivariate Analysis* **94**, 227–249.

Porter, T. & Duffle, T. (1984), 'Compositing digital images', *Computer Graphics* **18**, 253–25.

Pratt, W. K. (2001), *Digital Image Processing: PIKS Inside*, 3rd edn, John

Wiley and Sons, New York.

R Core Team (2017), *R: A Language and Environment for Statistical Computing*, R Foundation for Statistical Computing, Vienna, Austria. `http://www.R-project.org`.

RATP (2016), 'Qualité de l'air mesurée dans la station Châtelet', `https://data.iledefrance.fr/explore/dataset/qualite-de-lair-mesuree-dans-la-station-chatelet`. Régie autonome des transports parisiens – Département Développement, Innovation et Territoires. Accessed 2017-09-27.

Rényi, A. & Sulanke, R. (1963), 'Über die konvexe hülle von $n$ zufällig gewählten punkten', *Z. Wahrscheinlichkeitstheorie und Verw. Gebiete* **2**, 75–84.

Rodríguez-Casal, A. (2007), 'Set estimation under convexity type assumptions', *Annales de l'Institut Henri Poincaré: Probabilités et Statistiques* **43**, 763–774.

Rosenblatt, M. (1956), 'Remarks on some nonparametric estimates of a density function', *Annals of Mathematical Statistics* **27**, 832–837.

Rosenblatt, M. (1971), 'Curve estimates', *Annals of Mathematical Statistics* **42**, 1815–1842.

Rudemo, M. (1982), 'Empirical choice of histograms and kernel density estimators', *Scandinavian Journal of Statistics. Theory and Applications* **9**, 65–78.

Ruppert, D. & Cline, D. B. H. (1994), 'Bias reduction in kernel density estimation by smoothed empirical transformations', *Annals of Statistics* **22**, 185–210.

Rustagi, J. S., Javier, W. R. & Victoria, J. S. (1991), 'Trimmed jackknife kernel estimate for the probability density function', *Computational Statistics and Data Analysis* **12**, 19–26.

Saavedra-Nieves, P., González-Manteiga, W. & Rodríguez-Casal, A. (2014), 'Level set estimation', in M. Akritas, S. Lahiri & D. Politis, eds, *Topics in Nonparametric Statistics*, Springer Science+Business Media, New York, chapter 27, pp. 299–307.

Sacks, J. & Ylvisaker, D. (1981), 'Asymptotically optimum kernels for density estimation at a point', *Annals of Statistics* **9**, 334–346.

Sain, S. R. (2001), 'Bias reduction and elimination with kernel estimators', *Communications in Statistics: Theory and Methods* **30**, 1869–1888.

Sain, S. R. (2002), 'Multivariate locally adaptive density estimation', *Com-*

*putational Statistics and Data Analysis* **39**, 165–186.

Sain, S. R. & Scott, D. W. (1996), 'On locally adaptive density estimation', *Journal of the American Statistical Association* **91**, 1525–1534.

Samworth, R. J. & Wand, M. P. (2010), 'Asymptotics and optimal bandwidth selection for highest density region estimation', *Annals of Statistics* **38**, 1767–1792.

Savits, T. H. (2006), 'Some statistical applications of Faa di Bruno', *Journal of Multivariate Analysis* **97**, 2131–2140.

Schnabel, R. B., Koonatz, J. E. & Weiss, B. E. (1985), 'A modular system of algorithms for unconstrained minimization', *ACM Transactions on Mathematical Software* **11**, 419–440.

Schott, J. R. (1996), *Matrix Analysis for Statistics*, John Wiley and Sons, New York.

Schott, J. R. (2003), 'Kronecker product permutation matrices and their application to moment matrices of the normal distribution', *Journal of Multivariate Analysis* **87**, 177–190.

Schott, J. R. (2005), *Matrix Analysis for Statistics*, 2nd edn, John Wiley and Sons, New York.

Scott, D. W. (1979), 'On optimal and data-based histograms', *Biometrika* **66**, 605–610.

Scott, D. W. (1992), *Multivariate Density Estimation: Theory, Practice, and Visualization*, John Wiley and Sons, New York.

Scott, D. W. (2015), *Multivariate Density Estimation: Theory, Practice, and Visualization*, 2nd edn, John Wiley and Sons, New Jersey.

Scott, D. W. & Terrell, G. R. (1987), 'Biased and unbiased cross-validation in density estimation', *Journal of the American Statistical Association* **82**, 1131–1146.

Sheather, S. J. & Jones, M. C. (1991), 'A reliable data-based bandwidth selection method for kernel density estimation', *Journal of the Royal Statistical Society Series B* **53**, 683–690.

Silverman, B. W. (1986), *Density Estimation for Statistics and Data Analysis*, Chapman and Hall, London.

Simonoff, J. S. (1996), *Smoothing Methods in Statistics*, Springer-Verlag, New York.

Singh, A., Scott, C. & Nowak, R. (2009), 'Adaptive Hausdorff estimation of density level sets', *Annals of Statistics* **37**, 2760–2782.

Smith, P. J. (1995), 'A recursive formulation of the old problem of obtaining moments from cumulants and vice versa', *American Statistician* **49**, 217–218.

Stefansky, L. A. (1990), 'Rates of convergence of some estimators in a class of deconvolution problems', *Statistics and Probability Letters* **9**, 229–235.

Stefansky, L. A. & Carroll, R. J. (1987), 'Deconvoluting kernel density estimators', *University of North Carolina Mimeo Series* **1623**.

Stefansky, L. A. & Carroll, R. J. (1990), 'Deconvoluting kernel density estimators', *Statistics* **2**, 169–184.

Stuetzle, W. (2003), 'Estimating the cluster tree of a density by analyzing the minimal spanning tree of a sample', *Journal of Classification* **20**, 25–47.

Taylor, C. C. (1989), 'Bootstrap choice of the smoothing parameter in kernel density estimation', *Biometrika* **76**, 705–712.

Tenreiro, C. (2001), 'On the asymptotic behaviour of the integrated square error of kernel density estimators with data-dependent bandwidth', *Statistics and Probability Letters* **53**, 283–292.

Tenreiro, C. (2003), 'On the asymptotic normality of multistage integrated density derivative kernel estimators', *Statistics and Probability Letters* **64**, 311–322.

Terrell, G. R. (1990), 'The maximal smoothing principle in density estimation', *Journal of the American Statistical Association* **85**, 470–477.

Terrell, G. R. & Scott, D. W. (1992), 'Variable kernel density estimation', *Annals of Statistics* **20**, 1236–1265.

Thom, R. (1949), 'Sur une partition en cellules associée à une fonction sur une variété', *Comptes rendus des séances de l'Académie des Sciences* **228**, 973–975.

Triantafyllopoulos, K. (2003), 'On the central moments of the multidimensional Gaussian distribution', *Mathematical Scientist* **28**, 125–128.

Tsybakov, A. B. (2009), *Introduction to Nonparametric Estimation*, Springer, New York.

Tukey, J. W. (1977), *Exploratory Data Analysis*, Addison-Wesley, Reading.

USGS (2017), 'Earthquake glossary', `https://earthquake.usgs.gov/learn/glossary/?term=RingofFire`. United States Geological Survey. Accessed 2017-03-30.

Vieu, P. (1996), 'A note on density mode estimation', *Statistics and Probability Letters* **26**, 297–307.

Wand, M. P. (1992), 'Error analysis for general multivariate kernel estimators', *Journal of Nonparametric Statistics* **2**, 1–15.

Wand, M. P. (1994), 'Fast computation of multivariate kernel estimators', *Journal of Computational and Graphical Statistics* **3**, 433–445.

Wand, M. P. (1997), 'Data-based choice of histogram bin width', *The American Statistician* **51**, 59–64.

Wand, M. P. & Jones, M. C. (1993), 'Comparison of smoothing parameterizations in bivariate kernel density estimation', *Journal of the American Statistical Association* **88**, 520–528.

Wand, M. P. & Jones, M. C. (1994), 'Multivariate plug-in bandwidth selection', *Computational Statistics* **9**, 97–116.

Wand, M. P. & Jones, M. C. (1995), *Kernel Smoothing*, Chapman and Hall/CRC, London.

Wand, M. P., Marron, J. S. & Ruppert, D. (1991), 'Transformations in density-estimation', *Journal of the American Statistical Association* **86**, 343–353.

Wand, M. P. & Schucany, W. R. (1990), 'Gaussian-based kernels', *The Canadian Journal of Statistics* **18**, 197–204.

Wasserman, L. (2006), *All of Nonparametric Statistics*, Springer Science+Business Media, Inc., New York.

Wasserman, L. (2018), 'Topological data analysis', *Annual Review of Statistics and its Application* **5**.

Watson, G. S. & Leadbetter, M. R. (1963), 'On the estimation of the probability density, I', *Annals of Mathematical Statistics* **34**, 480–491.

Webb, A. R. & Copsey, K. D. (2011), *Statistical Pattern Recognition*, 3rd edn, John Wiley & Sons, Chichester.

World Bank Group (2016), 'World development indicators', `http://databank.worldbank.org/data/reports.aspx?source=world-development-indicators`. Accessed 2016-10-03.

Worton, B. J. (1989), 'Kernel methods for estimating the utilization distribution in home-range studies', *Ecology* **70**, 164–168.

Wu, T.-J., Hsu, C.-Y., Chen, H.-Y. & Yu, H.-C. (2014), 'Root *n* estimates of vectors of integrated density partial derivative functionals', *Annals of the Institute of Statistical Mathematics* **66**, 865–895.

Youndjé, E. & Wells, M. T. (2008), 'Optimal bandwidth selection for multivariate kernel deconvolution density estimation', *Test* **17**, 138–162.

Zeileis, A., Hornik, K. & Murrell, P. (2009), 'Escaping RGBland: Selecting colors for statistical graphics', *Computational Statistics and Data Analysis* **53**, 3259–3270.

# Index

9780367571733

## V